Surveys in Applied Mathematics

Proceedings of the First Los Alamos
Symposium on Mathematics
in the Natural Sciences

Surveys in Applied Mathematics

Essays dedicated to
S.M. Ulam

EDITED BY

N. METROPOLIS
S. ORSZAG
G.-C. ROTA

ACADEMIC PRESS, INC. New York San Francisco London 1976
A Subsidiary of Harcourt Brace Jovanovich, Publishers

COPYRIGHT © 1976, BY ACADEMIC PRESS, INC.
ALL RIGHTS RESERVED.
NO PART OF THIS PUBLICATION MAY BE REPRODUCED OR
TRANSMITTED IN ANY FORM OR BY ANY MEANS, ELECTRONIC
OR MECHANICAL, INCLUDING PHOTOCOPY, RECORDING, OR ANY
INFORMATION STORAGE AND RETRIEVAL SYSTEM, WITHOUT
PERMISSION IN WRITING FROM THE PUBLISHER.

ACADEMIC PRESS, INC.
111 Fifth Avenue, New York, New York 10003

United Kingdom Edition published by
ACADEMIC PRESS, INC. (LONDON) LTD.
24/28 Oval Road, London NW1

Library of Congress Cataloging in Publication Data

Los Alamos Symposium on Mathematics in the Natural
       Sciences, 1st, 1974
       Proceedings of the first Los Alamos Symposium on
Mathematics in the Natural Sciences.

       (Surveys in applied mathematics)
       1.   Mathematics—Congresses.   2.   Science—Congresses.
I.   Metropolis, Nicholas Constantine,   Date   II.   Ors-
zag, S.   III.   Rota, Gian Carlo,   Date   IV.   Series:
Surveys in applied mathematics (New York, 1976-   )
QA1.L588   1974   510   76-43107
ISBN 0-12-492150-7

PRINTED IN THE UNITED STATES OF AMERICA

*This volume
is affectionately dedicated to*

STANISLAW M. ULAM

who

*for many years
dedicated his efforts
to the well being of
The Los Alamos Scientific Laboratory*

CONTENTS

Contributors ix
Preface xi
Introduction xiii

ON THE ROLE OF APPLIED MATHEMATICS
   C.C. Lin   1

ON THE SHAPE OF A CURVE
   Raoul Bott   23

AUTOMORPHIC FORMS FOR SCHOTTKY GROUPS
   Lipman Bers   39

BIASED VERSUS UNBIASED ESTIMATION
   Bradley Efron   69

ALGORITHMS
   D.J. Kleitman   89

WHITNEY NUMBERS OF GEOMETRIC LATTICES
   Kenneth Baclawski   103

CONTINUED FRACTION EXPANSION OF ALGEBRAIC NUMBERS
   R.D. Richtmyer   117

RANDOM TIME EVOLUTION OF INFINITE PARTICLE SYSTEMS
   Frank Spitzer   123

BIFURCATIONS IN REACTION–DIFFUSION PROBLEMS
   Louis N. Howard   129

CONTENTS

SINGULAR PERTURBATION
J.D. Cole — 143

A SURVEY OF SOME FINITE ELEMENT METHODS PROPOSED FOR TREATING THE DIRICHLET PROBLEM
James H. Bramble — 157

THE MATHEMATICS OF QUANTUM FIELDS
James Glimm — 167

RENORMALIZATION GROUP METHODS
Kenneth G. Wilson — 179

REMARKS ON TURBULENCE THEORY
Robert H. Kraichnan — 197

ON AN EXPLICITLY SOLUBLE SYSTEM OF NONLINEAR DIFFERENTIAL EQUATIONS RELATED TO CERTAIN TODA LATTICES
M. Kac and Pierre van Moerbeke — 225

THREE INTEGRABLE HAMILTONIAN SYSTEMS CONNECTED WITH ISOSPECTRAL DEFORMATIONS
J. Moser — 235

ALMOST PERIODIC BEHAVIOR OF NONLINEAR WAVES
Peter D. Lax — 259

A CONSTRUCTIVE DEFINITION OF THE REAL NUMBERS
F. Faltin, N. Metropolis, B. Ross, and G.-C. Rota — 271

CONTRIBUTORS

KENNETH BACLAWSKI
Department of Mathematics
Harvard University
Cambridge, Massachusetts 02138

LIPMAN BERS
Department of Mathematics
Columbia University
New York, New York 10027

RAOUL BOTT
Department of Mathematics
Harvard University
Cambridge, Massachusetts 02138

JAMES H. BRAMBLE
Department of Mathematics
Cornell University
Ithaca, New York 14850

J. D. COLE
Department of Mathematics
University of California, Los Angeles
Los Angeles, California 90024

BRADLEY EFRON
Department of Mathematics
Stanford University
Stanford, California 94305

F. FALTIN
Department of Mathematics
Cornell University
Ithaca, New York 14850

JAMES GLIMM
Courant Institute of Mathematical
  Sciences
New York University
New York, New York 10012

LOUIS N. HOWARD
Massachusetts Institute of Technology
Cambridge, Massachusetts 02139

M. KAC
The Rockerfeller University
New York, New York 10021

D. J. KLEITMAN
Mathematics Department
Massachusetts Institute of Technology
Cambridge, Massachusetts 02139

ROBERT H. KRAICHNAN
Dublin, New Hampshire 03444

PETER D. LAX
Courant Institute of Mathematical
  Sciences
251 Mercer Street
New York, New York 10012

C. C. LIN
Massachusetts Institute of Technology
Cambridge, Massachusetts 02138

N. METROPOLIS
Los Alamos Scientific Laboratory
Los Alamos, New Mexico 87544

CONTRIBUTORS

PIERRE VAN MOERBEKE
University of Louvain
Louvain, Belgium

J. MOSER
Courant Institute of Mathematical Sciences
New York University
New York, New York 10012

R. D. RICHTMYER
T-Division
Los Alamos Scientific Laboratory
Los Alamos, New Mexico 87544

B. ROSS
Massachusetts Institute of Technology
Cambridge, Massachusetts 02139

G.-C. ROTA
Massachusetts Institute of Technology
Cambridge, Massachusetts 02139

FRANK SPITZER
Department of Mathematics
Cornell University
Ithaca, New York 14850

KENNETH G. WILSON
Laboratory of Nuclear Studies
Cornell University
Ithaca, New York 14850

PREFACE

Since its recent formation, the U.S. Energy Research and Development Administration has become involved in scientific programs whose complexity and impenetrability exceeds those of the past. It was foreseen that these programs would require mathematical sophistication well beyond that of even ten years ago.

The Los Alamos Scientific Laboratory decided to organize a workshop with two objectives. It was observed that some of the ongoing projects were not making full use of the latest mathematics and theoretical physics. Secondly, it was felt that a small group of outstanding scientists should be invited to Los Alamos and exposed to some of the applied mathematical aspects of important national problems.

This volume contains the collection of papers based upon most of the lectures delivered at the workshop. Discussions with Los Alamos staff members were also an integral part of the workshop. These consisted of panel discussions, wherein staff members along with invited speakers and visiting staff members participated. The staff made presentations of the mathematical aspects of problems and questions in the various areas of Laboratory interest; the invited speakers responded in general or particular terms depending upon the extent of overlap of interest. The reports of these discussions are not available for publication at present.

The Committee and all participants in the workshop thank Dr. Milton Rose for making it possible and lending his support.

N. Metropolis, Chairman
Steven A. Orszag
Gian-Carlo Rota

INTRODUCTION

C.C. Lin in "On the Role of Applied Mathematics" stresses the interaction between mathematical theory and the physical world. Lin argues the importance of improved communications between mathematicians and other scientists and engineers. We hope that the present volume helps to achieve this goal.

Raoul Bott in "On the Shape of a Curve" gives an introductory exposition of topology and algebraic geometry. Bott discusses the "shape" of the solution set to polynomial equations in several complex variables and explains the relationship of these results to modern developments in algebraic number theory including the famous Weil conjectures.

Lipman Bers in "Automorphic Forms for Schottky Groups" addresses the theory of complex variables via group theory. He provides an introduction to the theory of Kleinian groups and investigates their application to the deformation theory of Riemann surfaces.

Bradley Efron in "Biased versus Unbiased Estimation" shows that deliberate biasing of results can sometimes drastically improve statistical predictions. If the criterion is used that the statistical predictions should be very good most of the time but sometimes rather poor, it is possible to introduce statistical functions alternative to those that are best on the average. Efron investigates properties of these biased estimators.

D.J. Kleitman in "Algorithms" addresses the classification of problems that can and cannot be solved by efficient algorithms and illustrates the general results with a number of specific results concerning multiplication of binary numbers, testing a graph for planarity, and many other problems. This theory finds very useful application in the design of algorithms for modern computers.

Kenneth Baclawski in "Whitney Numbers of Geometric Lattices" uses sheaf theory to shed new light on the chromatic polynomials of graphs. He shows that the coefficients of these polynomials are Betti numbers.

R.D. Richtmyer in "Continued Fraction Expansion of Algebraic Numbers" presents some computer tests of the conjecture that irrational algebraic numbers of degree ≥ 3 have continued fraction expansion coefficients that satisfy Khintchine's law.

Frank Spitzer in "Random Time Evolution of Infinite Particle Systems" gives a synopsis of the latest developments in the theory of behavior of infinite coupled sets of dynamical equations. He states conditions for ergodic and nonergodic behavior.

INTRODUCTION

Louis N. Howard in "Bifurcation in Reaction–Diffusion Problems" surveys classical bifurcation theorems. A bifurcation theorem gives conditions such that stationary and periodic solutions of differential equations be stable to small perturbations. These theorems are important in such diverse fields as celestial mechanics, hydrodynamics, solid mechanics, etc. Howard presents some new general results and then applies the theory to the study of the Belousov–Zhabotinskii reaction, an oscillatory chemical reaction that propagates as a wave.

J.D. Cole in "Singular Perturbation" gives a heuristic introduction to the modern methods of perturbation theory. Cole gives representative applications of limit process expansions (boundary layer problems) and multiple-scale problems. Cole emphasizes the physical characteristics of the problems that are soluble by these methods.

James H. Bramble in "A Survey of Some Finite Element Methods Proposed for Treating the Dirichlet Problem" discusses several alternative methods for imposition of boundary conditions in finite element approximations to elliptic boundary value problems. Finite element methods are being actively applied to many kinds of boundary value problems arising in solid mechanics and elsewhere.

James Glimm in "The Mathematics of Quantum Fields" gives a terse, clear introduction to this intricate subject. The goal of quantum field theory is to obtain a fundamental description of elementary particles by combining quantum mechanics with special relativity in a self-consistent way. Glimm emphasizes what is now proved, what kinds of mathematics have been used to study quantum fields, and what the status is of various unsolved problems.

Kenneth G. Wilson in "Renormalization Group Methods" gives an elementary introduction to the renormalization group and its application to the solution of the Kondo problem. In recent years, Wilson and his co-workers have also used renormalization group techniques to solve critical phenomena problems in statistical mechanics. The technique is very powerful and may be useful for molecular bond problems in physical chemistry, turbulent flows in fluid dynamics, and quantum field theory.

Robert H. Kraichnan in "Remarks on Turbulence Theory" summarizes current theoretical ideas on homogeneous turbulence in fluids. Turbulence is perhaps the most widespread but least understood of all fluid phenomena. Kraichnan reviews attempts to apply renormalized perturbation theory and to compute associated eddy damping coefficients. He also discusses the nature of nonlinear energy cascade and intermittency and the prospects for obtaining useful information about turbulent flows by various theories.

M. Kac and Pierre van Moerbeke in "On an Explicitly Soluble System of Nonlinear Differential Equations Related to Certain Toda Lattices" discuss the exact solution of finite and infinite coupled lattice equations with exponential interactions between mass points. The problem is solved by the inverse scattering method which has proved so successful in the exact solution of model nonlinear evolution equations.

J. Moser in "Three Integrable Hamiltonian Systems Connected with Isospectral

INTRODUCTION

Deformations" shows that several recently studied exactly soluble problems are integrable Hamiltonian systems. This paper which is closely related to those of Lax and Kac and van Moerbeke develops the close relation between the inverse scattering method and spectral operator theory by an ingenious use of Jacobi matrices.

Peter D. Lax in "Almost Periodic Behavior of Nonlinear Waves" discusses recent work toward the proof of almost periodic behavior in three model nonlinear dynamical systems. This theory should be particularly useful in understanding the almost periodic behavior in time of spatially periodic solutions to the Korteweg-de Vries equation.

Finally, in a spirit of naïveté matched by ineffable delusions of grandeur, Faltin, Metropolis, Ross, and Rota propose a *soi-disant* constructive derivation of the real number system. Blissfully unaware of the unpleasant practicalities of arithmetic, whatever contribution—if any—they propose gets drowned in a verbiage of heavy-handed formalities.

Surveys in Applied Mathematics

On the Role of Applied Mathematics

C. C. LIN

Massachusetts Institute of Technology, Cambridge, Massachusetts

TO MY GOOD FRIEND STANISLAW MARCIN ULAM IN
HONOR OF HIS 65TH BIRTHDAY

> This paper concerns itself with the nature of applied mathematics, and how to make its role effective. Its differences from pure mathematics and from theoretical sciences are pointed out with the help of illustrative examples. Recommendations are made for supporting a program of education and research in applied mathematics to make it an independent scientific discipline and to create a community of applied mathematicians.

I. INTRODUCTION

1. *The Role of Applied Mathematicians*

It is a pleasure for me to be given this opportunity to speak to this distinguished audience on the role of applied mathematics—even though I would rather, like everybody else, speak about my own research work on stellar systems, on galaxies, on hydrodynamics and on turbulence—because I do have a few thoughts to share with you regarding the profession of applied mathematics as a whole. Every applied mathematician here knows what he is doing in his own line of specialty, and it would be presumptive for me to comment on that. However, I feel that there are things which are left unattended and which should be done by all applied mathematicians as a group to enhance the role of our profession. There are also several rather philosophical issues which, I feel, deserve our continued attention, because they would influence the long-term outlook of applied mathematics. It is toward these issues that I shall address my discussion. I hope that my remarks are also of interest to the pure mathematicians and to the theoretical and empirical scientists, with whom the applied mathematicians often develop close working relationships.

I should add that, like anybody else, my opinions will be biased by my own experience, which does not include any extensive stay in a laboratory such as this one. I hope, therefore, that I shall be excused and corrected if my remarks do not quite fit the conditions here.

It is almost tautological to say that the role of the applied mathematician is to promote the most effective use of mathematics in the natural and social sciences, in applied sciences, in engineering, in industry, in government, and in all other kinds of human activity. But this is not the complete list of his professional activities. As an intelligent scientist, he can also ask intelligent questions and originate new ideas which are only remotely or partly mathematical but which are nevertheless extremely important. I have in mind such ideas and practical inventions as those pioneered by our friend Stan Ulam at this laboratory.

However, in this talk, I shall restrict myself more narrowly to the effective use of mathematics in the sciences and to the stimulation of pure mathematical research through such efforts. One of the focal points of my discussion will be *how* to make this role effective.

In the course of this discussion, I shall comment on a few philosophical issues for which there seem to be, as yet, no generally accepted opinion. This is not surprising: philosophy is a subject in which it is easy to ask good questions and difficult to find good answers. The questions I would like to ask are the following: (a) Applicability: Why is mathematics so broadly applicable?[1] (b) Universality: Are the mathematics usable in one science also most likely to be usable in another? (c) Probability of applicability: (i) Can we identify a body of knowledge known as "applicable mathematics"? (ii) Is mathematics ever to be applicable to such subjects as history?

I shall attempt to give answers to some of these questions.

To continue with this discussion, we must first agree on an answer to the question "What is applied mathematics?" Different people will probably give answers with different emphases. I hope that my answer will be broad enough to receive general acceptance and be useful as a basis for our discussion.

---

[1] Cf. [1].

II. What is Applied Mathematics?

2. *The Nature and Scope of Applied Mathematics*

Generally speaking, applied mathematics is a disciplined activity which lies between the empirical sciences and pure mathematics. In essence, it is characterized by an attitude, an approach, a way of thinking. The principal theme is the *interdependence* of mathematics and the sciences. In common with the pure mathematician, the applied mathematician is interested in the stimulation of the development of new mathematics (see [2]),—but with primary emphasis on those aspects *directly* or at least very strongly motivated by scientific problems. In common with the theoretical scientists, the applied mathematician seeks knowledge and understanding of scientific facts and real world phenomena through the use of mathematical methods.

A particularly challenging type of activity is to find *new* ideas for the *application* of mathematics and indeed to develop mathematical theories in those scientific subjects which have not hitherto been subjected to systematic mathematical treatment. (Social sciences and biology are the oftcited examples.) These efforts may in turn lead to the creation of new mathematical ideas and theories (by abstraction, generalization, or otherwise), which are interesting in their own right as a part of pure mathematics. The recognition of this duality is essential to the spirit of applied mathematics. By stressing this duality, there is obviously a *difference in emphasis* between applied mathematics and either pure mathematics or the empirical sciences.

The scope of applied mathematics is very broad, and can be best described by borrowing the following words from Albert Einstein [3]:

> "Its realm is accordingly defined as that part of the sum total of our knowledge which is capable of being expressed in mathematical terms."

These words were originally used to define physics. Taken literally, the statement clearly includes the mathematical theories of economics, biology, communication, etc., and is perhaps a more adequate description of applied mathematics.

Clearly, the activity of every branch of science includes major efforts *not* deeply mathematical. For example, the *act* of observation of the behavior of sun spots and solar winds, isolation of radio-active elements, taking of opinion polls (not the analysis), and cultivation of bacteria. Thus, applied mathematics does not *include* all the sciences. It merely *overlaps* with all the sciences. The extent of this overlap may be large (as in physics) or small (as in biology, for the moment). It is indeed the extension of this overlap that should concern the applied mathematician. At the same time, the *wisdom* of the applied mathematician, generated by his intimate knowledge of mathematical methods (or "applicable mathematics," if you wish) and his experience with a *variety* of mathematical applications, puts him in the best position to judge in which areas or problems of science the power of a particular branch of mathematics would be most effective, would have only limited effectiveness, or would have no effectiveness at all. His efforts must obviously be devoted to those aspects of science where mathematical methods can be effectively used. The practitioner must *continue* to improve upon his own mathematical knowledge[2] and new mathematics must be created when necessary. Fortunately, our experience has shown that there is indeed simplicity and order in the fundamental aspects of the sciences. All fundamental laws of physics can be stated in mathematical form on one sheet of paper. At the same time, it appears to be true that human intelligence insists on conceptual simplicity in the formulation of the fundamentals of any scientific subject. *Thus, the mathematical approach is also the basic approach.* At the other extreme, mathematical methods also can be effectively used in detailed scientific or technological calculations.

To summarize, a well-educated applied mathematician must have a clear appreciation of the above-mentioned spirit of duality and a comprehensive knowledge of the total picture of the mathematical sciences. Let us now turn to a more specific discussion of his creative activities.

3. *Creative Activities of Applied Mathematicians, Part A* [3]

Applied mathematicians are often found to be engaged in the following efforts: (a) the formulation of scientific concepts and problems in mathematical terms, (b) the solution of the resultant mathematical problems, and (c) the discussion, interpretation and the evaluation of the results of his analysis, including the making of specific predictions. But the solution of specific problems often serves merely as a focus and

---

[2] See, for example, Theodore von Kármán's remarks at the age of 80 in [4, Preface].
[3] Cf. [5].

an aid in reaching a deeper understanding. The final goal of these efforts of an applied mathematician lies in (d) the creation of ideas, concepts, and methods that are of basic significance and general applicability to the subject in question, including the formulation of general principles. As mentioned above, these efforts may lead to (e) the creation of new mathematical ideas and theories. It might be added that, because of the origin of such mathematical theories, they are more likely to be applicable to other branches of the sciences.[4]

The basic difference in motivation between pure and applied mathematicians is reflected in the habits and practices of their activities. Although the applied mathematician understands and appreciates the nature of a rigorous demonstration, he cannot be made inactive by these considerations. In phase (b) of his activities, his primary emphasis is always directed toward the ultimate solution, and he frequently uses heuristic scientific reasoning to achieve this end. In this way, an analysis might be made more amenable to solution, approximations introduced, or arguments made plausible. His treatment is at all times responsible and disciplined, but he is not a deductive logician interested solely in the beauty of form and the power of abstraction. He should, however, have enough background in pure mathematics to be able to distinguish between rigorous proof, clear demonstration, plausible arguments, and hopeful speculation; all of which are used in the study of applied mathematics. In particular, when he is engaged in the creation of new mathematics, the applied mathematician should follow the practices (including the degree of rigor) used in pure mathematics in the *formulation* of his results. Because of his background, heuristic reasoning leading to the final form of the theory will doubtlessly be emphasized.

In many cases, the development of a rigorous form of the mathematical method or theory cannot be accomplished, yet the method has to be used with only plausible arguments to support its reliability. Most work in numerical solution of scientific problems belongs to this category.

In the other two phases of his activity, (a) and (c), the applied mathematician should follow the practices of a theoretical scientist. *The construction of an idealized mathematical model is indeed the most important and the most difficult phase,* especially in a new field of application of

---

[4] Cf. Hirsh Cohen [6]. In this article, he pointed out the similarity of the nature of these mathematical problems and their solution to those in mechanics and classical applied mathematics. Also, Lawrence Klein, in [7], stated ". . . but there are few examples in which it can be said that our subject has called forth its own branch of mathematics or given inspiration for great new mathematical discoveries."

mathematics. It usually requires a comprehensive knowledge and a deep understanding of the empirical facts related to the particular phenomenon under consideration as well as penetrating insight and mature judgement. These are also required in phases (c) and (d) of his activities. In these phases, the applied mathematician should examine his results to reach a deeper understanding of the problem at hand, and attempt to abstract the essentials to form concepts which are of more general applicability. At the same time, his conclusions must, of course, be checked against the existing body of knowledge. Any new inferences or predictions from his results are also subject to verification by further experimentation and observation, since their truth cannot be determined by purely logical means. With complementary theoretical and empirical efforts, a deeper and more penetrating understanding may be achieved.

Despite these similarities of activity, there are *subtle differences in attitude between a theoretical scientist and an applied mathematician*. The theoretical physicist, for example, has his primary interest in the discovery of new physical laws. Indeed, a correct theory can be formulated only after a great deal of trial and error. (Consider the current efforts in high-energy physics.) In contrast, the applied mathematician places more emphasis on the use of mathematical methods for the description of physical phenomena in terms of known physical laws. (Consider the current efforts in numerical weather forecasting.) He is also greatly interested in the stimulation of new mathematical ideas. One might say that the difference lies in the relative extent to which *inductive* and *deductive* reasoning are emphasized in each discipline.

However, these subtle differences in motivation can lead to substantial differences in the choice of the educational program and in the attitude developed in a person, which in turn color his activities. For example, classical mechanics (including particle mechanics and continuum mechanics) is seldom studied in detail by modern students of physics, yet it remains one of the principal subjects for the education of applied mathematicians. The fundamental concepts and the mathematical methods which have been developed through such studies still constitute the essential basis for the application of mathematics, whether the particular subject in question is atmospheric turbulence, radio astronomy, or demography. Indeed, the applied mathematician does not draw sharp lines along traditional boundaries of subject matter, and it is desirable for him to adapt his interests to the present and future vitality of the subject matter if he expects his research efforts to have an impact beyond the development of applicable mathematical methods (cf. Appendix).

The desire and ability to cut across traditional scientific disciplines, through the medium of mathematics, are perhaps the unique characteristics of an applied mathematician. Indeed, the education of an applied mathematician must provide him with a breadth of knowledge in both mathematics and the fundamentals of the sciences. He is then in the best position to advance a specific subject by the creation of suitable mathematical models, by his critical and precise thinking, and by transferring the mathematical knowledge gained through the study of another scientific discipline.

The above description of an applied mathematician almost compels him to stay at the general fundamental level of a braod spectrum of sciences. Of course, an applied mathematician could, if he so chooses, become a specialist in a given scientific subject. In doing this, he should perhaps subtly change his attitude and adopt entirely the spirit of the specialists in that field.

In order to educate applied mathematicians, a program of study must be provided both in the undergraduate and the graduate levels. Opportunities for post-doctoral research should also be available. In particular, there must be a course dealing with the fundamentals of the five-step processes described at the beginning of this section, with special emphasis on the first step, the construction of suitable mathematical models in a variety of problems. This last part is usually lacking in most of the curricula in mathematics.

4. *Creative Activities of Applied Mathematicians, Part B*

Applied mathematicians are sometimes also found in the resolution of mathematical issues that arise from their scientific problems. This interest they share with the willing pure mathematicians. Although this is a less frequent type of activity than that described above, it would be unfortunate to ignore this aspect of interaction between mathematics and the sciences. To quote from Chapter II of the COSRIMS Report [8], "Fourier considered mathematics as a tool for describing nature. But the impact of Fourier, important as it was in physics and engineering, was particularly felt in some of the purest branches of mathematics."

It is also to be recalled that when Fourier submitted his mathematical theory of heat to the French Academy for the Grand Prize, the referees,

Laplace, Lagrange, and Legendre pointed out that the mathematical treatment left much to be desired in the way of rigor, even though they admitted the novelty and importance of Fourier's work. More than a century was to elapse before all the subtle mathematical issues were satisfactorily resolved.[5] (To be sure, some of these issues were raised later.)

Here is a case from which much can be learned on the nature of the interaction. *The type of mathematical theory which is developed for the solution of certain scientific problems can often find application in other scientific problems.* For, when formulated in abstraction, i.e., in mathematical terms, they become similar or even identical. Secondly, in the early stages of the development of such work, its implication in pure mathematics is frequently unclear, and cannot be fully appreciated. The evaluation of such activities, therefore, must be made by the professionals as in any other discipline, to avoid unfair judgment.

In order to carry on with this tradition of Fourier, it is clear that the applied mathematician should be educated in the modern developments in pure mathematics. Even if his aim is less ambitious, he should be well informed in pure mathematics in order to make use of the latest accomplishments. In a university, special courses should be planned to introduce the essentials of pure mathematics to students in applied mathematics (cf. Sections 7 and 8 in Part III). As mentioned before, the applied mathematician also must be prepared to learn new mathematics throughout his career whenever it becomes necessary.

The contribution of pure mathematicians, who show interest in applications, can be extremely great in this area of interaction between science and mathematics. They can frequently contribute to the abstraction, consolidation, and generalization of special developments into a more elegant and a more powerful structure. Naturally, they would also contribute to *other mathematical developments which might appear to be useless at the beginning, but would later turn out to have important scientific applications.* However, I must disqualify myself from pursuing the discussion further along these lines. I hope that my friends in pure mathematics would expound on this theme further and furnish us with impressive examples.

From the standpoint of an applied mathematician, I would like, however, to add one comment. The theorems we need are often not the most general. We would be perfectly happy to have theorems proved

---

[5] See [9, 10].

under suitable restrictions. For Fourier analysis, piecewise smooth functions are usually sufficient. The extension to functions of bounded variation, Lebesque measurable functions, etc., are decidedly of lower importance. (For a modern problem in control theory, see [11]. Piecewise analytic functions are found to be most suitable.)

5. *Examples*

We shall now examine a few examples in physical mathematics to see how the above idealized description of the creative activities of applied mathematics fits realistic developments in specific instances.

(a) *Fourier Analysis, a Case of Unqualified Success*

First, consider the case of Fourier analysis. This is perhaps the best example of a complete "cycle" of development. A relatively simple class of problems in mathematical physics, heat conduction, was solved according to steps (a)–(c) outlined in Section 3. The mathematical theories of Fourier series, Fourier integral, and generalized harmonic analysis are very impressive developments. The theory of Lebesque measure is easily motivated by asking a simple question: Is the Converse to Parseval's theorem a true statement? Modern radio astronomy, with its multimillion dollar equipments and multimillion dollar annual budgets, depends on Fourier analysis for its very operation. The process of Fourier analysis is indeed built into the automatic online data processing equipment. The spectral analysis of the signals from distant astronomical objects is actually done by calculating the autocorrelation function and its Fourier transform (see [12, Fig. 5]).

The spatial Fourier analysis of the sky is also essential to radio astronomy. For this purpose, an array of radio telescopes is needed. A very large array, shaped in the form of the letter Y has been planned for completion in this country as a 10-year program, with a total budget exceeding one hundred million dollars. Figure 1 shows a linear array in Holland. It has a system of 12 telescopes spaced over a distance of 1600m. While the signal received by each of the telescopes can be analyzed in time by the method described above, the use of pairs of telescopes as interferometers enables the system to give a spatial Fourier analysis of the sky along the array. As the earth spins around its axis and revolves in its orbit, different directions in the sky may be covered. The Dutch astronomers are now in the process of doubling the length of the telescope system and the number of interferometer pairs in it.

Fig. 1. Part of a row of 12 radio-telescopes, each 25 m in diameter. The system, extending over a distance of 1600 m, provides 20 interferometers for the analysis of the structure of the radio sources. Present plans call for a doubling of the distance and the number of interferometers.

At microscales, the study of molecular structure is done by the study of X-ray diffraction patterns. The analysis is again based on Fourier methods. On the theoretical side, fundamental concepts in quantum mechanics, the wave-particle duality, can be best understood by using Fourier integrals.

(b) *Turbulence, an Incomplete Picture*

However, complete success of this type is rare. In many instances, success is only partial.

Consider the problem of turbulence. Fourier analysis is a necessary part of the theory, and indeed, provides the basic language for the subject. However, there are essential difficulties elsewhere. Turbulence is basically a nonlinear random process, in which the dissipative effect plays a paramount role as the stochastic process. There is as yet no mathematical theory for turbulence despite many heroic efforts over several decades.

Even on the level of a single field of fluid motion, which is governed by a well-established set of partial differential equations (the Navier-Stokes equations) the rigorous mathematical theory provides little help. The equations are quasilinear. The existence theory for this set of partial differential equations has been established only for very small values of the Reynolds number, a characteristic dimensionless parameter of the problem. To study the theory of turbulence, one must however deal with the fluid motion at extremely large Reynolds numbers.

In the more limited problem of hydrodynamic stability, where we may linearize the equations in the first approximation, rigorous mathematical theories have been found to be more helpful. There have also been very interesting new mathematical developments stimulated by the needs in the solution of the physical problem.

(c) *Numerical Analysis, a Case of Acquiescence*

The difficulty of dealing with nonlinear partial differential equations shows up even more clearly in the need to proceed with large-scale numerical calculations, despite the lack of mathematical theorems that would guarantee the convergence of such procedure. Examples that easily come to mind include much of the work done in this laboratory on the numerical solution of fluid dynamical problems and numerical weather forecasting. Logical-minded persons have to acquiesce to such decisions to go ahead for the lack of a better solution. Applied mathematicians are of course quite at home with such approaches, because there are various ways to ensure the plausible correctness of such procedures.

(d) *Plasma Physics and Stellar Dynamics, a Case of Mutual Support Without Firm Mathematical Foundation*

Plasma physics is a subject of great interest to this laboratory: The problem of plasma containment is still an unsolved problem, Professor Harold Grad would be better qualified than I to speak on the mathematical issues. But I am also not aware of any significant contributions to this science from new mathematical theorems. Most of the theoretical progress has been made in the spirit of applied mathematics. Indeed, in some of the exploratory calculations being made now, the procedures followed would seriously worry theoretical hydrodynamicists. The substantiation of such theoretical analysis depends heavily on experimental verification. At the same time, I am informed by my M.I.T. colleagues, who are theoretical plasma physicists working closely with experimental programs, that most of the desired experiments, with very few exceptions, are yet to be done. Here is a challenging opportunity for applied mathematicians to make substantial contributions in consolidating and improving upon the methods of analysis to make the predictions more reliable.

At the same time, these theoretical calculations can derive indirect support from distant places. The study of spiral structure in galaxies has led to the investigation of collective modes in stellar systems which are essentially plasmoidal in nature. These modes have been observed in terms of their effects on star formation. To be sure, the analysis in these cases has been done quite carefully, but even modes calculated with similar care in plasma physics have not yet been observed.

I mention this case to stress the importance of reasoning by analogy in the work of applied mathematicians. Confidence and insight can be gained in our understanding of plasma physics even though the work on galaxies contains no logical deduction in the context of electromagnetic plasmas.

Indeed, the concept of collective modes might be applicable (perhaps already applied) to social sciences. Each person is analogous to a star. At least in economics, a person interacts more with the institutions than with individuals. The latter kind of interaction plays a rather minor role to influence the economics of the country as a whole.

(e) *Transonic Flow, a Case of Mismatch*

When engineers began to design airplanes to fly at very high subsonic speeds, the problem of the existence of a local region of supersonic flow became important. People were concerned that shocks might develop to influence the behavior of the boundary layer, and thereby produce seriously increased drag. However, it was later realized that if reasonable and ingenious care were exercised to avoid strong shocks, the problem of highly increased resistance could be overcome.

In the meantime, the problem has stimulated a mathematical investigation of the possibility of obtaining smooth solutions of partial differential equations of mixed elliptic and hyperbolic types (parabolic on the line of transition). This is a very interesting mathematical development in itself, but its value for the practical problem is limited. The stringent mathematical conditions required for guaranteeing a smooth flow would exclude very *weak* shocks, whose occurrence does not really matter in the practical problem.

I cite this example to warn our colleagues against investing too much effort in problems posed by scientists and engineers before finding out the real nature and implications of the solution of the mathematical problem. In the example cited, the effort is not fruitless, because the mathematical problem is of interest in itself. In other cases, the loss could be almost total.

6. *Applied Mathematics as an Independent Discipline*

I now come to the central point of my talk, which is a plea for the establishment of a healthy community of applied mathematicians and for increased support of education and research in applied mathematics, especially the support of academic activities. We cannot expect the important role of applied mathematics to be fulfilled without a steady supply of applied mathematicians of very high caliber. It is not enough to have only pure mathematicians who would devote a small fraction of their efforts to applied mathematics. Neither is it sufficient to have only theoretical scientists who are unaware of mathematical methods used in other sciences. Applied mathematicians must be educated with a broad knowledge of applicable mathematics and an extensive exposure to the application of mathematics.

A common educational program for applied mathematicians is also needed to create a community spirit, without which applied mathematics cannot remain a healthy profession.

I think it is time that we should recognize applied mathematics as an independent discipline, fairly distinct from pure mathematics.

The difference in spirit between pure and applied mathematics is clear and often very great. In a very interesting and important article [13], Professor J. Schwartz warned mathematicians against the danger of single-mindedness, literal-mindedness, and simple-mindedness in dealing with scientific problems. The above examples also show clearly the difference between the approach of pure and applied mathematicians. Yet the need for the education of applied mathematicians in a comprehensive spirit described in this article has been generally neglected. It was recommended in [8, Recommendation 18]. Nevertheless, only a few schools have established such programs.

One might ask whether it is realistic to adopt this comprehensive approach. From my own examination of this matter, the answer is an unqualified "yes." I have recently read papers and books in ecology, mathematical theories of population, economics, combinatorial analysis, etc. I did not find the mathematics used to be *that* different one from the other. The mathematical methods used in biological sciences also do not differ from those in physical sciences (see [6]). Certainly, people use similar methods in numerical analysis and similar computing machines.

If the mathematical methods used in the various sciences were widely divergent, the establishment of a community spirit among applied mathematicians would of course be more difficult.

In this country, there had not been a tradition of applied mathematics. People now in this profession come from a variety of backgrounds. There is thus a definite need for practicing applied mathematicians to do something important for this community. This is to educate each other in the subject of his own special line of research in a well-motivated manner. This would create mutual understanding, establish mutual confidence, and build up the community spirit.

In the following sections, I present a discussion of the educational program and conclude with a number of recommendations. (In the oral presentation, this part was omitted.)

III. Education

7. *Education of Applied Mathematicians*

The ultimate aim of any liberal education program must include the following:

(1) the cultivation of an attitude, points of view, and value judgements;

(2) the acquirement of a vast body of coherent knowledge which deserves to be handed down from one generation to another, and which is best understood and special to this particular area of pursuit; and

(3) the development of certain talents, especially creative ability in the particular area in question.

From these, it is expected that the person will develop, through his personal experience in creative activity, a wisdom which enables him to act as an advisor to his co-workers and a teacher to the younger generation. In the latter capacity, he must be able to present a perspective of the total activity so that the younger generation will be able to decide how and where to devote their efforts, and to carry on their work with confidence. The lack of this perspective would often lead to wasteful efforts.

How can we arrange an educational program in applied mathematics to achieve the objectives described above?

First of all, the education must be started in the undergraduate years, during the formative period of the youth. If a person has formed an attitude to consider it far more valuable to prove an elegant theorem in algebra than to explain a phenomenon observed in the atmosphere or in the galaxies with the help of mathematical methods (or to consider the latter as "somebody else's business"), it would be very difficult to convert him to another set of beliefs. If a person has been impressed with a zeal to use his mathematical talent to elucidate a point in high energy physics, he might not be as enthusiastic in attacking a problem in biology or in economics.

The educational material (or the primary content of the knowledge of a student in applied mathematics) then must be based on an account of the typical impressive past accomplishments of mathematics selected from *all* the sciences. The method of case study must be used. At the same time, the students should also be made to understand that, while there would be a continuous opportunity for the applied mathematician in established areas (such as mechanics, or more broadly, physics) dramatic accomplishments of the future might lie in some other areas, where mathematics has not, as yet, been fully utilized. One must *avoid the danger of sterility by an overemphasis of traditional subject matter*.

It is the lack of survey courses of this broad nature that makes it difficult to educate future comprehensive applied mathematicians. A young student with an aspiration to become one usually finds himself in the almost impossible position of having to sift out for himself the mathematical aspects of the basic sciences from the courses that are available to him from the various scientific departments at a university. While this attempt itself is of great educational value, the total effort needed is overwhelming. The usual course of action for the student is then to attach himself to one of the traditional disciplines, either in mathematics or in one of the sciences. This specialization tends to defeat the major purpose of the education of comprehensive applied mathematicians.

The job of abstracting and synthesizing the total picture of the interplay between mathematics and the sciences must be done for these inspired young students by the experienced scientists. Indeed, this type of effort is the primary responsibility of the teachers in any profession. They must provide the introductory or survey courses to present the (undergraduate) students with a *total* picture of the *status quo* in the whole field so that future research efforts can be pursued with the greatest effectiveness.

Courses of another type, best taught by professional applied mathematicians, are those usually designated "methods of applied mathematics." These should be available at various levels. Besides these two types of "professional courses," the students must have a good foundation in basic pure mathematics, and some depth of education in at least one branch of science.

We may thus summarize a basic educational program in comprehensive applied mathematics[6] as follows:

(a) an education in the attitude of an applied mathematician,

(b) an education in the usual working ability (methods of applied mathematics),

(c) a survey in applied mathematics with emphasis on the total picture,

(d) a basic education in pure mathematics,

(e) An education in at least one branch of science in depth.

The subprograms (a), (b), and (c) should be normally taught by applied mathematicians; subprogram (d), by pure mathematicians; and subprogram (e), by professional scientists in various departments.

Graduate education should naturally be an extension and a continuation of the basic program discussed above; it must naturally include the development of the ability of the student to do research. The broad outlines of such abilities have been given in Section III. Here, as in all subjects, specialization is unavoidable, and the available lines of research at a given institution must depend on the faculty members available. It would, therefore, be useless to go into further details in this general discussions.

One should note the *central role of courses for subprogram* (c) (survey in applied mathematics), which should be approximately a two-year sequence in the junior and senior years. (It should be started in the

---

[6] In practice, this might mean three semester courses in pure mathematics, four semester courses in scientific subjects, and six semester courses in applied mathematics [(b) and (c)].

sophomore year, if possible.) Without these courses, we tend to educate only (i) pure mathematicians who might be occasionally interested in applications, or (ii) highly specialized applied mathematicians and theoretical scientists who would individually be very competent in a relatively narrow area of mathematics in his particular field of interest. These people do contribute to applied mathematics; but their presence alone is not sufficient for the healthy development of applied mathematics. The propagation of a comprehensive educational discipline of applied mathematics is essential to its success, since we must attract enterprising and brilliant young students in large numbers into the profession, and offer them a promising academic career. This last point is perhaps the most important point to the success of the development of applied mathematics. Applied mathematicians must be judged as such in terms of their total ability in promoting the interaction. They must also have the opportunity to educate their own future professionals.

Clearly, we do not advocate the education of applied mathematicians by first training them as pure mathematicians. Such a training has the danger of introducing a frame of mind which is disadvantageous to their creative activities. Those already educated in pure mathematics may wish to practice applied mathematics; but they should keep in mind the warnings of Professor Schwartz mentioned earlier in this paper.

8. *Some Practical Aspects of the Educational Program*

There are at least three important practical reasons for providing an undergraduate educational program in applied mathematics at a university.

(A) If only a graduate program is provided, one has to concentrate on education toward research ability in a certain specialized area; and there will not be time for education in breadth. This results in a splintering of activities in applied mathematics, making the total effect less effective. The usual sympton of difficulty is the small number of excellent students entering the graduate school to study applied mathematics.

(B) The maintenance of an undergraduate program would give the faculty in applied mathematics a common focus of interest and activity. The subject matter in an undergraduate program will evolve, but comparatively slowly (Newton mechanics still has its role, quantum mechanics will not be changed overnight). This permits a relatively stable joint effort on the part of the faculty. Joint effort creates stability.

In a graduate program, there is necessarily a diversification of interest among the faculty members. Furthermore, their interests often have to change quite rapidly in accordance with the current trends of scientific progress and the vitality of the subject matter. There cannot be real communication except among the specialists themselves. The complete lack of communication among colleagues is not healthy for any group.

(C) The courses in mathematical methods, and even the survey course, have tremendous educational value to many other students in pure mathematics, science, and engineering. Indeed, by a systematic offering of courses in mathematical methods, one may eliminate the need to provide special "service courses," which have often been the source of great discontent at a university.

A faculty member is usually required to fulfill a certain minimum teaching obligation. With the undergraduate program, the teaching obligations of an applied mathematician would be essentially to offer courses for the education of future applied mathematicians. The usefulness of these courses for the other purposes mentioned above is a natural beneficial side effect. If the applied mathematicians must teach mostly "service courses," they cannot help the feeling of being "second-class citizens" in the academic community.

IV. RECOMMENDATIONS

9. *Support of Academic Activities in Applied Mathematics*

To prevent mathematics from drifting further and further away from science and technology, there is a need to develop academic applied mathematics: There is a need to have highly qualified scientists specifically devoted to such activities. I wish to make the following recommendations toward accomplishing this goal.

Recommendation 1. The university administration and the scientific community, including the pure mathematicians, should be made keenly aware of the above-mentioned need.

Recommendation 2. Programs in academic applied mathematics should be offered in the universities, at both the undergraduate and graduate levels, aiming at the highest intellectual caliber.[7]

At first, because of the shortage of applied mathematicians, these programs could be offered by the cooperative efforts of scientists and engineers together with pure mathematicians who have an interest in applications. But it is clear that the service teaching mentality would be detrimental to the morale. Persons of highest scientific caliber, whose primary efforts are elsewhere and who must work in a highly competitive atmosphere, rarely can be depended upon to offer their time and energy to "other people's business" for an extended period of time. We therefore need professional applied mathematicians to educate future professional applied mathematicians, whose ability and contributions are to be judged in terms of the creative activities described above and in terms of promoting the interaction between mathematics and the sciences.

We therefore arrive at the following most important recommendation.

Recommendation 3. The universities should create faculty positions specifically in Applied Mathematics and appoint to these positions only people of the highest caliber.

Their research activities should be well respected in the scientific community. Judgement of their work should be based on their contributions to the effective interaction between mathematics and science.

Since the National Science Foundation is awarding research grants to support all branches of pure science, the academic aspect of applied mathematics (AAM) must not be an exception. Otherwise, due to the pressure of the needs of mission-oriented support,[8] the academic nature of such work, with all its long-range benefits, would be jeopardized. We therefore offer

Recommendation 4. The National Science Foundation should continue to expand its explicit support to academic activities in applied mathematics, as such (not under the guise of research in some subject matter). Those activities, whose usefulness is easily identifiable, may continue to be supported by mission-oriented agencies. A plurality of pattern of support should be maintained.

---

[7] A secondary beneficial effect of such programs is the elimination of the need to offer service courses of low caliber to students in engineering and science.

[8] It must be said that, at present, the mission-oriented agencies have been very generous in their support of basic research whose usefulness may not be directly apparent. But this is a policy susceptible to pressure from the U.S. Congress.

V. Appendix

Toward New Horizons with Great Traditions

While the newly developed and developing subjects in applied mathematics would naturally attract greater attention, the role of applied mathematics developed in the classical spirit should also be fully appreciated in future research, and especially as a vehicle for education. The usual danger of emphasizing all kinds of traditional effects whether in science or in other fields is to fall into a pit of sterility. It must be avoided by accepting new challenges with a flexible attitude. With this in mind, one can indeed make the best of a wealth of knowledge made available to us through efforts of the great minds of past generations.

(a) *Education*

While excitement is often created by new developments in any scientific subject, traditional subject matter, sifted through many generations of critical thinking and brought up to its modern form, can often serve as a core of knowledge on which future developments can be based. It is indeed these crystalized thoughts, ideas, and reasoning, which must be taught in the undergraduate years as a source of stimulation and as a basis for future research efforts. Indeed, a common background of undergraduate education would serve as a basis of establishing rapport among applied mathematicians who might later specialize in different lines of research.

Needless to say, the pattern of education and detailed choice of subject matter must evolve in the course of time as our perspective over the total picture of applied mathematics changes in view of new developments.

(b) *Research*

The nature of research efforts in the "traditional" line actually evolves rapidly and substantially in the course of time. These research efforts also bring out clearly unsolved classical problems whose implications are obviously important but yet not fully understood. As an example in the latter category, the study of "turbulence" in the context of hydrodynamics is still an unfinished task. It is now clear that the existing studies and the still unresolved problems both relate to the broader problem of "nonlinear random processes." In the case of incompressible viscous fluids, the physical processes are very well understood from a general point of view, and many detailed comparisons of theoretical and observed results have become possible; but the fundamental mathematical theory remains unformulated.

Stability and instability, wave motion, linear versus nonlinear processes, reversible versus irreversible processes, entropy, and one can name some more, are certainly classical topics whose implications are general. The concept of entropy has already found its new application in information theory. It is quite possible that the concept of wave motion and the description of the process of shock wave formation might turn out to be useful in the mathematical description of social processes. On the other hand, one can easily trace the change of direction and find new vitality in an old subject like fluid mechanics. Around the turn of the century, aerodynamics, as related to aeronautical engineering, was one of the great challenges of the times. It remained strong as a subject for research of applied mathematicians for a long time. Nowadays, aerodynamics is so well understood that it is no longer the major concern of the applied mathematician. In its place, other aspects of fluid mechanics, those related to geophysical and astrophysical problems, are attracting greater attention. Certain concepts developed primarily in aerodynamics, such as the boundary layer, however, remain important even in these newer studies.

Continuum mechanics is of course but one aspect of classical applied mathematics, although it does have a special position, at least for the following three reasons: (i) the physical concepts are simple, (ii) the mathematical problems involved are interesting and of wide applicability, and (iii) a variety of natural phenomena fall within its scope. Particle dynamics also occupied and continues to occupy an important position in applied mathematics. There is currently a great deal of activity based on the theory of dynamical systems (integrals of motion, adiabatic invariants, etc.) in the construction of models of galaxies of stars. Even the whole subject of mechanics and statistical mechanics is but one part of "classical" mathematical physics. There will no doubt remain many problems to be explored in the whole broad area of such activities for years to come. The solid foundation of traditional efforts will always help us to reach new horizons if we adopt the progressive point of view.

References

1. S. ULAM, The applicability of mathematics, *in* "Mathematical Sciences, a Collection of Essays," MIT Press, Cambridge, 1969.
2. H. WEYL, Relativity theory as a stimulus in mathematical research, *Proc. Amer. Phil. Soc.* **93** (1949), 535.

3. ALBERT EINSTEIN, "Out of my Later Years," Philosophical Library, New York, 1950.
4. T. VON KÁRMÁN AND C. C. LIN, On the existence of an exact solution of the equations of Navier–Stokes, *Comm. Pure Appl. Math.* **14** (1961), 645–655.
5. H. P. GREENSPAN, Applied mathematics as a science, *Amer. Math. Monthly* **68** (1961), 872–880.
6. H. COHEN, Mathematics and the biological sciences, *in* "Mathematical Sciences, A Collection of Essays," MIT Press, Cambridge, 1969.
7. L. KLEIN, The role of mathematics in economics, *in* "Mathematical Sciences, A Collection of Essays," MIT Press, Cambridge, 1969.
8. Report of the Committee on Support of Research *in* the Mathematical Sciences. National Academy of Sciences, Washington, D.C. 1969.
9. E. T. BELL, "Men of Mathematics," pp. 197–198. Simon and Schuster, New York, 1937.
10. G. BIRKHOFF, "Source Book in Classical Analysis," Harvard Univ. Press, Cambridge, 1974.
11. N. LEVINSON, Minmax, leapunov and "bang–bang," *J. Differential Equations* **2** (1966), 218.
12. R. WIELEBINSKI, The Effelsberg 100-m radio telescope, *Naturwissenschaften* **58** (1971), 109–116.
13. J. SCHWARTZ, The pernicious influence of mathematics on science, *in* "Proceedings of 1960 International Congress on Logic, Methodology, and Philosophy of Science," pp. 356–360, (Nagel, Suppes, and Tarske, Eds.), Stanford Univ. Press, Stanford, 1962.

On the Shape of a Curve

Raoul Bott

Department of Mathematics, Harvard University, Cambridge, Massachusetts 02138

DEDICATED TO STAN ULAM

The dubious honor of being your anti-keynote speaker[1] has no doubt fallen to me because I am an engineer and applied mathematician gone astray—a victim of the seductive wiles of pure mathematics.

Of course the days of penitence in matters of chastity are long gone, so do not expect me to be contrite. Rather my theme will be that the temptation was simply too great. However let me say this concerning the various branches of our subject. Subjectively I find no difference among any of them. When I worked on networks with Duffin I found them as fascinating as I found Lie groups later on in my work with Samelson, fixed point theory in my work with Atiyah, and as I find foliations in my work with Haefliger at the moment.

But to get on with it, let me cite an instance of the "Pure Temptation" in mathematics which was eminently successful:

We start from the "practical" question: How many solutions does a real polynomial equation of degree n have?

A little experimentation then shows that such an equation, say,

$$a_n x^n + \cdots + a_0 = 0, \qquad a_n \neq 0,$$

might well have no solutions, that it has at most n of them, and that there seems to be something going on behind the scenes which is hard to get at. On the other hand if we abandon the practical world of the real numbers and succumb to the "pure" by introducing the *complex numbers* $\mathbb{C}$, we find an infinitely more satisfying state of affairs which sheds light not only on our original question but eventually illuminates nearly every aspect of analysis. Over $\mathbb{C}$ the situation is of course the following.

---

[1] First Los Alamos Workshop on Mathematics in the Natural Sciences, June 12–18, 1974.

THEOREM 1. *The general polynomial equation of degree n has precisely n solutions.*

Here and in the following, "general" or "generic" is used to mean all polynomials except possibly those whose *coefficients* satisfy a finite number of polynomial equations.

For example, the general polynomials of degree two are all

$$p(x) \equiv ax^2 + bx + c,$$

with $a \neq 0$ and $b^2 - 4ac \neq 0$.

In particular the general polynomials are open and dense in the set of all polynomials and their complement is of measure zero. Thus the "special" polynomials in their complement are always limits of the generic ones so that even their properties can often be best understood as the limiting property of the generic ones. Thus Theorem 1 illuminates all polynomial equations.

Note finally that the cardinality is really the only invariant of a finite set, so that Theorem 1 does describe the shape of the generic solution set completely.

At this stage the pure temptation is of course to proceed further and ask the corresponding question for polynomials in more variables, and the moment one succumbs to it, one is introduced into the heart not only of topology and algebraic geometry but also of much of modern mathematics.

Let me pursue this path in the case of two variables, and to keep the discussion as concrete as possible, let us try first to determine the "shape" of the set S of points $(x, y) \in \mathbb{C}^2$ satisfying the equation

$$x^3 + y^3 = 1. \qquad (*)$$

A first result is the

PROPOSITION. *The set $S \subset \mathbb{C}^2$ is a smooth surface of real dimension two in $\mathbb{C}^2 = \mathbb{R}^4$.*

To prove this, proceed just as over the reals. These *real* points, $S_\mathbb{R}$, of S, are clearly the level sets where the function $f(x, y) = x^3 + y^3$ equals unity. It follows that

$$xf_x + yf_y = 3 \quad \text{on} \quad S_\mathbb{R}.$$

Hence the gradient of $f \neq 0$ along $S_\mathbb{R}$ and therefore, by the implicit function theorem, $S_\mathbb{R}$ is a smooth *curve* in $\mathbb{R}^2$. Over $\mathbb{C}$ this argument is equally valid, and that also explains why the algebraic geometer calls S a *curve* over $\mathbb{C}$. Geometrically, i.e., over $\mathbb{R}$, S is however given by two independent real conditions in $\mathbb{R}^4$ and hence has dimension two.

We next try to determine the "shape" of S. In particular, let us see how S fits into the classification of compact orientable surfaces which Lipman Bers already discussed earlier during our Conference. You will recall that the basic shapes of such objects are the sphere with g-handles (Fig. 1).

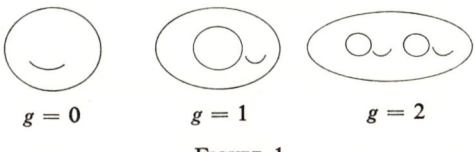

FIGURE 1

To investigate this problem, let us project S onto its first variable x. Clearly this projection, say π, maps S *onto* $\mathbb{C}$ and the inverse image of a point $x \in \mathbb{C}$ consists of all $y = (1 - x^3)^{1/3}$. Thus two types of situations arise:

(a) If $x^3 = 1$, then $\pi^{-1}(x)$ consists of one point.

(b) If $x^3 \neq 1$ then $\pi^{-1}(x)$ consists precisely of the three cube roots of $1 - x^3$.

In Fig. 2 I have tried to illustrate the situation.

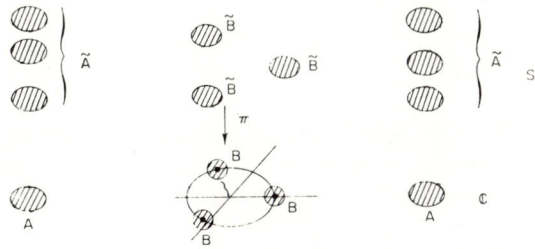

FIGURE 2

Over a region such as A, S has three disjoint copies of A, while over a region such as B, S has essentially just *one* copy of B. How this all fits together is admittedly a little mind-boggling but locally there is no difficulty. Near the branch points (the cube roots of 1) one argues as follows. Let ρ be such a cube root so that $\rho^3 = 1$. Then near ρ write $x = \rho + \epsilon$, so that

$$1 - x^3 = 1 - (\rho + \epsilon)^3 = -3\rho^2 \epsilon \cdot \{1 + (1/\rho)\epsilon + (1/3\rho^2)\epsilon^2\}.$$

Hence near ρ, y is given by

$$y = \sqrt[3]{\epsilon} \cdot \{a_0 + a_1\epsilon + \cdots\}$$

where $a_0 = (-3\rho^2)^{1/3}\rho \neq 0$ and the power series converges for small ϵ. Thus near $\pi^{-1}\rho$, the projection π looks essentially like the map $y \to y^3$ which sends the unit disk into itself. (In other respects the point $\pi^{-1}\rho$ is however quite undistinguished on S, for we already know that S has *no* singular points!)

Before assembling S out of the little disks of Fig. 1, we should first come to grips with the fact that S is not compact. Clearly as we march to ∞ in x, say along the integers $x = 1, 2, 3, ...$, every one of the three inverse images under π in S will also march to ∞ on S.

However a remedy is again close at hand. Just as the Gauss Sphere, that is the surface of genus zero, is obtained by adding "one point at ∞" to $\mathbb{C}$, so in order to obtain a closed surface $\bar{S}$ we expect to add three points to S. This turns out to be correct, but care has to be taken to establish this fact. Essentially we have to check that our point at ∞ is not "ramified relative to π" the way the cube roots $1, \rho, \rho^2$ were. For instance consider the equation

$$y^2 + x^3 = 1 \qquad (*')$$

instead of (*). Here the solution set S' clearly looks like *two* sheets spread out over the $\mathbb{C}$-plane, ramified at $1, \rho, \rho^2$ as before, but in this case I claim that one should add only *one* point at ∞, to obtain a smooth closed surface $\bar{S}'$.

Now how is all this to be understood and not only by the inspired but also by us, the simple minded?

The technique that reduces these questions to a routine computation again starts with an abstraction, the so-called projective space $\mathbb{C}P_2$, which plays the role of the Gauss Sphere in dimension two.

If in $\mathbb{C}^3$ we remove the origin and then identify two triples (x, y, z) and (x', y', z') of $\mathbb{C}^3 - 0$, if and only if there exists a $\lambda \in \mathbb{C} - 0$ with

$$(\lambda x', \lambda y', \lambda z') = (x, y, z),$$

the resulting space is called the complex projective space of dimension two and is notated $\mathbb{C}P_2$.

Notice that the map

$$(x, y) \to (x, y, 1) \tag{†}$$

imbeds $\mathbb{C} \times \mathbb{C}$ into $\mathbb{C}P_2$ and that the complement of this $\mathbb{C} \times \mathbb{C}$ in $\mathbb{C}P_2$ is given by the class of $(x, y, 0)$, that is, by a $\mathbb{C}P_1$, which in turn is precisely $\mathbb{C} \cup \infty$, i.e., our "Gauss Sphere." Thus the spaces $\mathbb{C}P_2$ etc., which are easily seen to be compact, are indeed the *natural higher dimensional analogs of the Gauss Sphere*.

Once we are aquainted with this concept it naturally suggests that we consider the set PS of points in $\mathbb{C}P_2$ on which the *homogenous equation*

$$x^3 + y^3 = z^3$$

is satisfied. Indeed this set is again easily seen to *be compact* and *obviously agrees with our S on* $\mathbb{C} \times \mathbb{C} \subset \mathbb{C}P_2$, *because there, z can be taken to be equal to unity*.

Thus our PS is certainly a "compactification" of S, but now the question arises in what sense PS is to be understood as a surface, whether it is smooth etc. In short the question one encounters here is, to *what extent and in what sense can one extend the calculus from* $\mathbb{C}^2$ *to* $\mathbb{C}P_2$.

Actually the answer is quite simple: Let

$$U_x = \{(x, y, z) \in \mathbb{C}P_2 \text{ with } x \neq 0\},$$
$$U_y = \{(x, y, z) \in \mathbb{C}P_2 \text{ with } y \neq 0\}, \text{ and}$$
$$U_z = \{(x, y, z) \in \mathbb{C}P_2 \text{ with } z \neq 0\}.$$

Then, by our earlier remark U_x, U_y, U_z are each isomorphic to $\mathbb{C} \times \mathbb{C}$ and *between them they cover all points of* $\mathbb{C}P_2$. *Thus to do the calculus on* $\mathbb{C}P_2$ *one simply has to work three times as hard as in* $\mathbb{C} \times \mathbb{C}$; one has to study all phenomena in *each* of the three coordinate patches:

U_x, U_y, and U_z. To illustrate: Using the plausible names for our variables, e.g. (y, z in U_x, etc.), the equation

$$x^3 + y^3 = z^3$$

is satisfied precisely on the sets

$$x^3 + y^3 = 1 \quad \text{in} \quad U_z,$$
$$x^3 + 1 = z^3 \quad \text{in} \quad U_y,$$
$$1 + y^3 = z^3 \quad \text{in} \quad U_x,$$

and, since each of these defines a smooth curve (by our old method of proof in $\mathbb{C} \times \mathbb{C}$), PS is indeed a smooth object. Furthermore the *points at ∞ of $x^3 + y^3 = 1$ now appear plainly in both U_y and U_x* as the points where $z = 0$, i.e., where x, respectively y, range over $-1, -\rho, -\rho^2$.

We test this procedure next for the equation

$$y^2 + x^3 = 1.$$

Here the homogeneous version is

$$y^2 z + x^3 = z^3$$

and its manifestations on U_x, U_y, U_z are

$$y^2 + x^3 = 1 \quad \text{on} \quad U_z,$$
$$z + x^3 = z^3 \quad \text{on} \quad U_y,$$
$$y^2 z + 1 = z^3 \quad \text{on} \quad U_x.$$

Hence the point at ∞ of $y^2 + x^3 = 1$ appears in U_y *alone* as the point $z = 0, x = 0$.

After this long detour, we may return to our curve S. What we have learned is that $PS \in \mathbb{C}P_2$ is a smooth compact surface of $\mathbb{C}P_2$ and that PS is obtained from S by adding three *points* at ∞. Thus:

$$PS = S \cup p_1^\infty \cup p_2^\infty \cup p_3^\infty.$$

In short, PS is $\bar{S}$, our closed version of S, and we can finally turn to our main global problem, which now can be formulated in the form:

What is the genus (number of handles) of $\bar{S}$?

Alas, we aren't quite there yet. First we need to check that $\bar{S}$ is orientable, that is, that $\bar{S}$ can be covered by little *real* coordinate disks so that on the overlap the Jacobian is positive. At this stage another magnificent quality of the complex numbers (and complex analytic functions) makes its appearance. The point is that, if

$$\omega = f(z)$$

is an analytic function of the complex variable z, then the Jacobian = $\det(\partial f_i/\partial x_j)$ of the coordinate transformation of real and imaginary parts of ω to the real and imaginary parts of f, is always ≥ 0, and >0 when $f'(z) \neq 0$.

I leave the proof of this fact to you, and, armed with it, we can now easily prove the orientability, by first finding a *good* complex coordinate near each point of S. Indeed x will do near all points which are not over branch points for π, and at these y will do. Finally, at ∞ in PS, z will do, but here one could easily argue that orientability is not affected by removing a few points, so that S orientable $\Rightarrow \bar{S}$ orientable. So then, $\bar{S}$ is orientable and our question concerning its genus is legitimate. Note further that along the way of proving this, we have actually noted that: (1) not only the calculus but really the whole theory of complex variables makes sense on PS, and (2) that orientability is automatic for such complex analytic beings.

But let me now really turn to the problem of identifying the shape, i.e., genus of $\bar{S}$, and here I will again take the most "uninspired" way out. I will not try to make you imagine *how* PS really looks, although it would be a good exercise to do so; rather I will apply to this problem the oldest *invariant* of topology, and as you will see this invariant will reduce the question to the simplest arithmetic. In a sense then this is actually the *most inspired* method, only it is not our inspiration that is needed, only that of Euler.

The invariant of a surface I want to use goes back to Euler, is called its "Euler number," and is computed as follows:

Consider a "triangulation" of a smooth surface: into α_2 curvilinear triangles, and let α_1 denote the number of edges and α_0 the number of vertexes.

For instance, for the Gauss Sphere (Fig. 3) we could take the triangulation with $\alpha_2 = 6$, $\alpha_1 = 9$, and $\alpha_0 = 5$. The number

$$e = \alpha_0 - \alpha_1 + \alpha_2$$

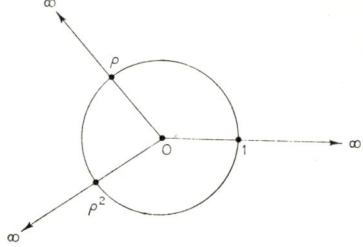

FIGURE 3

is then called the Euler number of the triangulation, and, as was already essentially known to Euler, the following theorem is valid:

THEOREM 2. *The Euler number of a triangulation of a surface X is independent of the triangulation. Thus $e(X)$ is well defined, and this invariant of X is related to the genus of X by the formula*

$$e(X) = 2 - 2g(x).$$

For instance for the Gauss Sphere we find by Fig. 3 that

$$e(\mathbb{C}P_1) = 5 - 9 + 6 = 2.$$

Therefore to determine the shape of $\bar{S}$ it is sufficient to compute its Euler number, and this is now easy to do in view of our geometric understanding of $\bar{S}$ spread over the Gauss Sphere (Fig. 2).

We triangulate $\mathbb{C}P_1 = \mathbb{C} \cup \infty$, say as indicated in Fig. 3, but any refinement of that triangulation will also do. Now since our branch points are vertexes in the triangulation, it is easy to see that the *inverse image of the triangulation under π is a triangulation of $\bar{S}$* and simply counting we see that the number $\tilde{\alpha}_i$ (of that triangulation) is related to the α_i of the initial triangulation of $\mathbb{C}P_1$ by the relation:

$$\tilde{\alpha}_2 = 3\alpha_2, \qquad \tilde{\alpha}_1 = 3\alpha_1, \qquad \tilde{\alpha}_0 = 3\alpha_0 - 3 \times 2.$$

The 3×2 of course corrects for the fact that the three branch points were covered only once rather than three times. These equations lead to

$$e(\bar{S}) = 3e(\mathbb{C}P_1) - 3 \times 2,$$

so that from $e(\mathbb{CP}_1) = 2$ we get

$$e(\bar{S}) = 0.$$

Thus we have shown that: $\bar{S}$ is of the type of a torus, and our original S looks like a torus with three points deleted.

Let us test this procedure, which goes back to Hurwitz, I believe, on a few more examples. For instance for the equation

$$y^2 + x^3 = 1, \qquad (*')$$

a triangulation count gives:

$$\tilde{\alpha}_2 = 2\alpha_2, \qquad \tilde{\alpha}_1 = 2\alpha_1, \qquad \tilde{\alpha}_0 = 2\alpha_0 - 4 \times 1$$

because there are now four branch points. Hence

$$e(\bar{S}') = 2e(\mathbb{CP}_1) - 4 \times 1$$
$$= 0.$$

Thus $\bar{S}'$ is again a torus while S is now a torus with *one point* removed. Actually much more can be said; namely, *the torus is the shape of the generic polynomial* of degree three in the *following sense*.

THEOREM 3. *Consider the polynomials in two variables $p(x, y)$ of degree three, whose associated homogeneous polynomial in three variables $\tilde{p}(x, y, z)$ vanishes on a nonsingular surface in $\bar{S}(p) \subset \mathbb{CP}_2$. These polynomials then form a generic set, and $\bar{S}(p)$ is of the shape of a torus.*

To prove this, one just has to show that (1) the nonsingularity of $\bar{S}(p)$ can be written down in terms of polynomials in the coefficients of p, and (2) that nearly all polynomials have *topologically* isomorphic shapes. Then it is easy to see that these polynomials form a *connected set*, and of course on a connected set the Euler number cannot change so that it is sufficient to compute it for *one* example.

In the same spirit we can now believe the following solution of our initial question.

THEOREM 4. *The generic polynomial in two variables of degree n, has for its compactified solution set in $\mathbb{CP}_2$, a surface of Euler number*

$$e(\bar{S}) = ne(\mathbb{CP}_1) - n(n-1)$$
$$= -n(n-3).$$

Indeed, all we really need to do is to apply our procedure to the equation

$$x^n + y^n = 1$$

which is only ramified over the nth roots of one, and therefore clearly leads to the equation for $e(\bar{S})$ given above. Q.E.D.

So much for my main example. I wanted to cover it in great detail, so that you would not only be exposed to some of the concepts and techniques of topology and algebraic geometry, but also learn a useful fact which isn't as generally known as it should be.

Let me point out, for instance, that the moment one tackles higher order partial differential equations, our theorem becomes very pertinent, and this is especially so in the study of hyperbolic equations à la Petrovsky [1].

But I will not try to propagandize for this fledgling result in topology here. Rather, I would like to show you an *application* of the full grown topology to number theory which is maybe the most spectacular achievement of recent mathematics.

To explain this development let me sketch in the extensions of those concepts we have already met which will be needed later on.

First of all the notion of a smooth surface has its natural extension in the notion of smooth n-manifold. For our purposes, let us make do with this definition of such an object.

A subset M_n of $\mathbb{R}^N$ is called a smooth n manifold, if in the vicinity of each point $p \in M_n$, M_n is the level set of $N - n$, smooth functions $\{f_\alpha\}$ whose gradients are independent at p in the sense that the matrix $\| \partial f_\alpha / \partial x_j \| j = 1,..., N$ has rank, $N - n$ at p.

On such M_n's one can locally describe points by n parameters, i.e., they are locally isomorphic to $\mathbb{R}^n$, but usually not globally so. Hence the calculus can be extended from $\mathbb{R}^n$ to M^n, but with care and usually more work, as global conditions have to be checked in many coordinate patches. These manifolds are also the natural habitat, in the view of the pure mathematician, of objects like tensor-fields, densities etc.

Now then, once we understand these n-manifolds, what about their "shape." Unfortunately already for $n = 3$ no complete list of shapes, such as in dim 2, can be constructed. We therefore have to get by with partial results, and it becomes even more vital than in dimension two to find a fairly good set of invariants of these shapes. A first candidate is the Euler number, and it is clear how to extend its definition. Namely, we cut M_n into α_n (*curvilinear triangles of* dim n) and let α_i, $i \leqslant n$ be the number of i dimensional curvilinear triangles (called i-simplexes in the jargon of topology) which occur in such a triangulation. Then we set

$$e(M) = \alpha_0 - \alpha_1 + \alpha_2 \pm \alpha_n$$

and hope that $e(M)$ turns our to be independent of the triangulation. Of course here it becomes even highly nontrivial whether such a triangulation exists, but it does, and indeed the Euler number is an invariant of M_n.

Next we might hope for more invariants as the dimension increases; and this is also true, though none of them are ever quite as tractable as Euler's. Let me just describe one of these invariants for it points the way for all of "cohomology theory."

I already remarked that the calculus carries over to M_n. In particular one can speak of differentials or line integrals, if you prefer, on M_n. Thus a differential ω on M_n induces a differential

$$\sum a_i(x) \, dx_i$$

on each coordinate patch.

Now just as in $\mathbb{R}^n$, we can also speak of *exact* differentials, i.e., one whose integral $\int_c \omega$ along the curve c depends only on the endpoints of c when c is perturbed a little bit. They satisfy the condition that for every one of their local manifestations

$$(\partial a_i / \partial x_j) - (\partial a_j / \partial x_i) = 0.$$

Again, just as in $\mathbb{R}^n$, any C^∞ function f on M_n has a differential df which is clearly *exact*. On the other hand, unlike the case in $\mathbb{R}^n$, it is not any more true that every exact form is df for some function. This fact leads to the vector space of *interesting line integrals*,

$$H^1(M^n) = \text{exact differentials/gradients of functions},$$

and a very instructive exercise for all of you not steeped in these matters would be to convince yourselves that on a surface

$$\dim H^1(M) = 2 \text{ genus } (M).$$

Thus for surfaces dim $H^1(M)$ is just as good an invariant as the Euler number was, except of course, *a priori*, harder to compute. Now returning to M_n, I think you can see how the land lies—one might hope to define for every $k \leqslant n$ vector spaces

$$H^k(M) = \text{exact } k\text{-dim volume integrals/trivially exact } k\text{-dim volume integrals}$$

and hope for finite dimensionality when M is compact. Well all this can be done, as was suggested by Eli Cartan and leads one to the so-called *de Rham Cohomology of* M_n. In terms of these slightly more elaborate invariants of shape, our old friend the Euler number is expressed as follows:

$$e(M_n) = \sum (-1)^i \dim H^i(M_n),$$

and this expression proves its invariance as no choices are involved in the right-hand side.

Let me next show you a marvelous phenomenon discovered by Lefschetz in the twenties which will play a fundamental role in the application to algebra and number theory I want to explain.

First we note that the change-of-variable formula in the calculus leads to an *induced homomorphism*

$$H^k(N) \xleftarrow{H^k(f)} H^k(M)$$

for every smooth function

$$N \xrightarrow{f} M.$$

(Note that the arrows are reversed, the characteristic contravariance of cohomology.) In particular then if $f: M \to M$ is a map of M into itself, then

$$H^k(f): H^k(M) \to H^k(M)$$

is a linear endomorphism and its trace is well defined whenever $H^k(M)$ is finite-dimensional, and, therefore in particular, when M is compact. Now what Lefschetz observed is the following remarkable fact:

$$\text{number of fixed points of } f = \sum (-1)^k \text{ trace } H^k(f)$$

Here of course $p \in M$ is fixed under f if $f(p) = p$, and the formula above is literally true only if the fixed points are "orientable" and "nondegenerate", in the sense that, at each fixed point p, we have

$$\det(1 - (\partial f_i/\partial x_j)(p)) > 0.$$

$\{x_\alpha\}$ being focal coordinates centered at p and f being locally expressed by the f_i. (Thus $x_i(f(p)) = f_i(x)$.)

Now this formula is a joy in itself but I really don't have time to explore it with you. Rather let me stop here with "topology" and return, in a sense, to our original problem of looking at the shape of the solution set to a polynomial, or a number of polynomial equations, *but this time more from the point of view of number theory.*

The number theorist is in the final analysis interested in the integer solutions of a set of equations, but over the years he has developed great insight into how to localize such problems to each prime and then put the local information together again. For him the field $\mathbb{Z}_p$, of the integers mod p, is therefore of profound interest, and he has by now a large body of algebraic information about it. Furthermore, much of our previous discussion makes sense over any field k; e.g., we can define the projective space kP_n of dim n for every field k, just as we did before by taking as points of kP_n the rays in k^{n+1}. Also, given a finite number of homogenous polynomials $p_1,\ldots,p_r$ in $(n+1)$ variables we can study the solution points over k:

$$S_k \subset kP_n \quad \text{where all} \quad p_i = 0;$$

and in algebraic geometry one tries to study this situation by analogy with what was found over the field $\mathbb{C}$. Of course the matter is delicate. Simple phenomena over $\mathbb{C}$ turn out to split into many cases in characteristic p. In fact the situation is reminiscent of the step from classical to quantum mechanics, so that what was found in the "real world of $\mathbb{C}$," is only a hint of what happens in characteristic p.

With this preamble we are about ready to explain the "Weil Conjectures" which have pointed the way for so much of this development in the last 30 years, and which were solved only last year by a young Belgian mathematician, Deligne, using all the machinery built over the past 20 years by Grothendieck and his school of algebraic geometry.

The problem deals with a set of homogenous polynomial equations

$$p_i(x_0,\ldots,x_n) = 0 \quad i = 1,\ldots,k \tag{**}$$

with coefficients in $\mathbb{Z}_p$, and the problem is again to describe the solutions of (**) over $\mathbb{Z}_p$ and its various extension fields.

Now $\mathbb{Z}_p$ has up to isomorphisms precisely one extension field $\mathbb{F}_k$ of order p^k for each k. Let us therefore set

$$S_k = \text{number of solutions of (**) in } \mathbb{F}_k P_n,$$

and combine these integers into the formal counting series,

$$Z(t) = \sum_{k=0}^{\infty} \frac{S_k t^k}{k}. \tag{R}$$

At first sight (and even at second!) there is very little that can be said about (R). After all, each S_k is computed by a difficult combinatorial procedure.

On the other hand if one knows a little algebra and a little topology and has a great imagination, as A. Weil had, then it is not hard to guess at some quite explicit properties of $Z(t)$.

The algebra we need to know is this:

(1) The role of $\mathbb{C}$ in the classical theory is always played by the *algebraic closure* of the field in which the equations have their coefficients.

(2) The algebraic closure, $\hat{\mathbb{Z}}_p$, has an automorphism $F: \hat{\mathbb{Z}}_p \to \hat{\mathbb{Z}}_p$, given by sending $x \to x^p$, which has the property, that the fixed set of F^k acting on $\hat{\mathbb{Z}}_p$ is isomorphic to the field $I\mathbb{F}_k$.

By the way, here is a nice example of the richness of characteristic p: The rather dull complex conjugation, taking us from $\mathbb{C}$ to $\mathbb{C}$ and having $\mathbb{R}$ as its fixed set is, in this framework, replaced by the much more flamboyant "Frobenius automorphism" F.

With this information at hand we can trivially reinterpret the number S_k in the following manner: Let $\hat{S}$ denote the set of solutions of (**) in $\hat{\mathbb{Z}}_p P_n$. Since the coefficients of the defining set of equations are in $\mathbb{Z}_p$ and hence fixed under F, this transformation acts on $\hat{S}$, and we clearly have:

$$S_k = \text{number of fixed points of } F^{(k)} \text{ acting on } \hat{S}.$$

Well, the cat is out of the bag now. In this formulation the topologist of course is reminded of the Lefschetz fixed point formula.

Indeed if F *were* an endomorphism of a *compact manifold* X, and if we *could* assume that all fixed points of all iterates are transverse and orientable, then by the Lefschetz formula we would have:

$$\text{number of fixed points } F^k = \sum (-1)^i \text{ trace } \{H^i(F)\}^k.$$

Hence if $\{\alpha_e\}$ and $\{\alpha_0\}$ run over the eigenvalues of $H^i(F)$: $H^i(X) \to H^i(X)$ with i even and i odd, respectively, then the above gives

$$\text{number of fixed points } F^k = \sum_e \alpha_e{}^k - \sum_0 \alpha_0{}^k,$$

and hence for the formal series in question:

$$Z(F, t) \equiv \sum \frac{(\text{number of fixed points } F^k) t^k}{k} = \log \frac{\prod(1 - t\alpha_e)}{\prod(1 - t\alpha_0)}.$$

It follows that in this topological situation we would have the relation:

$$e^{Z(F,t)} = \frac{\prod(1 - t\alpha_e)}{\prod(1 - t\alpha_0)}; \qquad (***)$$

i.e., the exponential of Z would be a rational function of t and the number of factors on top and on bottom correspond to the dimension of the even and odd cohomology of X, respectively.

Well, André Weil saw all this many years ago (see [4]) and conjectured that there should be finite dimensional vector spaces $H^i(\hat{S})$ associated to the set $\hat{S}$ with the Lefschetz property etc., and therefore that finally the counting series should have the desired property (***). In fact he went further. Using an even deeper analogy with the algebraic situation over $\mathbb{C}$, where X is a complex submanifold of $\mathbb{C}P_n$ and therefore has very special properties, he conjectured that the Eigenvalues α of F on $H^i(\hat{S})$ would have the bound

$$|\alpha| = p^{i/2}.$$

Well as I said, this assertion was proved only last year by P. Deligne in a tour de force of combined techniques. The rationality of $e^{Z(t)}$ had been proved earlier, and I refer you to [2] for a very short history of the subject.

Finally let me point out, again by earlier work of Deligne, that this solution of the Weil conjecture also implies the truth of the celebrated

"Ramanujan Conjecture" in number theory. This conjecture asserts that in the formal power series

$$\Delta = x \prod_{n=1}^{\infty} (1 - x^n)^{24}$$

the coefficient $\tau(p)$ of x^p (p a prime) satisfies the inequality

$$|\tau(p)| \leqslant 2p^{11/2}.$$

Of course it is a nontrivial and very beautiful story to see how this completely *elementary* conjecture is related to our main discussion, but I have neither the time nor really the expertise to make that connection for you. Rather I herewith rest my case for the fascination of pure mathematics, and, although all of this is far removed from turbulence, field theory and the many questions we have been hearing about, I am deeply convinced that uncanny relationships, such as the path from the Ramanujan Conjecture to the Lefschetz Formula, are still to be found in our subject and some of these may yet prove very useful in the more applied world. In fact, I would not be too surprised if discrete mod p mathematics and the p-adic numbers would eventually be of use in the building of models for very small phenomena.

References[1]

1. I. G. PETROVSKY, On the diffusion of waves and the lacunas for hyperbolic equations, *Math. USSR-Sb.* **17** (59)(1945), 289–370.
2. J.-P. SERRE, Valeurs propres des endomorphismes de Frobenius, Séminaire Bourbaki, No. 446, Feb. 1974.
3. C. L. SIEGEL, "Topics in Complex Function Theory," Vol. I, Wiley, New York, 1969.
4. A. WEIL, Number of solutions of equations in finite fields, *Bull. Amer. Math. Soc.* **55** (1949), 497–508.

---

[1] I cite here only four papers. C. L. Siegel's book on function theory is for those interested in following up the early part of my lecture. The A. Weil paper contains the original conjectures, and Serre's article gives an account of Deligne's arguments as well as an intelligent bibliography.

Automorphic Forms for Schottky Groups

LIPMAN BERS*

Department of Mathematics, Columbia University, New York, New York 10027

TO STANISLAW M. ULAM ON HIS SIXTY-FIFTH BIRTHDAY

Schottky groups stand out, among other Kleinian groups, by a particularly simple algebraic and geometric structure. They were invented (by Schottky) before Klein and Poincaré initiated the general theory, but some of their basic properties have been discovered only recently (cf. Chuckrow [15], Maskit [22], Hejhal [17], Marden [21]). A discussion of Schottky groups is a good introduction to Kleinian groups, and questions about Kleinian groups are sometimes best approached via the special case of Schottky groups. We do so here, for the following problem: how to represent, by Kleinian groups and Poincaré series, deformations of Riemann surfaces leading to the appearance of nodes. The full solution of the problem, outlined in [14], requires not Schottky groups but more complicated Kleinian groups. A detailed presentation of that solution will appear elsewhere.

This paper is a (revised and expanded) version of a lecture delivered at the Los Alamos Workshop on Mathematics in June, 1974, under the title *A Glimpse into Complex Analysis*. The first three sections are largely expository. A special case of the main result has been announced, without proof, in [11].

1. SCHOTTKY GROUPS

A *Kleinian group* G is a discrete group of Möbius transformations

$$z \mapsto \frac{az+b}{cz+d}, \quad a, b, c, d \in \mathbb{C}, \quad ad - bc = 1$$

which acts properly discontinuously on some open subset of the

---

* Work partially supported by the National Science Foundation.

Riemann sphere $\hat{\mathbb{C}} = \mathbb{C} \cup \{\infty\}$. The largest open set Ω on which G so acts is dense in $\hat{\mathbb{C}}$ and is called the *region of discontinuity* of G. The *limit set* of G is the complement $\hat{\mathbb{C}} \backslash \Omega$. A *fundamental region* for G is a measurable set $\omega \subset \Omega$ such that the boundary of ω in Ω has measure 0, no two distinct interior points of ω are G equivalent, and every point of Ω is G equivalent to some point of the closure of ω.

If G is a Kleinian group with region of discontinuity Ω and fundamental region ω, and γ is a Möbius transformation, then $\gamma G \gamma^{-1}$ is a Kleinian group with region of discontinuity $\gamma(\Omega)$ and fundamental region $\gamma(\omega)$.

The same is true if $z \mapsto \gamma(z)$ is a *quasiconformal selfmapping* of the Riemann sphere (cf. Ahlfors [1] and Lehto and Virtanen [19] for the theory of such mappings), provided that the Beltrami coefficient μ of γ,

$$\mu = (\partial \gamma / \partial \bar{z})/(\partial \gamma / \partial z), \tag{1.1}$$

satisfies the condition

$$\mu(g(z)) \overline{g'(z)}/g'(z) = \mu(z) \quad \text{for} \quad g \in G; \tag{1.2}$$

if also

$$\mu \mid \hat{\mathbb{C}} \backslash \Omega = 0, \tag{1.3}$$

then μ is called a Beltrami coefficient for G and

$$G \ni g \to \gamma \circ g \circ \gamma^{-1} \in \gamma G \gamma^{-1}$$

is called a *quasiconformal deformation* of G determined by μ.

Let $C_1, C_1', C_2, C_2', \ldots, C_p, C_p'$ be $2p$ disjoint Jordan curves on $\hat{\mathbb{C}}$, each of two-dimensional measure 0 which form the oriented boundary of a domain of connectivity $2p$ on the Riemann sphere Let $g_j, j = 1, \ldots, p$ be Möbius transformations such that g_j maps C_j onto C_j' reversing orientation, that is, maps the domain exterior to C_j onto the domain interior to C_j'. (Such g always exist if all C and C' are circles; Schottky himself considered only this case.) The group G generated by $g_1, \ldots, g_p$ is called a *Schottky group* (of genus p). A Schottky group G is a Kleinian group and a free group on p generators $g_1, \ldots, g_p$. Each nontrivial element γ of G is loxodromic. (This means that γ is conjugate, in the group of all Möbius transformations, to a mapping of the form $z \mapsto \lambda z$, $|\lambda| > 1$; the number λ is called the multiplier of γ.) The domain ω bounded by the curves $C_1, \ldots, C_p'$ is a fundamental region for G called a standard fundamental region belonging to the generators $g_1, \ldots, g_p$.

(Of course, ω is *not* determined by the generators.) The region of discontinuity Ω of G is connected. The limit set Λ of G is a perfect, totally disconnected set of measure 0 if $p > 1$, consists of two points if $p = 1$, and is empty if $p = 0$ and G trivial. From now on we assume that p is fixed and $p > 1$, unless the converse is stated specifically.

The results stated above are classical (see Appell, Goursat, and Fatou [6], Ford [16], Ahlfors and Sario [2]). In 1941 P. J. Myrberg [25] proved that Λ has positive logarithmic capacity. A refined study of the Hausdorff dimension of Λ has been carried out by Beardon [7] and by Akaza [3, 4, 5].

Maskit [22] proved that every finitely generated free Kleinian group all of which nontrivial elements are loxodromic is a Schottky group. This implies a result by Chuckrow: Every finitely generated subgroup of a Schottky group is a Schottky group. Chuckrow [15] also showed that to every set of free generators of a Schottky group there belongs a standard fundamental region. One can always construct standard fundamental regions bounded by analytic curves, but recently Marden [21] showed that there are Schottky groups for which *no* standard fundamental region, no matter what generators one chooses, is bounded by circles.

PROPOSITION 1. *Given a sequence $g_1,...,g_p$ of free generators of a Schottky group G, and a sequence of free generators $\hat{g}_1,...,\hat{g}_p$ of a Schottky group $\hat{G}$, there is a quasiconformal deformation of G onto $\hat{G}$ which takes g_j into $\hat{g}_j$, $j = 1,...,p$.*

Proposition 1 is a special case of an important theorem by Marden [20] which applies to an extensive class of Kleinian groups. Another proof will be found in a forthcoming paper by Maskit [24]. One can prove Proposition 1 quite simply if one is willing to use a result in the classical theory of Schottky groups, namely, the statement that

$$\Omega \in O_{AD}, \tag{1.4}$$

that is, that every holomorphic function on Ω, with finite Dirichlet integral, is a constant. This implies (since $\Omega \subset \hat{\mathbb{C}}$) that every holomorphic injection $\Omega \to \hat{\mathbb{C}}$ is a Möbius transformation. (Concerning the above statement, see Ahlfors and Sario [2, Chap. IV].)

To prove Proposition 1, we construct standard fundamental domains ω and $\hat{\omega}$, belonging to $g_1,...,g_p$ and to $\hat{g}_1,...,\hat{g}_p$, respectively, and having sufficiently smooth boundary curves. One can find a quasiconformal mapping γ of ω onto $\hat{\omega}$, which is a homeomorphism of the closure of ω onto that of $\hat{\omega}$ and satisfies the condition

$$\gamma(g_j(z)) = \hat{g}_j(\gamma(z)), \quad j = 1,..., p, \tag{1.5}$$

for z on the boundary of ω. This γ is the restriction to ω of a quasiconformal mapping, which we denote again by γ, of Ω onto $\hat{\Omega}$ (the region of discontinuity of $\hat{G}$) which satisfies (1.5) for all $z \in \Omega$. The Beltrami coefficient $\mu(z)$ of γ satisfies (1.2). Now set $\mu \mid \Lambda = 0$, and let (as usual) w^μ be the (unique normalized) quasiconformal selfmapping of $\hat{\mathbb{C}}$ with Beltrami coefficient μ, which fixes 0, 1 and ∞. Then $\check{G} = w^\mu G(w^\mu)^{-1}$ is a Schottky group with region of discontinuity $\check{\Omega} = w^\mu(\Omega)$, and the mapping $\alpha: \hat{\Omega} \to \check{\Omega}$ defined by $\alpha = w^\mu \circ \gamma^{-1}$ is a conformal bijection. By (1.4), α is the restriction of a Möbius transformation, which we denote by the same letter. Hence γ is the restriction to Ω of a quasiconformal self mapping $\alpha^{-1} \circ w^\mu$ of $\mathbb{C}$, and γ defines a quasiconformal deformation of G which takes g_j into $\hat{g}_j$.

2. Schottky Space and Augmented Schottky Space

A Schottky group is G *marked* by selecting a sequence $g_1,..., g_p$ of free generators. Two marked Schottky groups, $(G; g_1,..., g_p)$ and $(\hat{G}; \hat{g}_1,..., \hat{g}_p)$, are called *equivalent* if there is a Möbius transformation α such that $\alpha \circ g_j \circ \alpha^{-1} = \hat{g}_j$, $j = 1,..., p$. The set $\mathfrak{S}_p$ of all equivalence classes of marked Schottky groups of genus p is called the *Schottky space* (of genus p).

The Schottky space $\mathfrak{S}_p$ is made into a complex manifold by embedding it into $\mathbb{C}^{3p-3}$ as follows. If

$$\tau = (\tau_1, \tau_2,..., \tau_{3p-3}) \in \mathbb{C}^{3p-3},$$

set

$$a_1(\tau) = 0, \quad a_2(\tau) = \infty, \quad a_3(\tau) = 1, \quad a_{3+j}(\tau) = \tau_j, \quad j = 1,..., 2p-3$$

and

$$t_i(\tau) = \tau_{2p-3+i}, \quad i = 1,..., p.$$

In other words, set

$$\tau = (a_3(\tau),..., a_{2p}(\tau), t_1(\tau),..., t_p(\tau)) \in \mathbb{C}^{3p-3},$$
$$a_1(t) = 0, \quad a_2(t) = 1, \quad a_3(\tau) = \infty. \tag{2.1}$$

If $a_{2i-1}(\tau) \neq a_{2i}(\tau)$ and $0 < |t_i(\tau)| < 1$, then there is a loxodromic Möbius transformation $z \mapsto g_i(\tau, z)$ with repelling and attracting fixed points $a_{2i-1}(\tau)$ and $a_{2i}(\tau)$, and with multiplier $1/t_i$. More precisely,

$$g_1(\tau, z) = \frac{z}{t_1(z)},$$
$$\frac{g_j(\tau, z) - a_{2j}(\tau)}{g_j(\tau, z) - a_{2j-1}(\tau)} = t_j(z) \frac{z - a_{2j}(\tau)}{z - a_{2j-1}(\tau)}, \qquad j = 2 ..., p. \tag{2.2}$$

The set of those τ for which all $g_j(\tau)$, $j = 1,...,p$, are defined and are the free generators of a Schottky group G_τ, can be identified with $\mathfrak{S}_p$.

If $\tau \in \mathfrak{S}_p$, any standard fundamental region belonging to $g_1(\tau, \cdot),..., g_p(\tau, \cdot)$ will be called a τ-region.

PROPOSITION 2. *The Schottky space $\mathfrak{S}_p$ is a domain in $\mathbb{C}^{3p-3}$.*

Proof. Let $\tau \in \mathfrak{S}_p$ and let $C_1, C_1',..., C_p'$ be the boundary curves of a τ-region. For τ' close to τ, set $C_j'' = g_j(\tau', C_j)$. Then $C_1, C_1'', C_2, C_2'',..., C_p''$ bound a τ'-region so that $\tau' \in \mathfrak{S}_p$. Thus $\mathfrak{S}_p$ is open.

If τ and τ' belong to $\mathfrak{S}_p$, there is, by Proposition 1, a quasiconformal automorphism of $\mathbb{C}$ which induces a deformation of the group G_τ onto $G_{\tau'}$, which takes $g_j(\tau, \cdot)$ into $g_j(\tau', \cdot)$. In particular, this automorphism must take $0, 1, \infty$ into $0, 1, \infty$. Hence it is a normalized quasiconformal automorphism w^μ of $\mathbb{C}$ (this notation has been defined in the proof of Proposition 1), $\mu = \mu(z)$ being some Beltrami coefficient for G_τ. If s denotes a complex number with $|s| \leq 1$, then $s\mu(z)$ is a Beltrami coefficient for G_τ and, as is well known (cf. [2]), $w^{s\mu}(z)$ is a holomorphic function of s. Define $\xi = \xi(s)$ by the relations

$$g_j(\xi(s), \cdot) = w^{s\mu} \circ g_j(\tau, \cdot) \circ (w^{s\mu})^{-1}, \qquad j = 1,..., p.$$

Then $s \to \xi(s)$ is a holomorphic mapping, $\xi(0) = \tau$, $\xi(1) = \tau'$. Since a quasiconformal deformation of a Schottky group is a Schottky group, $\xi(s) \in \mathfrak{S}_p$ for $|s| \leq 1$, in particular for $0 \leq s \leq 1$. Hence τ and τ' can be joined by a curve in $\mathfrak{S}_p$. Thus $\mathfrak{S}_p$ is connected.

For $\tau \in \mathfrak{S}_p$, let Ω_τ be the region of discontinuity of G_τ. The *fiber space* $\mathfrak{F}\mathfrak{S}_p$ over $\mathfrak{S}_p$ is the set of points $(\tau, z) \in \mathbb{C}^{3p-2}$ with $\tau \in \mathfrak{S}_p$, $z \in \Omega_\tau$.

PROPOSITION 3. *The fiber space $\mathfrak{F}\mathfrak{S}_p$ is a domain in $\mathbb{C}^{3p-2}$.*

Proof. Let $\tau \in \mathfrak{S}_p$. The Schottky group G_τ is quasiconformally stable in the sense of [10], in view of the stability criterions tated there. Indeed, the proof of the criterion implies the following. There are $3p - 3$ Beltrami coefficients $\mu_1(z),..., \mu_{3p-3}(z)$ such that there is a holomorphic bijection $s \to \xi(s)$ of a neighborhood of the origin (in $\mathbb{C}^{3p-3}$) onto a neighborhood of τ with

$$g_j(\xi(s), \cdot) = w^{s \cdot \mu} \circ g_j(\tau, \cdot) \circ (w^{s \cdot \mu})^{-1}, \qquad j = 1,\ldots, p$$

where

$$s \cdot \mu = s_1 \mu_1(z) + \cdots + s_{3p-3} \mu_{3p-3}(z).$$

Since $w^{s \cdot \mu}(z)$ depends holomorphically on s, and since $\Omega_{\xi(s)} = w^{s \cdot \mu}(\Omega_\tau)$, we conclude that for every $z \in \Omega_\tau$ there is an $\epsilon > 0$ such that $z \in \Omega_{\tau'}$ for $|\tau - \tau'| < \epsilon$. This shows that $\mathfrak{F}\mathfrak{S}_p$ is open. It is connected since $\mathfrak{S}_p$ is.

Let $\partial \mathfrak{S}_p$ denote the *boundary* of $\mathfrak{S}_p$ in $\mathbb{C}^{3p-3}$, and let $\delta \mathfrak{S}_p$ denote the set of those points of $\tau \in \partial \mathfrak{S}_p$ at which either (i) two of the points $a_j(\tau)$, $j = 1,\ldots, p$, coincide, or (ii) one of the numbers $t_i(\tau)$, $i = 1,\ldots, p$, vanishes. Note that $\delta \mathfrak{S}_p$ is the intersection of $\partial \mathfrak{S}_p$ with finitely many complex hyperplanes, so that $\delta \mathfrak{S}_p$ has positive real codimension in $\partial \mathfrak{S}_p$.

For every $\tau \in \partial \mathfrak{S}_p \setminus \delta \mathfrak{S}_p$, the group G_τ is well defined. Chuckrow [15] proved the remarkable theorem that such a G_τ is always a free group on p generators. At some points of $\partial \mathfrak{S}_p \setminus \delta \mathfrak{S}_p$ the group G_τ may contain parabolic elements, but such τ lie on countably many analytic hypersurfaces. For all other τ, G_τ must be non Kleinian, for it either contains an elliptic element of infinite order and is therefore not discrete or it is purely loxodromic and if it were Kleinian it would be, by Maskit's theorem, a Schottky group. In the latter case, τ would belong to $\mathfrak{S}_p$ and not to $\partial \mathfrak{S}_p$.

In the present paper we are concerned primarily with certain boundary points of $\mathfrak{S}_p$ belonging to $\delta \mathfrak{S}_p$.

Let

$$I \subset \{1,\ldots, p\}, \qquad |I| = \text{number of elements in } I. \qquad (2.3)$$

Set $\delta^\phi \mathfrak{S}_p = \mathfrak{S}_p$, and for $I \neq \phi$, let $\delta^I \mathfrak{S}_p$ denote the set of those $\tau \in \mathbb{C}^{3p-3}$ for which (i) the $g_j(\tau, \cdot)$, $j \neq I$, are defined and are the free generators of a Schottky group (which we still denote by G_τ), (ii) the $t_i(\tau)$ with $i \in I$ vanish, (iii) the $2|I|$ points $a_{2j-1}(\tau)$, $a_{2j}(\tau)$, $j \in I$, are distinct, and (iv) lie in a suitably chosen standard fundamental region for G_τ belonging to $g_j(\tau, \cdot)$, $j \notin I$. (A fundamental region with the above property will be still called a τ-region. The region of discontinuity of G_τ will be still denoted by Ω_τ.)

Note that conditions (iv) are vacuously satisfied if $I = \{1,\ldots, p\}$, in this case $G_\tau = 1$, the trivial group and $\omega = \Omega = \hat{\mathbb{C}}$.

We denote by $\mathfrak{S}_p{}^I$ the union of all $\delta^L \mathfrak{S}_p$ with $L \subset I$, and by $\mathfrak{S}_p{}^*$ the set $\mathfrak{S}_p{}^I$ with $I = \{1,\ldots, p\}$. We call $\mathfrak{S}_p{}^*$ the *augmented* Schottky space.

PROPOSITION 4. *The augmented Schottky space $\mathfrak{S}_p{}^*$ is a domain in $\mathbb{C}^{3p-3}$, and a subset of $\mathfrak{S}_p \cup \partial \mathfrak{S}_p$.*

For each $I \subset \{1,\ldots, p\}$, $\delta^I \mathfrak{S}_0$ is a domain in $\mathbb{C}^{3p-3-|I|}$ and $\mathfrak{S}_p{}^I$ is a subdomain of $\mathfrak{S}_p{}^*$.

Proof. For some $I \neq \phi$, let $\tau \in \delta^I \mathfrak{S}_p$. Let C_j, C_j', $1 \leqslant j \leqslant p$, $j \notin I$ be the boundary curves of a τ-region. Let C_i, $i \in I$, be circles about the points $a_{2i-1}(\tau)$, of sufficiently small radius. If $\tau' \in \mathbb{C}^{3p-3}$ and $|\tau' - \tau|$ is sufficiently small, the set $K \subset \{1,\ldots, p\}$ of numbers k with $t_k(\tau') = 0$ satisfies $K \subset I$, and the Jordan curves C_r, $C_r'' = g_r(\tau', C_k)$, $r \notin K$, $1 \leqslant r \leqslant p$, are mutually disjoint and bound a standard fundamental region for the Schottky group freely generated by $g_r(\tau', \cdot)$, $r \notin K$; this region contains all points $a_{2k}(\tau')$, $a_{2k-1}(\tau')$, $k \in K$. This shows that $\tau \in \partial \mathfrak{S}_p$ and that τ is an interior point of $\mathfrak{S}_p{}^*$.

The proof of the second statement is left to the reader.

For $\tau \in \delta^I \mathfrak{S}_p$, let Ω_τ' be the complement in Ω_τ of all points of the form, $g[\tau, a_{2i-1}(\tau)]$, $g[\tau, a_{2i}(\tau)]$, $i \in I$, $g \in G_\tau$. Note that $\Omega_\tau' = \Omega_\tau$ for $\tau \in \mathfrak{S}_p$. The *fiber space* $\mathfrak{FS}_p{}^*$ over the augmented Schottky space $\mathfrak{S}_p{}^*$ is the set of all points $(\tau, z) \in \mathbb{C}^{3p-2}$ with $\tau \in \mathfrak{S}_p{}^*$, $z \in \Omega_\tau'$. We denote by $\mathfrak{FS}_p{}^I$ the restriction of $\mathfrak{FS}_p{}^*$ to $\mathfrak{S}_p{}^I$, by $\mathfrak{F}\delta^I \mathfrak{S}_p$ the restriction of $\mathfrak{FS}_p{}^*$ to $\delta^I \mathfrak{S}_p$.

PROPOSITION 5. *The fiber space $\mathfrak{FS}_p{}^*$ is a domain in $\mathbb{C}^{3p-2}$.*

For every $I \subset \{1,\ldots, p\}$, $\mathfrak{FS}_p{}^I$ is a subdomain of $\mathfrak{FS}_p{}^*$, and $\mathfrak{F}\delta^I \mathfrak{S}_p$ a domain in $\mathbb{C}^{3p-2-|I|}$.

The proof is an extension of the argument used in establishing Proposition 3, and is left to the reader. Note that Proposition 5 would be false had we used Ω_τ instead of Ω_τ'.

3. Schottky Groups and Riemann Surfaces

If G is a Kleinian group, with region of discontinuity Ω and a fundamental region ω, then the Hausdorff space $\Omega/G = \omega/G$ has a natural complex structure defined by the condition that the canonical surjection $\Omega \to \Omega/G$ be holomorphic. Every component of Ω/G is a Riemann surface. If G is a Schottky group of genus p ($G = G_\tau$, $\tau \in \mathfrak{S}_p$), then $S = \Omega/G$ is a closed ($=$ compact) Riemann surface of genus p, and the mapping $\Omega \to \Omega/G = S$ is an unramified Galois covering; one sees this by choosing for ω a τ-region. Furthermore, τ and the choice of ω define on S a sequence of p disjoint homologically independent simple closed curves (retrosections) $\Gamma_1,..., \Gamma_p$, Γ_j being the image under $\Omega \to \Omega/G$ of the boundary curves C_j, C_j' of ω identified by $g_j(\tau, \cdot)$.

This assignment of a closed Riemann surface with retrosections to a marked Schottky group with a standard fundamental region can be reversed. That is the content of the classical theorem:

RETROSECTION THEOREM. *Every closed Riemann surface S of genus p can be represented as Ω/G, G being a Schottky group with region of discontinuity Ω.*

More precisely, given a sequence of p disjoint, homologically independent, sufficiently smooth simple closed curves $\Gamma_1,..., \Gamma_p$ on S, one can choose G, and p generators $g_1,..., g_p$ of G, so that there is a standard fundamental region ω of G, bounded by curves $C_1, C_1',..., C_p'$ with $g_j(C_j) = C_j'$, such that Γ_j is the image of C_j under $\Omega \to \Omega/G$.

The marked Schottky group $(G; g_1,..., g_p)$ is determined uniquely by $(S; \Gamma_1,..., \Gamma_p)$ except for replacing $g_1,..., g_p$ by $\alpha \circ g_1^{\epsilon_1} \circ \alpha^{-1},..., \alpha \circ g_p^{\epsilon_p} \circ \alpha^{-1}$ where α is a Möbius transformation and $\epsilon_j = \pm 1$.

This theorem was first stated by Klein in 1883 (though Schottky may have had an inkling of it earlier) but a rigorous proof was given by Koebe only much later. The standard classical existence proof (Koebe, Courant) proceeds as follows.

One cuts S along the curves Γ_j to obtain a Riemann surface of genus 0 with $2p$ boundary curves Γ_i^+, Γ_i^-, $i = 1,..., 2p$, makes countably many copies $\Sigma_0, \Sigma_1, \Sigma_2,...$ of this cut surface, and joins them into the Schottky covering surface $\hat{S}$ of S by attaching to each boundary curve Γ_i^+ of a Σ_l a Γ_i^- on a Σ_m, two Σ's being joined by only one curve. One verifies that $\hat{S}$ is of genus 0 and appeals to Koebe's general uniformization principle (every Riemann surface of genus 0 is conformal to a subdomain of $\hat{\mathbb{C}}$) to identify $\hat{S}$ with domain in $\hat{\mathbb{C}}$. Now S appears as $\hat{S}/G$, G being a group of holomorphic self-mappings of $\hat{S}$. The surface Σ_0 may be chosen as a fundamental region for G and G is seen to be generated by those elements of G which identify the boundary curves Γ_i^+ and Γ_i^- of Σ_0, $i = 1,..., p$. It remains to show that the elements of G are Möbius transformations. This is accomplished by showing that $\hat{S} \in O_{AD}$ (cf. the proof of Proposition 1).

A different existence proof [9] uses quasiconformal mappings. Start with some marked Schottky group $\langle G_0, g_{01},..., g_{0p}\rangle$ with region of discontinuity Ω_0, and let ω_0 be a standard fundamental region with boundary curves $C_{01}, C'_{01},..., C'_{0p}$, belonging to the generators $g_{01},..., g_{0p}$. Given $(S, \Gamma_1,..., \Gamma_p)$, there is a quasiconformal map $\Omega_0/G_0 \to S$ such that the inverse image of Γ_j under $\Omega_0 \to \Omega_0/G_0 \to S$ contains C_{0j} and C'_{0j}. There is a Beltrami coefficient $\mu(z)$ for G_0 such that the mapping $\Omega_0 \to \Omega_0/G_0 \to S$ is conformal with respect to the Riemannian metric $|\,dz + \mu(z)\,d\bar{z}\,|$ on Ω_0. One then verifies that the Schottky group $w^\mu G(w^\mu)^{-1}$ with standard fundamental region $\omega = w^\mu(\omega_0)$ has the required properties.

In order to prove the uniqueness statement one must show that given two marked Schottky groups $\langle G; g_1,..., g_p\rangle$ and $\langle G_0, g_{01},..., g_{0p}\rangle$, with region of discontinuity Ω and Ω_0, and a conformal mapping $h: \Omega \to \Omega_0$ with $h \circ g_j = g_{0j} \circ h$, $j = 1,..., p$, h is (the restriction of) a Möbius transformation. This follows directly from (2.4), and can also be deduced from Proposition 1 and Maskit's extension theorem [23].

It follows from the retrosection theorem that the Schottky space $\mathfrak{S}_p$ is, in some sense, a space of all closed Riemann surfaces of genus p. Now for $p > 1$, $\dim_{\mathbf{C}} \mathfrak{S}_p = 3p - 3$, and already Klein noticed that this verifies Riemann's statement that the conformal type of a closed Riemann surface of genus p depends on $3p - 3$ "moduli." But $\mathfrak{S}_p$ is not the space of moduli, since distinct points of $\mathfrak{S}_p$ may well define conformally equivalent Riemann surfaces. Rather, the space of moduli (sometimes called the Riemann space) $\mathfrak{R}_p$ is the quotient of $\mathfrak{S}_p$ by a certain properly discontinuous group of holomorphic automorphisms. Also, the Schottky space is not simply connected; its universal covering space $\tilde{\mathfrak{S}}_p$ can be identified with the Teichmüller space T_p of closed Riemann surface of genus p. (See [10] for an outline of the theory of Teichmüller spaces and references to the literature.)

In view of Proposition 1, the equality

$$\tilde{\mathfrak{S}}_p = T_p \tag{3.1}$$

is a very special case of a general theorem on Kleinian groups (cf. Maskit [23], Kra [18], and [13]). A complete direct proof of (3.1) has been given by Hejhal [17]; he also deduced from (3.1) that $\mathfrak{S}_p$ is a domain of holomorphy.

We want now to interpret points $\tau \in \mathfrak{S}_p{}^*\backslash \mathfrak{S}_p$, as complex spaces. If $\tau \in \delta^I \mathfrak{S}_p$, with $\phi \neq I \subset \{1,..., p\}$, then Ω_τ/G_τ is a compact Riemann surface of genus $p - |I|$ on which there are $|I|$ distinguished pairs of points: p_i, p_i', $i \in I$, where p_i is the image of $a_{2i-1}(\tau)$ under the mapping $\Omega_\tau \to \Omega_\tau/G_\tau$, and p_i' the image of $a_{2i}(\tau)$. We denote by S_τ the Riemann surface Ω_τ/G_τ with the points p_i, p_i' *identified*, for every $i \in I$. This S_τ is an example of a closed Riemann surface with nodes.

In general, *a closed Riemann surface with nodes X* is a compact complex space each point Q of which has a neighborhood isomorphic either to a disk $|z| < 1$ in $\mathbb{C}$ (with Q corresponding to $z = 0$) or to the set $|z| < 1$, $|w| < 1$, $zw = 0$ in $\mathbb{C}^2$ (with Q corresponding to $z = w = 0$). In the latter case, Q is called a node. Every component of $X\backslash\{\text{nodes}\}$ is called a *part* of X; it is an (ordinary) Riemann surface. X is called *stable* if no part of X is a sphere punctured at 0, 1 or 2 points or a compact surface of genus 1. The genus p of X is defined by the formula

$$p = \sum p_j + k + 1 - r, \qquad (3.2)$$

where r is the number of parts, k the number of nodes, and $\sum p_j$ the sum of the genera of the parts.

For $\tau \in \mathfrak{S}_p{}^*$, S_τ is a surface of genus p, perhaps with nodes (more precisely, with $|I|$ nodes if $\tau \in \delta^I \mathfrak{S}_p$) and with one parts. The retrosection theorem implies that every closed Riemann surface of genus p, with or without nodes, and with one part only, can be represented as an S_τ, $\tau \in \mathfrak{S}_p{}^*$.

Intuitively one may think of a compact Riemann surface with l nodes as having been obtained from an ordinary closed Riemann surface of genus p by squeezing l disjoint homotopically independent simple closed curves into points. There will be one part only, if these curves are chosen as homologically independent. Theorems 1–4 stated in Sections 4 and 5 show that this intuitive idea has a solid mathematical content.

Remark. In order to construct spaces of compact Riemann surfaces with nodes and with more than one part, it is convenient to use Kleinian groups of a more complicated nature. These groups are described in [11].

4. Poincaré Metrics

If S is a Riemann surface and S is neither a sphere punctured at 0, 1 or 2 points nor a closed surface of genus 1, then S carries a Poincaré metric, the unique conformal (that is of the form $ds = \lambda_S(w) |dw|$,

w a local parameter) complete Riemannian metric of Gaussian curvature $k = -1$; the latter condition means that

$$\Delta \log \lambda_S = \lambda_S{}^2, \qquad \Delta = \partial^2/\partial u^2 + \partial^2/\partial v^2,$$

where $w = u + iv$. (The Poincaré metric on the upper half-plane $U = \{z = x + iy \mid y > 0\}$ is $y^{-1} \mid dz \mid$.) The Poincaré metric is invariant under all conformal automorphisms of S, and for $S_0 \subset S$ we have that $\lambda_S(w) \leqslant \lambda_{S_0}(w)$ if w is a local parameter defined near a point $p \in S_0$. If S is a Riemann surface with nodes, the Poincaré metric on S is defined as the Poincaré metric on (each component of) $S\backslash\{\text{nodes}\}$. One can compute that the Poincaré area of a stable closed Riemann surface with nodes, of genus p, is $4\pi(p-1)$.

For $\tau \in \mathfrak{S}_p{}^*$, set

$$\lambda_\tau(z) = \lambda_{\Omega_\tau{}'}(z) \qquad \text{for} \quad z \in \Omega_\tau{}'. \tag{4.1}$$

Clearly, λ_τ induces the Poincaré metric on S_τ.

THEOREM 1. *The number $\lambda_\tau(z)$ is a continuous function of $(\tau, z) \in \mathfrak{F}\mathfrak{S}_p{}^*$.*

Proof. As a solution of the elliptic partial differential equation

$$\Delta \log \lambda_\tau(z) = \lambda_\tau(z)^2, \qquad (\Delta = \partial^2/\partial x^2 + \partial^2/\partial y^2), \tag{4.2}$$

$\lambda_\tau(z)$ is real analytic in z for every fixed τ. If the disk $|z - z_0| \leqslant r$ belongs to $\Omega_\tau{}'$ for all τ with $|\tau - \tau_0| < \epsilon$, then

$$0 < \lambda_\tau(z) \leqslant 2r(r^2 - |z - z_0|^2)^{-1} \qquad \text{for} \quad |\tau - \tau_0| < \epsilon, \quad |z - z_0| < r, \tag{4.3}$$

since $2r(r^2 - |z - z_0|^2)^{-1} \mid dz \mid$ is the Poincaré metric on the disk $|z - z_0| < r$.

Let $\Lambda_\alpha(z) \mid dz \mid$ denote, for every $\alpha \in \mathbb{C}\backslash\{0\}$, the Poincaré metric on $\mathbb{C}\backslash\{0, \alpha\}$; this function can be expressed explicitly by the elliptic modular function, and depends continuously on α. Since $1 \notin \Omega_\tau{}'$,

$$0 < \Lambda_1(z) \leqslant \lambda_\tau(z) \qquad \text{for} \quad (\tau, z) \in \mathfrak{F}\mathfrak{S}_p{}^*. \tag{4.4}$$

We conclude from (4.3) and (4.4) that both $\lambda_\tau(z)$ and $\log \lambda_\tau(z)$ are locally uniformly bounded in $\mathfrak{F}\mathfrak{S}_p{}^*$. Using Eq. (4.2) and standard potential theory it is not difficult to conclude that the partial derivatives of $\lambda_\tau(z)$, with respect to $x = \operatorname{Re} z$ and $y = \operatorname{Im} z$, up to any fixed order, are also locally uniformly bounded. Thus Theorem 1 will be proved once we establish that $\lambda_\tau(z)$ is continuous in τ, for a fixed z.

Hence, let $\{\tau_\nu\}$ be a sequence in $\mathfrak{S}_p{}^*$, with

$$\lim \tau_\nu = \tau_0 \in \mathfrak{S}_p{}^*. \tag{4.5}$$

Selecting, if need be, a subsequence, we may assume, by the uniform boundedness statements made above and by the Arzela–Ascoli theorem, that

$$\lim \lambda_{\tau_\nu}(z) = \hat{\lambda}(z), \qquad z \in \Omega_{\tau_0}', \tag{4.6}$$

exists, and that the functions λ_{τ_ν} together with their x and y derivatives of all orders converge to $\hat{\lambda}$ and to its corresponding derivatives, uniformly on compact subsets of Ω_{τ_0}'. It will suffice to show that $\hat{\lambda}(z) = \lambda_{\tau_0}(z)$.

At any rate, $\hat{\lambda}(z) > 0$ since by (4.6) and (4.4), $\hat{\lambda}(z) \geqslant \Lambda_1(z)$. Also $\Delta \log \hat{\lambda}(z) = \hat{\lambda}(z)^2$, by (4.6) and (4.2). Since $\lambda_{\tau_\nu}(z) \mid dz \mid$ is G_{τ_ν} invariant, $\hat{\lambda}(z) \mid dz \mid$ is G_{τ_0} invariant, again by (4.6). Hence $\hat{\lambda}(z) \mid dz \mid$ induces on $\Omega_{\tau_0}'/G_{\tau_0}$ a conformal Riemannian metric of Gaussian curvature -1. We must show that this metric is complete.

Let $I \subset \{1,\ldots, p\}$ consist of those i for which $t_i(\tau_0) = 0$. Since $\Omega_{\tau_0}'/G_{\tau_0}$ is compact, except for $2 \mid I \mid$ punctures, completeness of the $\hat{\lambda}$-metric will follow if we show that

$$\int_\Gamma \hat{\lambda}(z) \mid dz \mid = +\infty \tag{4.7}$$

for every rectifiable path joining a point z_0 in Ω_{τ_0}' to one of the points $a_{2i}(\tau_0)$, $a_{2i-1}(\tau_0)$ with $i \in I$, and lying, except for one endpoint, in Ω_{τ_0}'. We shall do so, assuming that $4 \in I$ and the endpoint of Γ is $a = a_4(\tau_0)$. Other cases can be treated similarly.

Since $\Omega_{\tau_\nu}' \subset \mathbb{C} \backslash \{0, a_4(\tau_\nu)\}$, we have that

$$\lambda_{\tau_\nu}(z) \geqslant \Lambda_{a_4(\tau_\nu)}(z), \qquad z \in \Omega_{\tau_\nu}'.$$

Since $\lim a_4(\tau_\nu) = a$, (4.6) implies that

$$\hat{\lambda}(z) \geqslant \Lambda_a(z) \qquad \text{for} \quad z \in \Omega_{\tau_0}'. \tag{4.8}$$

Since, as is well known,

$$\Lambda_a(z) \sim \mid z - a \mid^{-1} (-\log \mid z - a \mid)^{-1}, \qquad z \to a, \tag{4.9}$$

(4.8) implies that (4.7) holds.

Remark. Presumably Theorem 1 can be strengthened to the statement: $\lambda_\tau(z)$ is a real analytic function of (τ, z).

5. Automorphic Forms and Poincaré Series

Let $q > 0$ be an integer. A *holomorphic q-differential F* on a Riemann surface X (without nodes) is a holomorphic form of type $(q, 0)$; locally $F = f(w)\,dw^q$ where w is a local parameter and f a holomorphic function. If $P \in X$ and w is a local parameter with $w = 0$ at P, defined on a domain D with $P \in D \subset X$, then a holomorphic q-differential F on $D\backslash\{P\}$ may be written, near P, in the form

$$F = \sum_{n=-\infty}^{+\infty} a_n w^n \, dw^q.$$

The number a_{-q} is called the *residue* at P; it does not depend on the choice of w provided that $a_n = 0$ for $n < -q$ (such a differential is said to have at P a pole of order at most q).

Let X be a stable closed Riemann surface with nodes. A *regular q-differential* on X is, by definition, an assignment of a holomorphic q-differential to each part of S, with the provision that (α) at punctures corresponding to nodes there are poles of order at most q, and (β) at two punctures joined in a node the residues are equal (if q is even) or opposite (if q is odd).

Now let X be compact, of genus p. The number $\delta = \delta(p, q)$ of linearly independent regular q-differentials is computed by the Riemann–Roch theorem:

$$\delta = \begin{cases} p & \text{if } q = 1, \\ (2q-1)(p-1) & \text{if } q > 1. \end{cases} \tag{5.1}$$

In other words, δ does not depend on the presence or the number of nodes.

Let $F_1, \ldots, F_\delta$ be linearly independent regular q-differentials on X. They determine a so called *q-canonical holomorphic mapping* of X into the (complex) projective space $\mathbb{P}_{\delta-1}$. If $P \in X$ is not a node, w is a local parameter with $w = 0$ at P, and near P, $F_j = f_j(w)\,dw^q$, then the image of P has homogeneous coordinates

$$\{f_1(0), f_2(0), \ldots, f_\delta(0)\}.$$

If P is a node, the image of P has homogeneous coordinates

$$(a_1, a_2, ..., a_\delta)$$

where a_j is the residue of F_j at a puncture (on a part of X) corresponding to P. Using known results about q-canonical mappings of Riemann surfaces without nodes, one verifies that, for $q \geqslant 3$, the q-canonical mapping is an embedding. The image is, by Chow's theorem, an algebraic curve; its only singularities are simple nodes (corresponding to the nodes of X).

Returning to the augmented Schottky space $\mathfrak{S}_p^*$, one would like to associate to every $\tau \in \mathfrak{S}_p^*$ a q-canonical mapping $S_\tau \to \mathbb{P}_{\delta-1}$, depending holomorphically on τ. I could obtain, however, only a somewhat weaker result. Before stating it, we recall some definitions.

Let G be a Kleinian group, with region of discontinuity Ω. If $\Omega' \subset \Omega$ a domain such that Ω' is G invariant and $\Omega \backslash \Omega'$ discrete, $\varphi(z)$, $z \in \Omega'$, is a holomorphic function, and

$$\varphi(g(z))g'(z)^q = \varphi(z) \quad \text{for} \quad g \in G,$$

φ is called a holomorphic *automorphic form* of weight $(-2q)$, on Ω', or, briefer, a *holomorphic q-form*. Such a form induces a holomorphic q-differential on each component of Ω/G. The following method for constructing holomorphic q-forms is classical.

Let $G_0 \subset G$ be a subgroup, and let $\gamma_0, \gamma_1, ...$ be all right coset representatives of G modulo G_0 so that

$$G = G_0\gamma_0 + G_0\gamma_1 + G_0\gamma_2 + \cdots.$$

Let $\Phi(z)$, $z \in \Omega'$, be a holomorphic function satisfying

$$\Phi(\gamma(z))\,\gamma'(z)^q = \Phi(z), \quad \gamma \in G_0,$$

and set

$$\varphi(z) = \sum_{n=0}^{\infty} \Phi(\gamma_n(z))\,\gamma_n'(z)^q.$$

The series is called a Poincaré (theta) series, a relative Poincaré series of $G_0 \neq \mathrm{id}$. (The nth term of the series depends only on the coset of γ_n modulo G_0.) If the series converges uniformly and absolutely on compact subsets, then φ is a holomorphic q-form for G.

Now let $\tau \in \mathfrak{S}_p{}^*$. A holomorphic q-form for G_τ on $\Omega_\tau{}'$ will be called *regular* if it induces a regular q-differential on the Riemann surface with nodes S_τ.

THEOREM 2. *Let $p > 1$, $q > 1$ be integers. There exist $\delta = (2q - 1)(p - 1)$ holomorphic functions $\varphi_j(\tau, z)$, $(\tau, z) \in \mathfrak{F}\mathfrak{S}_p{}^*$, and an analytic subvariety $Z \subset \mathfrak{S}_p{}^*$ such that (1) Z is either empty or of pure codimension one, (2) Z avoids all points τ which lie on any set $\delta^I \mathfrak{S}_p$ with $I \subset \{1,...,p\}$, $|I| \geq p - 1$, (3) for each $\tau \in \mathfrak{S}_p{}^*$ the functions $\varphi_j(\tau, z)$ are regular q-forms for G_τ, and (4) these functions are linearly independent if and only if $\tau \notin Z$.*

I do not know whether Z can be made empty. If it could, the next theorem would be superfluous.

THEOREM 3. *Let $p > 1$, $q > 1$ be integers, $I \subset \{1,...,p\}$, $|I| < p - 1$, $\tau_0 \in \delta^I \mathfrak{S}_p$. There exists an analytic subvariety $Z \subset \mathfrak{S}_p{}^I$ and $\delta = (2p - 1)(q - 1)$ holomorphic functions $\varphi_j(\tau, z)$, $(\tau, z) \in \mathfrak{F}\mathfrak{S}_p{}^I$, such that (1) Z is either empty or of pure codimension 1, (2) $\tau_0 \notin Z$, (3) for each $\tau \in \mathfrak{S}_p{}^I$, the functions $\varphi_j(\tau, z)$ are regular q-forms for G_τ, and (4) are linearly independent if and only if $\tau \in Z$.*

The functions $\varphi(\tau, z)$ occurring in these theorems will be constructed, in the next section, as Poincaré series.

For $q = 1$ we state

THEOREM 4. *Let $p > 1$ and $\tau_0 \in \mathfrak{S}_p{}^*$. There is a neighborhood N of τ_0 in $\mathfrak{S}_p{}^*$, and p holomorphic functions $\varphi_j(\tau, z)$, $\tau \in N$, $z \in \Omega_\tau{}'$, such that, for each $\tau \in N$ the p functions of z, $\varphi_j(\tau, z)$, are linearly independent regular 1-forms for G_τ.*

The proof will be given in Section 6. The reason the functions in Theorems 2 and 3 are defined for all $\tau \in \mathfrak{S}_p{}^I$ and are linearly independent in a "Zariski neighborhood" (= complement of an analytic subvariety) of τ_0, while those in Theorem 4 are so only in an "ordinary neighborhood" of τ_0, seems to be related to two facts: (a) Poincaré series do not, in general, converge for $q = 1$ (though they do for a wide class of Schottky groups as was noticed many decades ago by Schottky and by Burnside), and (b) a marking of a Schottky group G does not define, unambiguously, a homology basis on Ω/G.

6. Proofs of Theorems 2 and 3

The proofs are, unfortunately, rather long and technical. They will be presented in a series of lemmas. We shall often use the (essentially known, cf. [24]).

PROPOSITION 6. *Let $\Delta \subset \mathbb{C}^r$ be a domain, Γ a properly discontinuous group of holomorphic self-mappings*

$$\Delta \ni \zeta = (\zeta^1,..., \zeta^r) \mapsto \gamma(\zeta) = (\gamma^1(\zeta),..., \gamma^r(\zeta)) \in \Delta,$$

$\Gamma_0 \subset \Gamma$ *a subgroup, $\{\gamma_0, \gamma_1 ,...\}$ a complete list of right coset representatives of Γ modulo Γ_0, and $q \geqslant 1$ an integer. Also, let $\rho(\zeta)$, $\zeta \in \Delta$, be a positive continuous function such that*

$$\rho(\gamma(\zeta)) |\operatorname{jac}_\gamma(\zeta)| = \rho(\zeta), \qquad \gamma \in \Gamma, \tag{6.1}$$

where

$$\operatorname{jac}_\gamma(\zeta) = \frac{\partial(\gamma^1,..., \gamma^r)}{\partial(\zeta^1,..., \zeta^r)}, \tag{6.2}$$

and let $\Phi(\zeta)$, $\zeta \in \Delta$, be a holomorphic function such that

$$\Phi(\gamma(\zeta)) \operatorname{jac}_\gamma(\zeta)^q = \Phi(\zeta), \qquad \gamma \in \Gamma_0, \quad \text{and} \tag{6.3}$$

$$\iint_{\Delta/\Gamma_0} \rho(\zeta)^{2-q} |\Phi(\zeta)| \, dV_\zeta < +\infty, \tag{6.4}$$

where

$$dV_\zeta = d\xi^1 \, d\eta^1 \, d\xi^2 \cdots d\eta^r \qquad (\zeta^j = \xi^j + \sqrt{-1}\, \eta^j).$$

Set

$$\varphi(\zeta) = \sum_n \Phi(\gamma_n(\zeta)) \operatorname{jac}_{\gamma_n}(\zeta)^q, \qquad z \in \Delta. \tag{6.5}$$

Then the series above converges absolutely and uniformly on compact subsets of Δ, $\varphi(\zeta)$ is holomorphic in Δ,

$$\varphi(\gamma(\zeta)) \operatorname{jac}_\gamma(\zeta)^q = \varphi(\zeta), \qquad \gamma \in \Gamma, \quad \text{and} \tag{6.6}$$

$$\iint_{\Delta/\Gamma} \rho(\zeta)^{2-q} |\varphi(\zeta)| \, dV_\zeta \leqslant \iint_{\Delta/\Gamma_0} \rho(\zeta)^{2-q} |\Phi(\zeta)| \, dV_\zeta. \tag{6.7}$$

Proof. In (6.4) and (6.7), Δ/Γ_0 stands for any fundamental region ω_0 for Γ_0 and Δ/Γ for any fundamental region ω for Γ. (The definition of a fundamental region is the same as the one given in Section 1 for $r = 1$; the existence of a fundamental region is tantamount to the proper discontinuity of the group.) Relations (6.1) and (6.3) imply that $|\Phi(\zeta)| \rho(\zeta)^{2-q} dV_\zeta$ is a Γ_0-invariant volume element, and relation (6.6) implies the Γ-invariance of $|\varphi(\zeta)| \rho(\zeta)^{2-q} dV_\zeta$. Hence the choice of the fundamental region of integration, in (6.4) or in (6.7), is irrelevant.

Choose an ω and set

$$\omega_0 = \gamma_0(\omega) \cup \gamma_1(\omega) \cup \gamma_2(\omega) \cup \cdots.$$

Now

$$\sum_n \iint_\omega |\Phi(\gamma_n(\zeta)) \, \mathrm{jac}_{\gamma_n}(\zeta)|^q \rho(\zeta)^{2-q} dV_\zeta$$

$$= \sum_n \iint_\omega |\Phi(\gamma_n(\zeta))| \rho(\gamma_n(\zeta))^{2-q} |\mathrm{jac}_{\gamma_n}(\zeta)|^2 dV_\zeta$$

$$= \sum_n \iint_{\gamma_n(\omega)} |\Phi(\zeta)| \rho(\zeta)^{2-q} dV_\zeta = \iint_{\omega_0} |\Phi(\zeta)| \rho(\zeta)^{2-q} dV_\zeta.$$

This implies that, for every compact $K \subset \omega$,

$$\sum_n \max_{\zeta \in K} |\Phi(\gamma_n(\zeta))| \, \mathrm{jac}_{\gamma_n}(\zeta)^q \, dV_\zeta < +\infty.$$

This inequality, in turn, implies the statements about the convergence of (6.5), the holomorphicity of φ, the functional equation (6.6) and the inequality (6.7).

For the rest of this section we introduce the following terminology. Let $\hat{G}$ denote an abstract free group on p free generators $\hat{g}_1, ..., \hat{g}_p$. For every $I \subset \{1, ..., p\}$, let $\hat{G}_I$ be the subgroup of $\hat{G}$ generated by the $p - |I|$ elements $\hat{g}_j$, $1 \leq j \leq p$, $j \notin I$. For $\tau \in \mathfrak{S}_p{}^I$ there is defined a canonical isomorphism of $\hat{G}_I$ onto G_τ which sends $\hat{g}_j$ into $g_j(\tau, \cdot)$, $j \notin I$. The image of a $\hat{\gamma} \in \hat{G}_I$ under this isomorphism will be denoted by

$$z \to \gamma_\tau(z);$$

in particular $g_{j,\tau} = g_j(\tau, \cdot)$.

LEMMA 1. *Every γ_τ depends holomorphically on τ.*

The proof is clear.

From now on we fix the numbers $p > 1$ and $q > 1$. We shall define three kinds of functions of $3p - 2$ variables.

For $j = 1,\ldots, p$ and $(\tau, z) \in \mathfrak{F}\mathfrak{S}_p{}^*$, set

$$\Psi_j(\tau, z) = \left\{\frac{a_{2j-1} - a_{2j}}{(z - a_{2j-1})(z - a_{2j})}\right\}^q \tag{6.8}$$

where

$$a_{2j-1} = a_{2j-1}(\tau), \qquad a_{2j} = a_{2j}(\tau), \tag{6.9}$$

and for $j = 1$ one interprets Ψ_j as

$$\Psi_1(\tau, z) = 1/z^q. \tag{6.10}$$

Let $G_\tau{}^j$ denote the group generated by $g_j(\tau, \cdot)$ if $t_j(\tau) \neq 0$, the trivial group if $t_j(\tau) = 0$, and let $\gamma_0 = \text{id}$, $\gamma_1, \gamma_2,\ldots$ be a complete list of distinct right coset representatives of G_τ modulo $G_\tau{}^j$. We set

$$\psi_j(\tau, z) = \sum_{n=0}^{\infty} \Psi_j[\tau, \gamma_n(z)]\, \gamma_n{}'(z)^q. \tag{6.11}$$

Denote by $\mathfrak{R}_\tau$, $\tau \in \mathfrak{S}_p{}^*$, the vector space of rational functions of z which have poles of order at most $q - 1$ at the points $0, 1, a_4(\tau),\ldots, a_{2p}(\tau)$, and no other singularities, and which vanish at ∞ of order at least $q + 1$. Denote by Π the vector space of polynomials in z of degree at most $2(p - 1)(q - 1) - 2$. Then every element of $\mathfrak{R}_\tau$ is of the form

$$\Psi_\pi(\tau, z) = \frac{\pi(z)}{\{z(z - 1)[z - a_4(\tau)] \cdots [z - a_{2p}(\tau)]\}^{q-1}} \tag{6.12}$$

with $\pi \in \Pi$. We set

$$\psi_\pi(\tau, z) = \sum_{\gamma \in G_\tau} \Psi_\pi[\tau, \gamma(z)]\, \gamma'(z)^q \tag{6.13}$$

for $\pi \in \Pi$ and $(\tau, z) \in \mathfrak{F}\mathfrak{S}_p{}^*$. We note for later reference that

$$\dim \mathfrak{R}_\tau = \dim \Pi = 2(p - 1)(q - 1) - 1. \tag{6.14}$$

For $I \subset \{1,\ldots, p\}$, let Σ_I denote the set of sequences σ of $2q - 1$ distinct nontrivial elements $\hat{\gamma}_1,\ldots, \hat{\gamma}_{2q-1}$ of $\hat{G}_I$ with the property: the attracting fixed points $\alpha_1(\tau),\ldots, \alpha_{2q-1}(\tau)$ of $\gamma_{1,\tau},\ldots, \gamma_{2q-1,\tau}$ are distinct for one (and hence for all) $\tau \in \mathfrak{S}_p{}^I$. For $\sigma \in \Sigma_I$, and $(\tau, z) \in \mathfrak{F}\mathfrak{S}_p{}^I$, we set

$$\Psi_\sigma(\tau, z) = \frac{1}{[z - \alpha_1(\tau)][z - \alpha_2(\tau)] \cdots [z - \alpha_{2q-1}(\tau)]}, \qquad (6.15)$$

$$\psi_\sigma(\tau, z) = \sum_{\gamma \in G_\tau} \Psi_\sigma[\tau, \gamma(\tau)]\, \gamma'(\tau)^q. \qquad (6.16)$$

We proceed to study the properties of the functions ψ_j, ψ_π and ψ_σ.

LEMMA 2. *The series* (6.11), (6.13), (6.16) *converge absolutely. The function* $\psi_j(\tau, z)$ *and* $\psi_\pi(\tau, z)$ *are regular q-differentials for* G_τ, *for every fixed* $\tau \in \mathfrak{S}_p{}^*$; *so are the functions* $\psi_\sigma(\tau, z)$, *for* $\sigma \in \Sigma_I$, $\tau \in \mathfrak{S}_p{}^I$.

Proof. For ψ_j and $\tau \in \mathfrak{S}_p{}^*$, $t_j(\tau) \neq 0$, we apply Proposition 6 with $r = 1$, $\zeta = z$, $\Delta = \Omega_\tau'$, $\Gamma = G_\tau$, Γ_0 the subgroup $G_\tau{}^j$ generated by $g_j(\tau, \cdot)$, $\rho(z) = \lambda_\tau(z)$ and $\Phi = \Psi_j$, $\varphi = \psi_j$.

For ψ_j and $t_j(\tau) = 0$, or for ψ_π and any τ, or for ψ_σ, $\sigma \in \Sigma_I$ and $\tau \in \mathfrak{S}_p{}^I$, we apply Proposition 6 as before, except that we set $\Gamma_0 = 1$ and, of course, $\Phi = \Psi_j$, $\varphi = \psi_j$ or $\Phi = \Psi_\pi$, $\varphi = \psi_\pi$, or $\Phi = \Psi_\sigma$, $\varphi = \psi_\sigma$.

In all cases, the only question is of verifying condition (6.4). We observe that $q \geqslant 2$ and that

$$\lambda_\tau(z) \geqslant \Lambda_A(z), \qquad (6.17)$$

where A is any finite set of fixed points of nontrivial elements of G_τ and $\Lambda_A(z)$ is the Poincaré metric of the domain $\mathbb{C}\backslash A$. Recall that, as is well known,

$$\Lambda_A(z) \sim \frac{1}{|z - \alpha|(-\log|z - \alpha|)}, \qquad z \to \alpha \in A \qquad (6.18)$$

and

$$\Lambda_A(z) \sim \frac{|z|}{\log|z|}, \qquad z \to \infty. \qquad (6.19)$$

The desired conditions follow easily.

LEMMA 3. *The functions* $\psi_j(\tau, z)$, $\psi_\pi(\tau, z)$ *are holomorphic functions of* $(\tau, z) \in \mathfrak{F}\delta^I\mathfrak{S}_p$ *for every* $I \subset \{1, ..., p\}$. *The same is true of* $\psi_\sigma(\tau, z)$ *provided that* $\sigma \in \Sigma_I$.

Proof. We first assume that $j \notin I$ and apply Proposition 6 to the case $r = 3p - 2 - |I|$,

$$\Delta = \{\zeta = (\tau, z) \mid \tau \in N, z \in \Omega_\tau'\},$$

where N is a relatively compact subdomain of $\delta^I \mathfrak{S}_p$ containing a given point $\tau_0 \in \delta^I \mathfrak{S}_p$, Γ the group of mappings

$$\Delta \ni \zeta = (\tau, z) \mapsto \gamma(\zeta) = (\tau, \gamma_\tau(z)), \qquad \hat{\gamma} \in \hat{G},$$

Γ_0 the subgroup of Γ generated by $\zeta = (\tau, z) \to (\tau, g_j(\tau, z))$,

$$\rho(\zeta) = \rho(\tau, z) = \lambda_\tau(z),$$

and, $\Phi(\zeta) = \Psi_j(\tau, z)$, $\varphi(\zeta) = \psi_j(\tau, z)$. Note that $\rho(\zeta)$ is continuous, by Theorem 1, and that Γ is a group of holomorphic self-mappings of Δ, by Lemma 1. Since

$$\mathrm{jac}_\gamma(\zeta) = \gamma_\tau'(z) \qquad \text{for} \quad \zeta = (\tau, z)$$

relations (6.1) and (6.3) follow.

If we choose N sufficiently small, the group Γ_0 will have in Δ a fundamental region

$$\omega_0 = \{(\tau, z) \mid \tau \in N, z \in \Omega_\tau \cap K_\epsilon(\tau)\}$$

where $K_\epsilon(\tau)$ is an annular region bounded by a fixed Jordan curve C around $a_{2j-1}(\tau_0)$ and by $C' = g_j(\tau, C)$. One sees easily that

$$\int_{\omega_0} |\Psi(\tau, z)| \lambda_\tau(z)|^{2-q} \, dV_\tau \, dx \, dy < +\infty$$

where dV_τ is the Euclidean $(3p - 3 - |I|)$-dimensional volume element in $\delta^I \mathfrak{S}_p$ and $z = x + iy$. This means that condition (6.4) in Proposition 6 is satisfied. We conclude, by the Proposition, that $\psi_j(\tau, z) = \varphi(\zeta)$ is holomorphic in Δ and, since τ_0 was arbitrary, holomorphic for $\zeta = (\tau, z) \in \mathfrak{F} \delta^I \mathfrak{S}_p$.

The holomorphicity of $\Psi_j(\tau, z)$ for $(\tau, z) \in \mathfrak{F} \delta^I \mathfrak{S}_p$ in the case when $j \in I$ is proved in the same way, except that we set $\Gamma_0 = 1$ and use for Δ the domain

$$\Delta = \{(\tau, z) \mid \tau \in N, z \in \Omega_\tau' \backslash D_\epsilon(\tau)\}$$

where $D_\epsilon(\tau)$, for a sufficiently small $\epsilon > 0$, is the union of a closed disk of radius ϵ about $a_{2j-1}(\tau)$, a closed disk of radius ϵ about $a_{2j}(\tau)$ [or the "disk" $|z| \geqslant 1/\epsilon$ if $j = 1$], and the images of these disks under the group G_τ. The inequality

$$\int_{\Delta/\Gamma} |\Psi_j(\tau, z)| \lambda_\tau(z)^{2-q} \, dV_\zeta < +\infty$$

is easily established using Theorem 1, the inequality

$$(2p-2) \lambda_\tau(z) \geqslant \sum_{\nu=3}^{2p} \Lambda_{a_\nu(\tau)}(z) \tag{6.20}$$

(cf. relation (4.9) in Section 4), and a fundamental region ω for Γ in Δ of the form

$$\omega = \{(\tau, z) \mid \tau \in N, z \in \omega(\tau) \setminus [\omega(\tau) \cap D_\epsilon(\tau)]\}.$$

Here $\omega(\tau)$ is a τ-region bounded by $p - |I|$ fixed Jordan curves C_k, $k \in \{1,..., p\} \setminus I$ and $p - |I|$ curves $C_k' = g_k(\tau, C_k)$. Such a fundamental region exists if N and ϵ are sufficiently small.

For ψ_π and ψ_σ the proofs are similar, but simpler.

LEMMA 4. *Let* $I \subset \{1,..., p\}$, $|I| < p$, *and let* $k \in \{1,..., p\} \setminus I$, $K = I \cup \{k\}$. *Let* φ *be one of the functions* ψ_j, ψ_π *or* ψ_σ, *with* $\sigma \in \Sigma_K$. *If*

$$\{\tau_\nu\} \subset \delta^I \mathfrak{S}_p, \lim_{\nu \to \infty} \tau_\nu = \tau_0 \in \delta^K \mathfrak{S}_p, \tag{6.21}$$

then

$$\lim_{\nu \to \infty} \varphi(\tau_\nu, z) = \varphi(\tau_0, z) \quad \text{for} \quad z \in \Omega'_{\tau_0}. \tag{6.22}$$

Proof. We consider first the case when $\varphi = \psi_\pi$. To simplify writing, assume that $k \neq 1$.

For $\tau \in \mathfrak{S}_p^K$ let $G_{\tau K}$ denote the subgroup of G_τ generated by all $g_j(\tau, \cdot), j \notin K$, that is, the image of $\hat{G}_K$ in G_τ and let $\Omega'_{\tau K}$ denote the region of discontinuity of $G_{\tau K}$, with the points $a_i(\tau)$, $1 \leqslant i \leqslant 2p$, and all their images under $G_{\tau K}$ removed. Arguing as in Section 2 one verifies that the set $\mathfrak{F}^* \mathfrak{S}_p^K$ of pairs (τ, z) with $\tau \in \mathfrak{S}_p^K$, $z \in \Omega'_{\tau K}$ is a domain in $\mathbb{C}^{3p-2}$. Now set, for $(\tau, z) \in \mathfrak{F}^* \mathfrak{S}_p^K$,

$$\hat{\Psi}(\tau, z) = \sum_{\gamma \in G_{\tau K}} \Psi_\pi[\tau, \gamma(z)] \, \gamma'(z)^q. \tag{6.23}$$

Of course, if $|K| = p$, the series in (6.23) consists of only one term, with $\gamma = \mathrm{id}$. On the other hand, if $\tau \in \delta^K \mathfrak{S}_p$ then $G_{\tau K} = G_\tau$ and $\hat{\Psi}(\tau, z) = \psi_\pi(\tau, z) = \varphi(\tau, z)$. In particular,

$$\hat{\Psi}(\tau_0, z) = \varphi(\tau_0, z). \tag{6.24}$$

One verifies, by repeating the arguments used in proving Lemmas 2 and 3, that the series in (6.23) converges absolutely and that

$$\hat{\Psi}(\tau, z) \text{ depends holomorphically on } (\tau, z) \in \mathfrak{F}^* \mathfrak{S}_p{}^K. \tag{6.25}$$

We also have, for $(\tau, z) \in \mathfrak{F} \delta^I \mathfrak{S}_p$ an absolutely convergent expansion

$$\varphi(\tau, z) = \psi_\pi(\tau, z) = \sum_{n=0}^\infty \hat{\Psi}[\tau, \gamma_n(z)] \gamma_n'(z)^q \tag{6.26}$$

where $\mathrm{id} = \gamma_0, \gamma_1, \gamma_2, \ldots$ is a complete list of distinct right coset representatives of G_τ modulo $G_{\tau K}$. One sees this by substituting (6.23) into (6.26) whereupon the series in (6.26) becomes identical with that in (6.13).

Now set, for $\nu = 1, 2, \ldots$, and $z \in \Omega'_{\tau_\nu K}$

$$R_\nu(z) = \sum_{n=1}^\infty \hat{\Psi}[\tau_\nu, \gamma_n(z)] \gamma_n'(z)^q. \tag{6.27}$$

In view of (6.26), (6.24) and (6.25), assertion (6.22) will be shown once we show that

$$\lim_{\nu \to \infty} R_\nu(z) = 0 \quad \text{for} \quad z \in \Omega'_{\tau_0}. \tag{6.28}$$

For large ν, we may assume that the τ-region ω_ν is bounded by $p - |I| - 1$ fixed curves C_j, surrounding $a_{2j-1}(\tau)$, $j \in \{1, \ldots, p\} \setminus K$, $p - |I| - 1$ curves $C'_{j\nu} = g_j(\tau_\nu, C_j)$, a curve $C_{k\nu}$ surrounding $a_{2k-1}(\tau)$, of diameter $|C_{k,\nu}| = o(1)$, $\nu \to \infty$, and the curve $C'_{k\nu} = g_k(\tau_\nu, C_{k\nu})$. The $2p - 2|I| - 2$ curves C_j, $C'_{j\nu}$, $j \in \{1, \ldots, p\} \setminus K$, bound a standard fundamental region $\hat{\omega}_\nu$ for $G_{\tau_\nu K}$ belonging to the generators $g_j(\tau_\nu, \cdot)$, $j \in \{1, \ldots, p\} \setminus K$. We note that $\lim C'_{j\nu} = C'_{j0}$, where C_j, C'_{j0} are $2p - 2|I| - 2$ boundary curves of a τ_0 region ω_0.

Using Proposition 6 and estimates (6.20) we verify that

$$\iint_{\hat{\omega}_\nu} |\hat{\Psi}(\tau_\nu, z)| \lambda_{\tau_\nu}(z)^{2-q} \, dx \, dy = O(1), \quad \nu \to \infty. \tag{6.29}$$

This implies that for a properly chosen fixed circle c, surrounding $a_{2k-1}(\tau_\nu)$, and for all large ν,

$$\max_{z \in c} |\hat{\psi}(\tau_\nu, z)| = O(1), \qquad \nu \to \infty. \tag{6.30}$$

Now, $\hat{\Psi}(\tau_\nu, z)$ is either holomorphic in the domain interior to c, or has there only one singularity, a pole at $a_{2k-1}(\tau_\nu)$ of some fixed order, at most $q - 1$; what happens depends on the choice of π but not on ν. Hence we conclude from (6.30) and (6.20) that there is constant M_0 and M such that, for large ν and z interior to c,

$$|\hat{\Psi}(\tau_\nu, z)| \leqslant \frac{M_0}{|z - a_{2k-1}(\tau_\nu)|^{q-1}}$$

so that

$$|\hat{\Psi}(\tau_\nu, z)| \lambda_{\tau_\nu}(z)^{2-q} \leqslant \frac{M}{|z - a_{2k-1}(\tau_\nu)|^{3/2}}; \tag{6.31}$$

here M_0 and M are constants independent of ν.

We may assume that the coset representatives $\gamma_1, \gamma_2, \ldots$ in (6.27) are chosen so that each $\gamma_n(\omega_\nu)$ lies either in the domain interior to $C_{k\nu}$ or in the domain interior to $C'_{k\nu}$. (This can be achieved by premultiplying γ_n be a properly chosen element of $G_{\tau K}$, which does not change the coset of γ_n.) Since $\gamma_n(\omega_\nu) \cap \omega_\nu = \phi$ for $n > 0$ and $\gamma_n(\omega_\nu) \cap \gamma_m(\omega_\nu) = \phi$ for $n \neq m$, we obtain that

$$\gamma_1(\omega_\nu) \cup \gamma_2(\omega_\nu) \cup \cdots \subset \hat{\omega}_\nu \backslash \omega_\nu.$$

Therefore

$$\iint_{\omega_\nu} |R_\nu(z)| \lambda_{\tau_\nu}(z)^{2-q} \, dx \, dy$$

$$\leqslant \sum_{n=1}^{\infty} \iint_{\omega_\nu} |\hat{\Psi}[\tau_\nu, \gamma_n(z)]| \, |\gamma_n'(z)|^q \lambda_{\tau_\nu}(z)^{2-q} \, dx \, dy$$

$$= \sum_{n=1}^{\infty} \iint_{\gamma_n(\omega_\nu)} |\hat{\Psi}(\tau_\nu, z)| \lambda_{\tau_\nu}(z)^{2-q} \, dx \, dy$$

$$\leqslant \iint_{\hat{\omega}_\nu/\omega_\nu} |\hat{\Psi}(\tau_\nu, z)| \lambda_{\tau_\nu}(z)^{2-q} \, dx \, dy. \tag{6.32}$$

Since $\hat{\omega}_\nu \backslash \omega_\nu$ is contained in the union of the domains interior to

$C_{k\nu}$ and $C'_{k\nu}$, and these domains have diameters converging to 0, we conclude by (6.31) and (6.32) that

$$\lim_{\nu \to \infty} \iint_{\omega_\nu} |R_\nu(z)| \lambda_{\tau_\nu}(z)^{2-q} \, dx \, dy = 0. \tag{6.33}$$

This implies that (6.28) holds.

The proof for the case when $\varphi = \psi_\sigma$ is very similar and will not be carried out in detail. We proceed to prove the lemma for $\varphi = \psi_j$. We distinguish three cases: $j \in I$, $j \notin K$ and $j = k$. The proof for case $j \in I$ is very similar to the one given above for ψ_π; we shall not carry out the details.

We consider next case $j \notin K$. To simplify writing we assume that $1 \neq j$ and $1 \neq k$. We use the symbols $G_{\tau K}$, $\Omega'_{\tau K}$, $\mathfrak{F}^* \mathfrak{S}_p{}^K$ with the same meaning as above, and for $(\tau, z) \in \mathfrak{F}^* \mathfrak{S}_p{}^K$ we define

$$\hat{\Psi}(\tau, z) = \sum_{n=0}^{\infty} \Psi_j[\tau, \hat{\gamma}_n(z)] \hat{\gamma}_n'(z)^q, \tag{6.34}$$

where Ψ_j is defined by (6.8) and $\hat{\gamma}_0 = \text{id}$, $\hat{\gamma}_1, \hat{\gamma}_2, \ldots$ is a complete list of distinct right coset representatives of $G_{\tau K}$ modulo $G_\tau{}^j$. For $\tau = \tau_0$, $G_{\tau_0 K} = G_{\tau_0}$, so that (6.24) still holds. The absolute convergence of (6.34) and the fact that (6.25) still holds, follow by repeating the arguments used in proving Lemmas 2 and 3. One verifies also that (6.31) still holds, so that the proof of (6.22) reduces to showing that the function R_ν defined by (6.27) satisfies (6.33).

We define, as in the proof for the case $\varphi = \psi_\pi$, the fundamental regions $\hat{\omega}_\nu$ and ω_ν, and obtain inequality (6.32). We also note that the two curves, C_j and $C'_{j\nu}$, bound a fundamental region $\check{\omega}_\nu$ for the group $G_{\tau_\nu}^j$. Now, by Proposition 6,

$$\iint_{\hat{\omega}_\nu} |\hat{\Psi}(\tau_\nu, z)| \lambda_{\tau_\nu}(z)^{2-q} \, dx \, dy \leq \iint_{\check{\omega}_\nu} |\Psi_j(\tau_\nu, z)| \lambda_{\tau_\nu}(z)^{2-q} \, dx \, dy.$$

For $\nu \to \infty$ the right side converges to

$$\iint_{\check{\omega}_0} |\Psi_j(\tau_0, z)| \lambda_{\tau_0}(z)^{2-q} \, dx \, dy < \infty$$

where $\check{\omega}_0$ is a fundamental region for the group $G_{\tau_0}^j$. Since $\hat{\Psi}(\tau_\nu, z)$ is holomorphic at $a_{2k-1}(\tau_\nu)$ and at $a_{2k}(\tau_\nu)$, we conclude that, for large ν and for z close to $a_{2k-1}(\tau_0)$ or to $a_{2k}(\tau_0)$, $|\hat{\Psi}(\tau_\nu, z)| \lambda_{\tau_\nu}(z)^{2-q}$ is uniformly bounded. Now we may use, as for $\varphi = \psi_\pi$, the inequality (6.31) to show that (6.32) holds.

It remains to consider the case $j = k$; we assume again that $j = k \neq 1$. Recalling the definition of $\psi_j(\tau, z)$ we write, for $\nu = 1, 2,...$

$$\psi_j(\tau_\nu, z) = \hat{\Psi}(\tau_\nu, z) + R_\nu(z),$$

where

$$\hat{\Psi}(\tau, z) = \sum_{\gamma \in G_{\tau K}} \Psi_k[\tau, \gamma(z)] \gamma'(z)^q,$$

$$R_\nu(z) = \sum_{n=1}^{\infty} \Psi_k[\tau_\nu, \check{\gamma}_n(z)] \check{\gamma}_n'(z)^q$$

where $\check{\gamma}_1, \check{\gamma}_2, ...$ is a complete list of those right coset representatives of G_τ modulo G_τ^j which cannot be represented by elements of $G_{\tau K}$.

Except for the new definition of Ψ and R_ν, we retain the notations introduced in the previous arguments. Relations (6.24) and (6.25) still hold, so that we must prove (6.28).

For each fixed $\nu = 1, 2,...$, the domains $\check{\gamma}_n(\omega_\nu)$, $n = 1, 2,...$, are disjoint and lie in the union $\check{\omega}_n$ of the domains into which the Möbius transformations $g_j(\tau_\nu, \cdot)$, $g_j^{-1}(\tau_\nu, \cdot)$, $j \in \{1,...,p\}\backslash K$, map the two components of $\hat{\omega}_\nu \backslash \omega_\nu$. We note that

$$\lim_{\nu \to \infty} \check{\omega}_\nu = \check{\omega}_0 \tag{6.35}$$

is a finite set consisting of the images of $a_{2k-1}(\tau_0)$ and $a_{2k}(\tau_0)$ under $g_j(\tau_0, \cdot)$ and $g_j^{-1}(\tau_0, \cdot)$, $j \in \{1,...,p\}\backslash K$. In a neighborhood of $\check{\omega}_0$ the functions $\Psi_k(\tau_\nu, z) \lambda_{\tau_\nu}(z)^{2-q}$ are uniformly bounded, for large ν.

A calculation analogous to one carried out before leads to the inequality

$$\iint_{\omega_\nu} |R_\nu(z)| \lambda_{\tau_\nu}(z)^{2-q} \, dx \, dy \leqslant \iint_{\check{\omega}_\nu} |\Psi_k(\tau_\nu, z)| \lambda_{\tau_\nu}(z)^{2-q}. \tag{6.36}$$

In view of (6.35) and the boundedness statement made above, the right side in (6.34) is $o(1)$ for $\nu \to \infty$. This implies (6.28).

LEMMA 5. *The functions $\psi_j(\tau, z)$ and $\psi_\pi(\tau, z)$ are holomorphic functions of $(\tau, z) \in \mathfrak{F} \mathfrak{S}_p{}^*$. The functions $\psi_\sigma(\tau, z)$, with $\sigma \in \Sigma_I$, are holomorphic functions of $(\tau, z) \in \mathfrak{F} \mathfrak{S}_p{}^I$.*

Proof. We establish only the assertion concerning ψ_j, the other assertions are proved similarly.

Let X_i, $i = 1,..., p$, be the analytic hyperplane in $\mathfrak{F}\mathfrak{S}_p{}^*$ defined by $t_i(\tau) = 0$. Set $X = X_1 \cup X_2 \cup \cdots \cup X_p$, let Y be the union of all intersections $X_i \cap X_j$, $i \neq j$, and set $X_0 = X \backslash Y$. By Lemma 3, ψ_j is holomorphic in $\mathfrak{F}\mathfrak{S}_p = \mathfrak{F}\mathfrak{S}_p{}^* \backslash X$. By Lemma 4, ψ_j has a continuous extension to X_0. Hence ψ_j is holomorphic in $\mathfrak{F}\mathfrak{S}_p \cup X_0 = \mathfrak{F}\mathfrak{S}_p{}^* \backslash Y$. Since $\dim Y < \dim \mathfrak{F}\mathfrak{S}_p{}^* - 1$, ψ_j has a holomorphic extension to Y. Using Lemma 4 once more, we conclude that this extension coincides with ψ_j.

LEMMA 6. *Let* $\pi_1,..., \pi_m$ *be* $m = 2(p-1)(q-1) - 1$ *linearly independent polynomials of degree at most* $2(p-1)(q-1) - 2$. *The* $(2q-1)(p-1)$ *functions* $\psi_1(\tau, z),..., \psi_p(\tau, z), \psi_{\pi_1}(\tau, z),..., \psi_{\pi_m}(\tau, z)$ *are linearly independent for every* $\tau \in \delta^I \mathfrak{S}_p$, *provided that* $|I| \geq p - 1$.

Proof. If $|I| = p$, then, for $\tau \in \delta^I \mathfrak{S}_p$, $\psi_j = \Psi_j$, $\psi_\pi = \Psi_\pi$ and the linear independence is evident.

If $|I| = p - 1$, assume, for the sake of simplicity, that $I = \{2,..., p\}$. For $\tau \in \delta^I \mathfrak{S}_p$, the p functions ψ_j are

$$1/z^q \quad \text{and} \quad \sum_{n=-\infty}^{+\infty} \left(\frac{(a'-a)\lambda^n}{(\lambda^n z - a)(\lambda^n z - a')} \right)^q$$

where

$$\lambda = 1/t_1(\tau), \qquad a = a_{2k-1}(\tau), \qquad a' = a_{2k}(\tau), \qquad k = 2,..., p;$$

whereas the m functions ψ_{π_i} may be chosen as

$$\frac{1}{z^{q-1}} \sum_{n=-\infty}^{+\infty} \frac{\lambda^n}{(\lambda^n z - 1)(\lambda^n z - a)},$$

$$\lambda = 1/t_1(\tau), \qquad a = a_s(\tau), \qquad s = 4, 5,..., 2p,$$

and

$$\frac{1}{z^{q-1}} \sum_{n=-\infty}^{+\infty} \frac{\lambda^n}{(\lambda^n z - a)^{1+r}},$$

$$\lambda = 1/t_1(\tau), \qquad a = a_i(\tau), \qquad i = 3, 4,..., 2p, \qquad r = 2, 3,..., q - 2.$$

The linear independence follows by considering the behavior of these functions at the points $a_i(\tau)$, $i = 2, 3,..., p$.

LEMMA 7. *Given a point* $\tau_0 \in \delta^I \mathfrak{S}_p$, *there are* $(2q-1)(p-1)$

holomorphic functions $\varphi_j(\tau, z)$, $(\tau, z) \in \mathfrak{F}\mathfrak{S}_p{}^I$, each a linear combination of functions ψ_j, ψ_π, ψ_σ, which are linearly independent functions of z for $\tau = \tau_0$.

Proof. In view of Lemma 6 we need only to consider the case $[I \mid < p - 1$. There are $(2q - 1)(p - 1)$ holomorphic q-differentials on the Riemann surface with nodes S_{τ_0}. Given any of those, say F, we can find a linear combination of the functions ψ_j, ψ_π which defines on S_τ a regular holomorphic q-differential F_0 with the same poles at the $|I|$ nodes. Then $F - F_0$ is represented in Ω'_{τ_0} by a regular q-form $\varphi(z)$ which satisfies the condition: $|\varphi(z) \lambda_{\tau_0}(z)^{2-q}| \leqslant$ const. Such a φ is called a cusp form (for G_{τ_0}) and is, by virtue of a general theorem proved in [12], a linear combination of functions of the form ψ_σ.

LEMMA 8. *Let $\varphi_j(\tau, z)$, $j = 1,..., (2q - 1)(p - 1)$, be linear combinations of the functions ψ_j, ψ_π, ψ_σ, all holomorphic in some $\mathfrak{F}\mathfrak{S}_p{}^I$. Let these functions be linearly independent, as functions of z, for some $\tau_0 \in \mathfrak{S}_p{}^I$. Then they are so for τ near τ_0.*

Proof. Let $z_0 \in \Omega'_{\tau_0}$ and let ϵ be a sufficiently small positive number. The Gram determinant

$$\det \iint_{|z-z_0|<\epsilon} \varphi_j(\tau, z) \overline{\varphi_k(\tau, z)} \, dx \, dy$$

is positive for $\tau = \tau_0$. Hence it is so for τ near τ_0.

LEMMA 9. *The set $Z \subset \mathfrak{S}_p{}^I$ such that the $(2q - 1)(p - 1)$ functions $\varphi_j(\tau, z)$ from Lemma 7 are linearly dependent if and only if $\tau \in Z$ is either empty or an analytic subvariety of pure codimension 1.*

Proof. We already know, by Lemma 8, that Z is closed. Let $\tau_0 \in Z$, and let $z_0 \in \Omega'_{\tau_0}$ be not a "τ_0-point," that is, not a "Weierstrass point" for regular q-forms on G_{τ_0}. This means that if $\hat{\varphi}_j(z)$, $j = 1,...,$ $(2q - 1)(p - 1)$ is any basis of regular q-forms for G_{τ_0}, the Wronskian determinant

$$\det \hat{\varphi}_j^{(k-1)}(z_0), \qquad j, k = 1,..., (2q - 1)(p - 1)$$

does not vanish. Such a z_0 exists, since τ_0-points form a discrete set. By Lemma 8, z_0 is also not a τ-point for τ close to τ_0. Now set

$$W(\tau) = \det \frac{\partial^{k-1} \varphi_i(\tau, z)}{\partial z^{k-1}} \bigg|_{z=z_0}, \qquad j, k = 1,..., (2q - 1)(p - 1).$$

A point $\tau \in \mathfrak{S}_p{}^I$ close to τ_0 belongs to Z if and only if $W(\tau) = 0$. Since $W(\tau)$ is holomorphic in a neighborhood of τ_0, the assertion of the lemma follows.

Lemmas 2, 5, 6, 7, and 9 contain Theorems 2 and 3.

7. Proof of Theorem 4

Let $\tau_0 \in \delta^I \mathfrak{S}_p$. To simplify writing we assume that the set I is of the form $\{r, r+1,..., p\}$; the further simplification resulting in the cases $r = 1$ and $I = \phi$, will be self-evident.

Let C_i, $i = 1, 2,..., r-1$ and $C_i' = g_i(\tau_0, C_i)$ be the boundary curves of a τ_0-region, and let C_j, $j = r, r+1,..., p$ be small circles about $a_{2r-1}(\tau_0), a_{2r+1}(\tau_0),..., a_{2p-1}(\tau_0)$. If τ is close to τ_0 the curves C_i and $g_i(\tau, C_i)$, with $t_i(\tau) \neq 0$, are the boundaries of a τ-region.

From the theory of Abelian differentials on closed Riemann surfaces we conclude that there is a uniquely determined basis $\varphi_j(\tau, z)$, $j = 1,..., p$, of regular 1-forms for G_τ, τ close to τ_0, determined by the conditions

$$\int_{C_j} \varphi_i(\tau, z)\, dz = \begin{cases} 1 & \text{if } i = j, \\ 0 & \text{if } i \neq j. \end{cases}$$

Theorem 4 will be proved if we show that $\varphi_i(\tau, z)$ is a holomorphic function of (τ, z) for $\tau \in N$ (sufficiently small neighborhood of τ_0) and $z \in \Omega_\tau'$.

Let $\check{\phi}(\tau, z)$ be a linear combination of the functions from Theorem 3 belonging to $q = 2$ and to the point τ_0, and such that the regular 2-form $\check{\phi}(\tau_0, z)$ has a double pole at every point $a_{2k-1}(\tau)$ if $t_k(\tau) = 0$. Then $\check{\phi}(\tau_0, z)$ has exactly

$$4(p - |I| - 1) + 2 \cdot 2 |I| = 4p - 4$$

zeros in Ω_{τ_0} which are not G_{τ_0} equivalent. We can choose $\check{\phi}$ so that these zeros are all simple. There is a τ_0-region ω_{τ_0} such that $\check{\phi}(\tau_0, z)$ vanishes at $4p - 4$ distinct interior points of ω_{τ_0}. We conclude that there exist $4p - 4$ holomorphic functions of τ (defined in a neighborhood of τ_0) $\zeta_1(\tau),..., \zeta_{4p-4}(\tau)$, whose values are distinct interior points of ω_{τ_0}, no two of which are equivalent under G_τ, such that

$$\check{\phi}(\tau, \zeta_i(\tau)) = 0, \quad i = 1,..., 4p - 4. \tag{7.1}$$

Next, let $\hat{\varphi}_1(\tau, z), \ldots, \hat{\varphi}_{5p-5}(\tau, z)$ be the functions from Theorem 3 belonging to $q = 3$ and the point τ_0. We want to represent every $\varphi_j(\tau, z)$ in the form

$$\varphi_j(\tau, z) = \frac{1}{\check{\varphi}(\tau, z)} \sum_{k=1}^{5p-5} A_k{}^j(\tau)\, \hat{\varphi}_k(\tau, z). \qquad (7.2)$$

For a fixed j, the coefficient $A_k{}^j$ must satisfy first of all the conditions

$$\sum_{k=1}^{5p-5} A_k{}^j(\tau)\, \hat{\varphi}_k(\tau, \zeta_i(\tau)) = 0, \qquad i = 1, \ldots, 4p - 4 \qquad (7.3)$$

and also the conditions

$$\sum_{k=1}^{5p-5} A_k{}^j(\tau) \int_{C_i} \frac{\hat{\varphi}_k(\tau, z)}{\check{\varphi}(\tau, z)}\, dz = \begin{cases} 1 & \text{if } j = i \\ 0 & \text{if } j \neq i \end{cases}, \quad i = 1, \ldots, p. \qquad (7.4)$$

These are $5p - 4$ conditions for $5p - 5$ "unknowns." Assume, however, that, for some τ, (7.3) holds for $i = 1, \ldots, 4p - 5$. Then $\varphi_j(\tau, z)$ induces on the Riemann surface Ω_τ/G_τ an Abelian differential which has at most simple poles at the points corresponding to $\zeta_{4p-4}(\tau)$ and to the points $a_{2k-1}(\tau)$, $a_{2k}(\tau)$ with $t_k(\tau) = 0$. But the residues at the images of $a_{2k-1}(\tau)$ and $a_{2k}(\tau)$ cancel each other. (This is so since $\check{\varphi}$ and all $\hat{\varphi}_k$ are regular forms.) Hence the residue at the image of $\zeta_{4p-4}(\tau)$ is 0. Thus the first $4p - 5$ equations (7.3) imply the last.

One verifies that the homogeneous system corresponding to the equations (7.3) and (7.4) has only the trivial solution. Hence the nonhomogeneous system is uniquely solvable, and the solutions $A_k{}^j(\tau)$ are holomorphic functions of τ.

This established the holomorphic dependence of $\varphi_j(\tau, z)$ on $\tau \in N$.

References

1. L. V. AHLFORS, "Lectures on Quasiconformal Mappings," Van Nostrand, Princeton, NJ, 1966.
2. L. V. AHLFORS AND L. SARIO, "Riemann Surfaces," Princeton University Press, Princeton, NJ, 1960.
3. T. AKAZA, (3/2)-dimensional measure of singular sets of some Kleinian groups, *J. Math. Soc. Japan* 24 (1972), 448–464.
4. T. AKAZA, Local property of the singular sets of some Kleinian groups, *Tôhoku Math. J.* 25 (1973), 1–22.

5. T. AKAZA AND T. SHIMAZAKI, The Hausdorff dimension of the singular sets of combination groups, *Tôhoku Math. J.* **25** (1973), 61–68.
6. P. APPELL AND E. GOURSAT, "Theorie des Fonctions Algébriques et de leurs Intégrales," Vol. II, Gauthiers-Villars, Paris, 1930.
7. A. F. BEARDON, The Hausdorff dimension of singular set of properly discontinuous groups, *Amer. J. Math.* **88** (1966), 722–736.
8. L. BERS, Completeness theorems for Poincaré series in one variable, *in* "Proceedings of the International Symposium on Linear Spaces," pp. 88–100, Jerusalem, 1960.
9. L. BERS, Uniformization by Beltrami equation, *Comm. Pure Appl. Math.* **14** (1961), 215–228.
10. L. BERS, Uniformization, moduli, and Kleinian groups, *Bull. London Math. Soc.* **4** (1972), 257–300.
11. L. BERS, Spaces of degenerating Riemann surfaces, *in* "Discontinuous Groups and Riemann Surfaces" (L. Greenberg, Ed.), *Annals of Math. Studies* **79**, pp. 43–55, 1974.
12. L. BERS, Poincaré series for Kleinian groups, *Comm. Pure Appl. Math.* **26** (1973), 667–672.
13. L. BERS, On moduli of Kleinian Groups, *Uspehi Mat. Nauk.* **29** (1974), 86–102.
14. L. BERS, On spaces of Riemann surfaces with nodes, *Bull. Amer. Math. Soc.* **80** (1974), 1219–1222.
15. V. CHUCKROW, On Schottky groups with applications to Kleinian groups, *Ann. of Math.* **88** (1968), 47–61.
16. L. R. FORD, "Automorphic Functions," Second edition, Chelsea, New York, 1951.
17. D. HEJHAL, On Schottky and Teichmüller spaces, *Advances in Math.* **15** (1975), 133–156.
18. I. KRA, On spaces of Kleinian groups, *Comment. Math. Helv.* **47** (1972), 53–69.
19. O. LEHTO AND K. I. VIRTANEN, "Quasikonforme Abbildungen," Springer-Verlag, Berlin, 1965.
20. A. MARDEN, The geometry of finitely generated Kleinian groups, *Ann. of Math.* **99** (1974), 383–462.
21. A. MARDEN, Schottky groups and circles, *in* "Contributions to Analysis" (L. V. Ahlfors *et al.*, Eds.), Academic Press, New York, 1974.
22. B. MASKIT, A characterization of Schottky groups, *J. Analyse Math.* **19** (1967), 227–230.
23. B. MASKIT, Selfmaps on Kleinian groups, *Amer. J. Math.* **93** (1971), 840–856.
24. B. MASKIT, to appear.
25. P. J. MYREBERG, Die Kapazität der singulären Menge der linearen Gruppen, *Ann. Acad. Sci. Fenn.* **10** (1941), 1–19.

Biased Versus Unbiased Estimation

BRADLEY EFRON

Stanford University, Stanford, California 94305

DEDICATED TO STAN ULAM

> Statisticians have begun to realize that certain deliberately induced biases can dramatically improve estimation properties when there are several parameters to be estimated. This represents a radical departure from the tradition of unbiased estimation which has dominated statistical thinking since the work of Gauss. We briefly describe the new methods and give three examples of their practical application.

1. INTRODUCTION

Two young statisticians were interviewing for the same job. "Suppose," said the would-be employer, "that you observed n independent normally distributed random variables with mean θ and variance 1, say

$$x_i \stackrel{\text{ind}}{\sim} \mathcal{N}(\theta, 1) \qquad i = 1, 2, ..., n,$$

and on the basis of $\mathbf{x} \equiv (x_1, x_2, ..., x_n)$ I asked you to estimate the unknown parameter θ. What would you do?"

The first statistician, a quick fellow, answered "I would use the unbiased estimation rule

$$\delta^0(\mathbf{x}) \equiv \bar{x} \equiv \sum_{i=1}^{n} x_i/n.$$

It has minimum variance among all unbiased estimators (those satisfying $E_\theta \delta(\mathbf{x}) = \theta$ for all θ, "E_θ" indicating expectation when θ is the parameter value), and likewise among all translation invariant estimators ($\delta(\mathbf{x} + (c, c, c, ..., c)) = \delta(\mathbf{x}) + c$). Moreover it is minimax (minimizes the maximum expected squared error), admissable (no competing estimation rule has smaller expected squared error for all values of θ), and it is the maximum likelihood estimator (choosing $\theta = \bar{x}$ maximizes, among all values of θ, the probability of obtaining the value of $\mathbf{x}$ actually observed)."

Naturally the employer was impressed, but a sense of fairness compelled him to give the second statistician his chance. "Well, sir," he responded after an embarrassing silence, "I think I might try

$$\delta^1(\mathbf{x}) \equiv \bar{x} - \Delta(\bar{x})$$

where $\Delta(-x) = -\Delta(x)$ and for $x \geq 0$,

$$\Delta(x) \equiv \frac{1}{2\sqrt{n}} \min(\sqrt{n}\, x, \Phi(-\sqrt{n}\, x)),$$

where of course

$$\Phi(t) \equiv \frac{1}{\sqrt{2\pi}} \int_{-\infty}^{t} e^{-(1/2)s^2}\, ds$$

represents the normal distribution function as usual."

After the laughter had died down the employer asked him how he could justify such a bizarre recommendation. The second statistician admitted that δ^1 had none of the nice properties of δ^0. It wasn't unbiased, wasn't invariant, wasn't minimax, and wasn't even admissable. "On the other hand my guess will be closer than my competitor's to the true value of θ more than half of the time, no matter what θ is!" After an easy computation, which the reader may want to do for himself, it turned out that indeed

$$\text{Prob}_\theta\{|\,\delta^1(\mathbf{x}) - \theta\,| < |\,\delta^0(x) - \theta\,|\} > \tfrac{1}{2}$$

for all θ. He got the job of course. As the employer put it, "Why should I settle for second best?"

Now it is possible to give a good argument that the first statistician deserved the job, and that δ^0 is really a better estimator than δ^1. Nevertheless our shaggy statistician story has a serious point: in more complicated situations involving the estimation of several parameters at the same time, statisticians have begun to realize that biased estimation rules have definite advantages over the usual unbiased estimators. This represents a radical departure from the tradition of unbiased estimation which has dominated statistical thinking since Gauss' development of the least squares method. A brief description of the new theory is given in Section 2 followed in Section 3 by examples of its application to three data analysis problems.

My purpose in writing this article is to whet the interest of non-statisticians in the use of these new estimators. An equivalent of the Surgeon-General's warning may be in order: these methods are not perfectly understood yet, and are still the subject of heated controversy among statisticians (see the discussion following [7]). Most of the material presented here is abstracted from a series of articles by Carl Morris and myself [3–8]. The unsatisfied reader may wish to read [6] for more theoretical background and [8] for a fuller description of data analysis procedures.

2. The James–Stein Estimator

Suppose we wish to estimate several parameters $\theta_1, \theta_2, ..., \theta_k$, and for each one we observe n independent random variables,

$$x_{ij} \stackrel{\text{ind}}{\sim} \mathcal{N}(\theta_i, \sigma^2) \quad \begin{matrix} i = 1, 2, ..., k, \\ j = 1, 2, ..., n. \end{matrix} \tag{2.1}$$

Here σ^2 is the common variance of the x_{ij}, which for convenience of presentation we assume known although this isn't necessary for the theory which follows. A sufficient statistic (one which contains all the information) for θ_i is

$$y_i \equiv \bar{x}_i \equiv \sum_{j=1}^{n} x_{ij}/n, \tag{2.2}$$

the more exact statement being that the vector $\mathbf{y} = (y_1, y_2, ..., y_k)$ is sufficient for the parameter vector $\mathbf{\theta} = (\theta_1, \theta_2, ..., \theta_k)$. The y_i are independently normally distributed with mean θ_i and variance $D \equiv \sigma^2/n$,

$$y_i \stackrel{\text{ind}}{\sim} \mathcal{N}(\theta_i, D) \quad i = 1, 2, ..., k. \tag{2.3}$$

From the vector $\mathbf{y}$ we wish to infer the value of $\mathbf{\theta}$. It is customary to do so by means of an estimation rule

$$\mathbf{\delta}(\mathbf{y}) \equiv (\delta_1(\mathbf{y}), \delta_2(\mathbf{y}), ..., \delta_k(\mathbf{y})),$$

$\delta_i(\mathbf{y})$ being the estimator of θ_i.

Model (2.3) is the simplest case of Gauss' "linear model"

$$\mathbf{y} = \mathbf{\theta M} + \mathbf{\epsilon} \tag{2.4}$$

where **y** is a $1 \times r$ vector of observed variables, **θ** a $1 \times k$ vector of unknown parameters, and **M** a known $k \times r$ "structure matrix" which we assume to be of rank k with $k \leq r$ to avoid some nasty details. (For (2.3), $r = k$ and **M** = **I**, the $k \times k$ identity matrix.) The noise vector $\epsilon = (\epsilon_1, \epsilon_2, ..., \epsilon_r)$ is assumed to have independent normal components with mean 0 and common variance, that being D in (2.3).

Gauss suggested the "least squares estimator" for this situation,

$$\delta^0(\mathbf{y}) \equiv \mathbf{y}\mathbf{M}^T(\mathbf{M}\mathbf{M}^T)^{-1},$$

which is also "maximum likelihood" and "minimum variance unbiased" for **θ**. (Among unbiased estimators, $E_\theta \delta(\mathbf{y}) = \theta$ for all **θ**, δ^0 has minimum variance component by component, and also for any linear combination of the components.)

The linear model is used extensively in all the sciences, δ^0 usually being the estimator of choice. It turns out that (2.4) can be reduced to (2.3) by suitable linear transformations, so that the simpler structure (2.3) actually has all the statistical content of the full linear model. The least squares estimator δ^0 is simply

$$\delta^0(\mathbf{y}) = \mathbf{y} \tag{2.5}$$

in this case. This says the obvious: y_i is the best unbiased estimator of θ_i in the model (2.3).

In 1960[1] James and Stein [9] suggested another estimator for **θ**,

$$\delta^1(\mathbf{y}) \equiv \left[1 - \frac{(k-2)D}{S}\right]\mathbf{y}, \tag{2.6}$$

where

$$S \equiv \|\mathbf{y}\|^2 \equiv \sum_{i=1}^{k} y_i^2. \tag{2.7}$$

(*Here and henceforth we assume* $k \geq 3$.) At first sight δ^1 appears ridiculous since the estimate of θ_i,

$$\delta_i^1 = \left[1 - \frac{(k-2)D}{S}\right] y_i,$$

depends not only on y_i but on the seemingly irrelevant values of y_j, $j \neq i$. Ah, but beware!

---

[1] Stein suggested a similar estimator in his 1955 paper [11].

Suppose one measures the performance of an estimator δ by its expected sum of squared error risk,

$$R(\theta, \delta) \equiv E_\theta L(\theta, \delta(\mathbf{y})) \equiv E_\theta \sum_{i=1}^{k} (\delta_i(\mathbf{y}) - \theta_i)^2$$
$$\equiv E_\theta \| \delta - \theta \|^2. \tag{2.8}$$

(In the usual parlance $R(\theta, \delta)$ is the *risk function* of δ under the *loss function* $L(\theta, \delta) = \| \delta - \theta \|^2$.) For $\delta^0 = \mathbf{y}$ we have from (2.3) that

$$R(\theta, \delta^0) = kD$$

for all θ. We know that δ^0 minimizes $R(\theta, \delta)$ among all unbiased estimators δ. However James and Stein showed that

$$R(\theta, \delta^1) < kD$$

for all θ, so that in terms of total squared error risk δ^1 is uniformly preferable to δ^0!

The advantage enjoyed by δ^1 is by no means trivial. Figure 1 compares

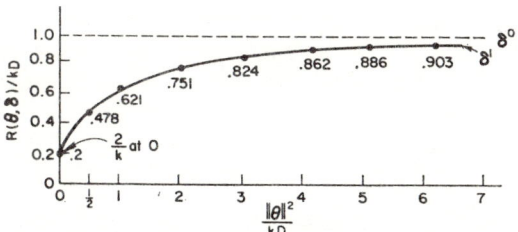

FIG. 1. Risk of the James–Stein rule δ^1 compared to that of the unbiased estimator δ^0, for $k = 10$.

$R(\theta, \delta^1)$ and $R(\theta, \delta^0)$, for $k = 10$, as a function of $\| \theta \|^2/kD$. We see that the most favorable case is $\theta = 0$ for which $R(\theta, \delta^1)/R(\theta, \delta^0) = 2/10$. (For dimension k, $R(0, \delta^1)/R(0, \delta^0) = 2/k$.)

Why haven't statisticians and users of statistics rushed to embrace this considerable improvement over the method of least squares? (They haven't—it's barely been used at all.) Several reasons can be given.

(i) *Blind stupid prejudice.* Unbiased estimators have been used on literally millions of real problems, with generally satisfactory results. The new estimators haven't. Their theoretical superiority has yet to be tested in the rigors of wide-spread application. Prejudice of this sort isn't really blind or stupid, just conservative.

(ii) *Counter-intuition.* We would expect that changing the origin and scale of our measurements would change any reasonable estimator in the obvious way,

$$\delta(b\mathbf{y} + (c, c, c, ..., c)) = b\delta(\mathbf{y}) + (c, c, ..., c)$$

for any numbers b and c. δ^1 does not have this *invariance* property. As mentioned before δ^1 is also statistically counter-intuitive in that it uses data other than y_i to estimate θ_i. If the different θ_i refer to obviously disjoint problems (e.g., θ_1 is the speed of light, θ_2 is the price of tea in China, θ_3 is the efficacy of new treatment for psoriasis, etc.) combining the data can produce a definitely uncomfortable feeling in the statistician.

(iii) *Bias.* The very name "unbiased" suggests the appeal of this concept to notions of scientific objectivity. All values of the parameter vector $\boldsymbol{\theta}$ are treated with an equal hand by δ^0. Not so with δ^1, which biases estimators toward the origin, the more so the closer $\mathbf{y}$ itself is to the origin. As a matter of fact (2.6) shows that for $S < (k-2)$, $\delta^1(\mathbf{y})$ is actually pulled past the origin in the direction opposite $\mathbf{y}$. It can be shown that if we legislate out this behaviour by refusing to go past the origin, that is we use the "plus-rule" estimator

$$\delta^{1+}(\mathbf{y}) \equiv \left[1 - \frac{(k-2)D}{S}\right]_+ \mathbf{y}, \qquad (2.9)$$

where $[x]_+ \equiv \max(0, x)$ as usual, then $R(\boldsymbol{\theta}, \delta^{1+}) < R(\boldsymbol{\theta}, \delta^1)$ for all $\boldsymbol{\theta}$, giving a reduction in risk for all $\boldsymbol{\theta}$. (The plus-rule and its variations are actually the proposed competitors to δ^0 in Section 3, but of course unless $S < k-2$ you can't tell the difference between δ^{1+} and δ^1.)

There is no need to give $\mathbf{0}$ the preferred position in the definition of δ^1 or δ^{1+}. For any vector $\boldsymbol{\mu} = (\mu_1, \mu_2, ..., \mu_k)$ define

$$\delta_i^{1+} \equiv \mu_i + \left[1 - \frac{(k-2)D}{S}\right]_+ (y_i - \mu_i), \qquad i = 1, 2, ..., k, \qquad (2.10)$$

where now

$$S \equiv \sum_{i=1}^{k} (y_i - \mu_i)^2 = \|\mathbf{y} - \boldsymbol{\mu}\|^2.$$

This rule also uniformly dominates δ^0 no matter what the choice of $\boldsymbol{\mu}$. By choosing $\boldsymbol{\mu}$ in different ways we can make $\delta^{1+}(\mathbf{y})$ take on any value we want within a sphere of radius $(k-2)^{1/2}$ centered at **y**. Even assuming we play fair and choose **µ** before observing **y** we are still biasing our results toward **µ**, a value of **θ** that may have vested interest for the statistician.

(iv) *Sum of squared-error loss.* What if we change the loss function from $L(\theta, \delta) = \|\delta - \theta\|^2$ to, say, $L(\theta, \delta) = \sum_{i=1}^{k} |\delta_i - \theta_i|$. Is δ^1 still preferable to δ^0? The answer is probably yes, though the mathematics speak less clearly than for $\|\theta - \delta\|^2$. Roughly speaking, present indications are that for any "ensemble" loss function, that is one that sums errors over all the coordinates, δ^1 or some variation of it will dominate δ^0. Even the assumption of normal distributions doesn't seem to be very important, see [2], [6].

Another example of an ensemble loss function is the multi-parameter analogue of that employed by the statistician who got the job in Section 1. It is possible to show that $\text{Prob}_\theta\{\|\delta^1 - \theta\| < \|\delta^0 - \theta\|\}$ is greater than $\frac{1}{2}$ for all values of **θ**. An explicit formula, which won't be derived here, is

$$\text{Prob}_\theta\{\|\delta^1 - \theta\| < \|\delta^0 - \theta\|\} = \text{Prob}\left\{\chi_k^2\left(\frac{\|\theta\|^2}{4D}\right) \leq \frac{\|\theta\|^2}{4D} + \frac{k-2}{2}\right\}, \quad (2.11)$$

where $\chi_k^2(\sum_1^k \sigma_i^2)$ is the distribution of $\sum_1^k w_i^2$, w_i being independent normal random variables with mean σ_i and variance 1. For k reasonably large (say ≥ 8) and $\|\theta\|^2/kD$ reasonably small (say ≤ 3), (2.11) gives quite high probabilities of $\|\delta^1 - \theta\|$ being less than $\|\delta^0 - \theta\|$, on the order of 90%.

It is actually the "sum" in "sum of squared errors" that is the crucial assumption. If we really aren't interested in $\theta_2, \theta_3, ..., \theta_k$, just θ_1, then $L_1(\theta, \delta) = (\delta_1 - \theta_1)^2$ is a more reasonable loss function than $\|\delta - \theta\|^2$. It is *not* true that $E_\theta(\delta_1^1 - \theta_1)^2 < E_\theta(\delta_1^0 - \theta_1)^2$ for all **θ**. As a matter of fact $E_\theta(\delta_1^1 - \theta_1)^2/E_\theta(\delta_1^0 - \theta_1)^2$ can be as large as about $\sqrt{k}$ for certain configurations of $\theta_1, \theta_2, ..., \theta_k$, (namely $\theta_1 = \sqrt{k}$, $\theta_2 = \theta_3 = \cdots = \theta_k = 0$).

There are the beginnings of a paradox here: we expect $|\delta_i^1 - \theta_i|$ to be smaller than $|\delta_i^0 - \theta_i|$ for a majority of the coordinates θ_i. (For example if $\|\theta\|^2/kD$ is near 1 and no one of the θ_i is enormously large then for large k it is possible to show that δ_i^1 will be closer for about 65% of the coordinates.) On the other hand for any one coordinate we can't guarantee we are doing better, even in expectation, and we may do considerably worse.

Morris and I have confronted those criticisms, at least the last 3, in our long series of papers. Some of our answers will be clear from the examples of Section 3. The truth is that δ^1 is less automatic than δ^0, does involve more judgmental factors in its use, and can lead to greater disasters if misused. Nevertheless, as the examples show, *not* using δ^1 or some close cousin can make one a very inefficient statistician, a disaster in its own right.

3. THREE EXAMPLES OF BIASED ESTIMATION IN PRACTICE

(i) 18 *Baseball Players*. Table I, column 1, shows the batting averages of 18 major league players after their first 45 times at bat in

TABLE I

1970 Batting Avergages for 18 Major League Players

i	Unbiased estimate $\delta_i^0 = y_i$	Parameter value θ_i	James–Stein est. δ_i^1		At bats, remainder of 1970
1	0.400	0.346	0.293	$(0.334)^a$	367
2	0.378	0.298	0.289	$(0.312)^a$	426
3	0.356	0.276	0.284	$(0.290)^a$	521
4	0.333	0.221	0.279		276
5	0.311	0.273	0.275		418
6	0.311	0.270	0.275		467
7	0.289	0.263	0.270		586
8	0.267	0.210	0.265		138
9	0.244	0.269	0.261		510
10	0.244	0.230	0.261		200
11	0.222	0.264	0.256		277
12	0.222	0.256	0.256		270
13	0.222	0.304	0.256		434
14	0.222	0.264	0.256		538
15	0.222	0.226	0.256		186
16	0.200	0.285	0.251		558
17	0.178	0.319	0.247	$(0.243)^a$	405
18	0.156	0.200	0.242	$(0.221)^a$	70
Player number	Batting average after 45 at bats $\bar{y} = 0.265$	Batting average remainder of season	$\bar{y} + [1 - (k-3)D/S](y_i - \bar{y})$ $= 0.265 + 0.212(y_i - \bar{y})$		Average remainder at bats $= 369.3$

^{a} Limited translation estimate. All other values agree with δ_i^1.

the 1970 season. These can be considered as unbiased estimates y_i of θ_i, the true probability of Player i getting a hit. Column 2 gives a much better estimate of θ_i, the batting average for Player i during the remainder of the season, based on an average of about 370 more at bats. We consider these to be the acutal θ_i, though a more careful analysis would include the sampling error in these numbers.

We now apply the James–Stein estimator in form (2.10), choosing all the μ_i equal to $\bar{y} \equiv \sum_1^k y_i/18$. Letting the data choose the μ_i in this way effectively removes one dimension ("one degree of freedom" in statistical jargon) from the problem, leading to the estimator

$$\delta_i^1 = \bar{y} + \left[1 - \frac{(k-3)D}{S}\right](y_i - \bar{y}),$$

$$S \equiv \sum_1^k (y_i - \bar{y})^2.$$

(3.1)

Notice that the lost dimension has changed $k-2$ to $k-3$. For this problem D is unknown (actually depending on θ_i) but from the properties of the binomial distribution we can estimate it by

$$D = \frac{\bar{y}(1-\bar{y})}{45} = 0.004332.$$

The resulting estimation rule, $\delta_i^1 = 0.265 + 0.212(y_i - \bar{y})$, is given in column 3. The losses are

$$\sum_1^{18} (\delta_i^0 - \theta_i)^2/D = 17.68, \quad \sum_1^{18} (\delta_i^1 - \theta_i)^2/D = 5.05 \qquad (3.2)$$

so δ_i^1 outperforms δ_i^0 by a factor of 3.50.

It is shown in [4] that if we follow the estimation rule δ_i^1 as closely as possible *subject to the constraint that no estimated value be more than one standard deviation D away from the unbiased estimate δ_i^0*, then (a) we still get most of the sum of squared errors risk reduction associated with δ_i^1; and (b) the maximum possible risk for individual components is reduced substantially. This "limited translation rule," an attempt to have our cake and eat it too vis-a-vis objection (iv) of Section 2, is also given in Table I. Notice that it does appear to protect Player 1 (Roberto Clemente!) from over shrinking toward the common mean.

A more careful treatment of this data is given in [8], but it is reassuring to see δ^1 working as predicted in a situation where no attempt has made to exactly satisfy the model (2.1).

(ii) *Ten Reaction Time Experiments.*[2] Each of ten subjects was asked to perform a certain task under seven different conditions. Let x_{ij} indicate the (natural) log reaction time of subject i under condition j, $i = 1, 2, \ldots, 10$, $j = 1, 2, \ldots, 7$. The two way analysis of variance model ("ANOVA")

$$x_{ij} = \mu + \alpha_i + \beta_j + \epsilon_{ij}, \qquad \sum_1^{10} \alpha_i = \sum_1^n \beta_j = 0, \qquad (3.3)$$

was used to analyze the data. Here μ is the overall mean, α_i the main effect for Subject i, β_j the main effect for Condition j, and ϵ_{ij} the random noise, assumed to be independent normal with mean 0 and variance σ^2,

$$\epsilon_{ij} \stackrel{\text{ind}}{\sim} \mathcal{N}(0, \sigma^2) \qquad i = 1, 2, \ldots, 10, \qquad j = 1, 2, \ldots, 7. \qquad (3.4)$$

For example a value of $\alpha_1 = 0.12$ means that Subject 1 would average 0.12 greater on the log scale than all 10 subjects together, in the absence of the noise from the ϵ_{ij}—that is he would be about 12% slower than the group average. In the discussion below we are only interested in estimating the patient main effects α_i from the data x_{ij}.

An unbiased estimate of α_i is

$$y_i \equiv \sum_{j=1}^7 x_{ij}/7 - \sum_{i=1}^{10} \sum_{j=1}^7 x_{ij}/70 \qquad (3.5)$$

for which a simple calculation gives

$$y_i \sim \mathcal{N}(\alpha_i, D), \qquad D = (9/70)\sigma^2. \qquad (3.6)$$

The y_i are not independent since they, like the α_i, must sum to zero. The applicable version of the James–Stein estimator is

$$\delta_i^1 = \left[1 - \frac{(k-3)D}{S}\right] y_i, \qquad S = \sum_1^k y_i^2, \qquad k = 10. \qquad (3.7)$$

This is really the same rule as (3.1) except here we are not interested in putting the overall mean back into the estimates.

---

[2] I am grateful to Dr. R. Angel of the Stanford Medical School for allowing me to abstract this data from a larger experiment he is conducting.

σ^2, hence D, is unknown to us, but can be estimated in the usual way from model (3.3). This gives an unbiased estimate $\hat{\sigma}^2$ independent of y_i, and we can take $\hat{D} = (9/70)\hat{\sigma}^2$. (For those familiar with ANOVA, $\hat{\sigma}^2$ is biased on a chi-square variable with 58 degrees of freedom. Actually it can be shown that a slightly biased estimator for D is preferable, but we shall ignore this small improvement.)

This whole experiment was repeated 10 times, yielding a total of 700 observations—10 Subjects, 7 conditions, 10 experiments. We can use these repetitions as a check on how a given estimator performed in any given experiment. Each column of Table 2 refers to an experiment analyzed separately from the others. The upper number in each box is $\delta_i^0 = y_i$, the unbiased estimate of α_i for that experiment. For example experiment 1 has $y_3 = -.16$, indicating that Subject 1 reacted about 16% faster than the average of the 10 subjects for that experiment. The lower number in each box is δ_i^1 as given in (3.7), with $\hat{D}$ substituted for D. $\hat{D}$ is given at the bottom of each column, along with the shrinkage factor $[1 - (k-3)\hat{D}/S]$ in (3.7).

For each i we can average the values of δ_i^0 over the 10 experiments to obtain a much more accurate unbiased estimate of α_i. We take these to be the true α_i values even though they still have some sampling variability in them. They are listed as the top numbers in the "combined" column. The last column of Table 2 compares δ_i^0 with δ_i^1 over the 10 experiments. If $(\delta_{ie}^0, \delta_{ie}^1)$ represent the two estimates for α_i in experiment e, $e = 1, 2, ..., 10$. Then the two numbers given are

$$\sum_{e=1}^{10} (\delta_{ie}^0 - \alpha_i)^2 \quad \text{and} \quad \sum_{e=1}^{10} (\delta_{ie}^1 - \alpha_i)^2. \tag{3.8}$$

The first of these is greater than the second for eight out of the ten Subjects. Overall

$$\sum_{i=1}^{10} \sum_{e=1}^{10} (\delta_{ie}^0 - \alpha_i)^2 = 1.025, \quad \sum_{i=1}^{10} \sum_{e=1}^{10} (\delta_{ie}^1 - \alpha_i)^2 = 0.759 \tag{3.9}$$

indicating that δ_i^1 was about 25% more accurate than δ_i^0 over all 10 experiments.

Notice that the two Subjects who do worse under δ_i^1 then δ_i^0 are the slowest (Subject 6) at the fastest (Subject 8, tied with Subject 3). This underscores the point that δ_i^1 can have high risk on the more unusual components θ_i. The limited translation modification mentioned previously can be used effectively here.

TABLE II

Ten Reaction-Time Experiments. Upper Number is Unbiased Estimate—
Lower Number is James-Stein Estimate on that Column's Data

Subject	Experiment →										Combined	$\sqrt{\Sigma_e (\delta_{ie}^0 - \alpha_i)^2}$ $\sqrt{\Sigma_e (\delta_{ie}^1 - \alpha_i)^2}$
	1	2	3	4	5	6	7	8	9	10		
1	−0.03	0.06	−0.08	−0.04	−0.05	0.15	0.05	0.03	−0.10	−0.06	−0.00	0.0545
	−0.01	0	−0.05	−0.02	−0.03	0.13	0.04	0.01	−0.07	−0.02	−0.00	0.0278
2	0.08	0.24	0.17	0.14	−0.06	0.37	0.20	0.11	0.26	−0.07	0.14	0.1704
	0.03	0	0.10	0.08	−0.04	0.32	0.14	0.04	0.18	−0.02	0.13	0.1389
3	−0.16	0.07	−0.32	−0.30	−0.28	−0.18	−0.19	−0.25	−0.08	−0.09	−0.18	0.1300
	−0.06	0	−0.18	−0.18	−0.19	−0.15	−0.13	−0.10	−0.06	−0.03	−0.17	0.0936
4	−0.04	0.02	0.02	0.04	−0.05	−0.05	−0.16	−0.02	0.13	−0.12	−0.03	0.0615
	−0.01	0	0.01	0.02	−0.03	−0.04	−0.11	−0.01	0.09	−0.04	−0.03	0.0256
5	0.08	0.01	0.03	−0.00	0.14	0.06	−0.22	−0.00	−0.23	−0.10	−0.02	0.1307
	0.03	0	0.02	−0.00	0.10	0.05	−0.15	−0.00	−0.16	−0.03	−0.02	0.0612

Subject												Total
6	0.25	0.15	0.17	0.35	0.21	0.24	0.29	0.02	0.28	0.22	0.22	0.0718
	0.09	0	0.10	0.21	0.14	0.21	0.20	0.01	0.19	0.07	0.20	0.1542
7	0.09	−0.10	−0.06	0.09	0.30	0.09	0.09	0.11	0.07	0.29	0.10	0.1431
	0.02	0	−0.03	0.05	0.20	0.08	0.06	0.04	0.05	0.09	0.09	0.0525
8	−0.13	−0.14	−0.16	−0.23	−0.23	−0.34	−0.18	−0.08	−0.15	−0.10	−0.18	0.0524
	−0.05	0	−0.09	−0.14	−0.16	−0.29	−0.13	−0.03	−0.10	−0.03	−0.17	0.1214
9	−0.09	−0.13	0.17	−0.03	0.04	−0.20	0.03	−0.08	−0.17	0.05	−0.04	0.1183
	−0.02	0	0.10	−0.02	0.03	−0.17	0.02	−0.03	−0.12	0.02	−0.04	0.0572
10	−0.04	−0.16	0.07	−0.02	−0.03	−0.14	0.09	0.17	−0.00	−0.02	−0.01	0.0898
	−0.01	0.00	0.04	−0.01	−0.02	−0.12	0.06	0.07	−0.00	−0.01	−0.01	0.0266
$\hat{D}$	0.011	0.020	0.012	0.014	0.011	0.007	0.010	0.010	0.011	0.016	0.00122	1.025/0.759
$\left[1 - \dfrac{7\hat{D}}{S}\right]$	0.35	−0.02	0.57	0.60	0.68	0.86	0.70	0.40	0.69	0.31	0.929	
$F_{9,54}$	1.25	0.82	1.93	2.09[a]	2.64[a]	6.01[b]	2.74[b]	1.38	2.66[a]	1.21	3.42[b]	

[a] Sig at 0.05 level.
[b] Sig at 0.01 level.

In experiment 2 the shrinkage factor comes out negative, and we use the plus-rule δ_i^{1+} to estimate all $\alpha_i = 0$. One biased estimation procedure that has been widely used is to run the usual F test for the hypothesis "all $\alpha_i = 0$," estimating α_i by 0 if the test accepts the hypothesis, and by δ_i^0 if the test rejects the hypothesis. This amounts to estimating α_i by

$$[1 - I(F)] y_i,$$

where $I(F)$ equals 1 or 0 as F is less or greater than some conventional value, usually the 95th percentile of F under the hypothesis all $\alpha_i = 0$. δ^{1+} is a smoother version of this same idea, being expressable as

$$\delta_i^{1+} = \left[1 - \frac{8.33}{10} \frac{1}{F}\right]_+ y_i \qquad (3.10)$$

in the case at hand. It is shown in [10] that δ^{1+} uniformly dominates the F-test procedure in terms of sum of squared error risk. The F value for each experiment is listed as "$F_{9,54}$" (indicating the proper degrees of freedom) in Table II.

Actually the 10 experiments were run under somewhat different conditions. The widely varying values of F in Table II suggest that the α_i themselves may have changed from experiment to experiment. The reader may wish to propose an estimate for α_{ie}, the ith subject main effect in experiment e, from the data in Table II. Hint: in each row we can shrink the values δ_{ie}^0, $e = 1, 2,..., 10$, toward their mean value α_i.

The lower numbers in the combined column were obtained by applying δ^1 to the α_i. There isn't much shrinkage effect because the α_i are individually quite accurate themselves.

(iii) *The Sunspot Data.* Figure 2 is a graph of the log spectrogram for the number of sunspots occurring annually from 1947–1924. The plotted point at each frequency is an unbiased estimate of the log power at the frequency.[3] The successive frequencies plotted are separated from each other by approximately 0.0057 cycles/year (0.0057 = 1/176 years). Assuming the sunspot generating mechanism is stationary over time, there are theoretical reasons for believing that the plotted estimates will be independent of each other with mean say θ_i, the log

---

[3] Those numbers are obtained as log $S_p + 0.573$ where S_p is the Schuster's spectrogram for the sunspot data taken from Table A.3.2 of [1]. Adding 0.573 compensates for the bias induced by taking logarithms.

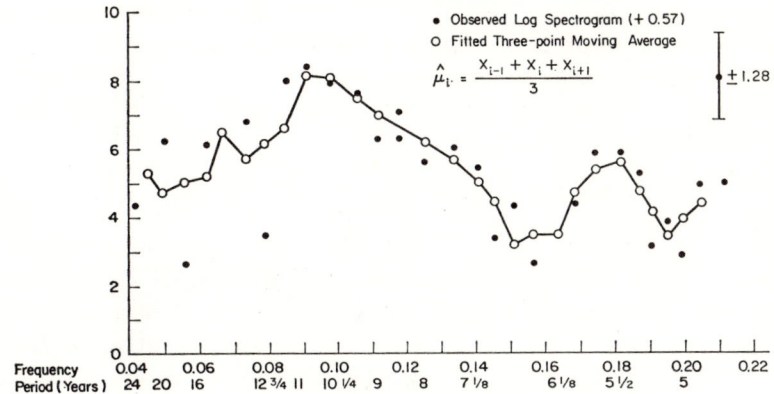

FIG. 2. Log Spectrogram for Schuster's sunspot data.

power at the ith frequency, and variance $\sigma^2 = (1.28)^2$. The sampling distribution is nonnormal (actually being the log of an exponential random variable), but this will not affect the analysis which follows.

We would like to estimate the "spectrum," that is the θ_i values. Spectral estimation is one area of statistics where biased estimates have traditionally played a preferred role. Smoothing the sample spectrogram by some type of moving average process is usually recommended. If successive θ_i are not too different from one another smoothing reduces sampling variance while inducing only mild bias, thus reducing the expected mean square error. Figure 2 shows the results of a three point moving average. If x_i is the observed value of the spectrogram at frequency i, the smoothed value is defined to be

$$\mu_i \equiv \frac{x_{i-1} + x_i + x_{i+1}}{3}. \tag{3.11}$$

The residuals

$$y_i \equiv x_i - \mu_i \tag{3.12}$$

have variance

$$D = \tfrac{1}{3}\sigma^2 = 1.09$$

and standard deviation $\sqrt{1.09} = 1.05$. Looking at Fig. 2 with this in mind it is clear that we have oversmoothed the spectrogram, at least near the low end of the frequency scale. If we are serious about estimating the ϕ_i, rather than just getting an idea of the spectrum's general appearance, we can use Stein-type ideas to compromise between the two extremes of complete smoothing and no smoothing.

A digression is necessary here. Suppose $\boldsymbol{\theta} = (\theta_1, ..., \theta_k)$ is a parameter vector and $\mathbf{y} = (y_1, ..., y_k)$ an observed vector such that y_i has mean θ_i and variance some known value D. Define

$$A \equiv \sum_1^k \theta_i^2/k. \tag{3.13}$$

It is easy to show that among all estimates of the form $\delta_i = (1-b)y_i$, the choice of the constant b which minimizes the expected total squared error is $B \equiv D/(A+D)$. (This result does not depend on normality or even independence of the y_i.) In practice we will not know the value of A, so we cannot use the optimum "linear shrinking rule" $\delta_i = (1-B)y_i$. However, $E_\theta \|y\|^2 = k(A+D)$, so $\|y\|^2/k$ is an unbiased estimator of $A+D$. This suggests the estimator $\hat{B} \equiv kD/\|\mathbf{y}\|^2$ for B, and therefore the estimation rule

$$\hat{\delta}_i \equiv \left(1 - \frac{kD}{\sum_1^k y_i^2}\right) y_i, \tag{3.14}$$

which should behave like the optimum linear estimator if $\hat{B}$ is near B. Actually the theory is quite forgiving, and it can be shown that using even a rough guess of B is quite likely to improve on the unbiased estimator, which uses $b = 0$, [6].

The similarity of (3.14) with (2.6) is one of the best arguments for robustness of δ^1 to changes of the underlying assumptions. The choice of the constant $k-2$ instead of k in (2.6) can be shown to produce a uniform improvement, but this is not the case for the plus-rule version (2.10), see [6]. We won't worry about this fine point any more.

Returning to the sunspot data, let

$$\mu_i \equiv \frac{\phi_{i-1} + \phi_i + \phi_{i-1}}{3}, \qquad \theta_i \equiv \phi_i - \mu_i \tag{3.15}$$

so

$$\phi_i = \theta_i + \mu_i.$$

The $\hat{\mu}_i$ are unbiased estimators for the μ_i, the y_i unbiased estimators for the θ_i. The unbiased, completely unsmoothed, estimator of ϕ_i is $x_i = \hat{\mu}_i + y_i$, which effectively estimates θ_i by y_i. The completely smoothed estimator estimates θ_i by zero and hence ϕ_i by $\hat{\mu}_i$. We can use (3.14) to obtain estimates of $\hat{\theta}_i$ between 0 and y_i; these can then be employed to improve the estimates of the ϕ_i.

Table III shows the data and the calculations. We are attempting to estimate ϕ_i for the 28 frequencies shown in column 2, beginning with 0.0417 and ending with 0.2051. Column 4 gives the value of x_i for these 28 values, plus the two end values necessary to calculate $\hat{\mu}_1$ and

TABLE III

Compromise Estimates of the Log Power Spectrum, Sunspot Data

i	Freq.	Period	x_i	$\hat{\mu}_i$	$y_i = \hat{\theta}_i$	$\hat{\delta}_i = (1-\hat{B})y_i$	$\hat{\phi} = \hat{\mu}_i + \hat{\delta}_i$
	0.0417	24	4.4				
1	0.0455	22	5.3	5.3	0	0	5.3
2	0.0500	20	6.3	4.8	1.5	0.4	5.2
3	0.0556	18	2.8	5.1	−2.3	−0.7	4.4
4	0.0625	16	6.2	5.2	1.0	0.3	5.5
5	0.0667	15	6.6	6.6	0	0	6.6
6	0.0727	$13\frac{3}{4}$	6.9	5.7	1.2	0.3	6.0
7	0.0784	$12\frac{3}{4}$	3.6	6.2	−2.6	−0.8	5.4
8	0.0843	$11\frac{6}{7}$	8.1	6.7	1.4	0.4	7.1
9	0.0909	11	8.5	8.3	0.2	0.1	8.4
10	0.0976	$10\frac{1}{4}$	8.2	8.2	0	0	8.2
11	0.1053	$9\frac{1}{2}$	7.8	7.4	0.4	0.1	7.5
12	0.1111	9	6.4	7.1	−0.7	−0.2	6.9
13	0.1176	$8\frac{1}{2}$	7.2	6.4	0.8	0.2	6.6
14	0.1250	8	5.7	6.3	−0.6	−0.2	6.1
15	0.1333	7.5	6.1	5.8	0.3	0	5.8
16	0.1403	$7\frac{1}{8}$	5.6	5.1	0.5	0	5.1
17	0.1455	$6\frac{7}{8}$	3.5	4.5	−1.0	0	4.5
18	0.1509	$6\frac{5}{8}$	4.5	3.3	0.8	0	3.3
19	0.1568	$6\frac{3}{8}$	2.5	3.6	−1.1	0	3.6
20	0.1633	$6\frac{1}{8}$	3.7	3.6	0.1	0	3.6
21	0.1686	5.93	4.6	4.8	−0.2	0	4.8
22	0.1739	$5\frac{3}{4}$	6.0	5.5	0.5	0	5.5
23	0.1818	$5\frac{1}{2}$	6.0	5.8	0.2	0	5.8
24	0.1860	$5\frac{3}{8}$	5.4	4.9	0.5	0	4.9
25	0.1905	$5\frac{1}{4}$	3.3	4.2	−0.9	0	4.2
26	0.1951	$5\frac{1}{8}$	4.0	3.5	0.5	0	3.5
27	0.2000	5	3.1	4.1	−1.0	0	4.1
28	0.2051	$4\frac{7}{8}$	5.1	4.5	0.6	0	4.5
	0.2105	$4\frac{3}{4}$	5.2	↑	↑	↑	↑
				$\hat{\mu}_i=(x_{i-1}+x_i+x_{i+1})/3$	$y_i=x_i-\hat{\mu}_i$	$\hat{B}=kD/\|y\|^2$	Compromise Estimate

$\hat{\mu}_{28}$. The values of $\hat{\mu}_i$ and y_i are given in columns 5 and 6. The θ_i have been estimated from the y_i by the rule (3.14). This has been done separately for the first fourteen coordinates, $i = 1, 2, ..., 14$, and for the last fourteen coordinates, $i = 15, 16, ..., 28$. The first 14 coordinates give $\hat{B} = 0.71$, hence $\hat{\delta}_i = 0.29 y_i$. The second 14 coordinates give an estimate $\hat{B}$ greater than 1. This would lead to a rule which shrinks past the origin, so as before we replace this by $\hat{\delta}_i = 0$. These values are listed in column 7 as $\hat{\delta}_i$. Finally, column 8 gives $\hat{\phi}_i = \hat{\mu}_i + \hat{\delta}_i$, our presumably improved estimates of the ϕ_i.

In this case we have no way of checking whether they really are improvements. Almost certainly they improve on the unbiased estimates x_i, but in this case they are so similar to the completely smoothed estimates it is doubtful whether they are much better than them, except on the obviously badly fitted coordinates $i = 3$ and $i = 7$. (For these two the limited translation rule should also be invoked.)

Writing ϕ_i as $\theta_i + \mu_i$ illustrates a technique for extending the usefullness of the James–Stein estimator. The usual unbiased estimators are used on part of the problem, the μ_i here, and then δ^1 is used to mop up what is left over, the θ_i here. The hope is that the θ_i, being residuals from a smooth model, will be small in magnitude and hence estimated very efficiently by δ^1.

It is usually unwise to pool too many estimation problems together. In [7] we recommend 10–12 as the best value for k. Here we have split the 28 coordinates into 2 groups of 14 in the most obvious way, but more subtle and potentially more advantageous methods of separating and recombining estimation problems are possible, see [5] and [7]. Overpooling relates to objection (ii) of Section 2, the combination of unrelated problems. The usual penalty for doing so is to reduce δ^1 to δ^0, S getting so large it nullifies the $(k-2)D/S$ term in (2.6).

References

1. T. W. Anderson, "The Statistical Analysis of Time Series," John Wiley, New York, 1971.
2. L. Brown, On the admissability of invariant estimators of one or more location parameters, *Ann. Math. Statist.* **37** (1966), 1087–1136.
3. B. Efron and C. Morris, Limiting the risk of Bayes and empirical Bayes estimators—Part I: the Bayes case, *J. Amer. Statist. Assoc.* **66** (1971), 807–815.
4. B. Efron and C. Morris, Limiting the risk of Bayes and empirical Bayes estimators—Part II: the empirical Bayes case, *J. Amer. Statist. Assoc.* **67** (1972), 130–139.

5. B. Efron and C. Morris, Empirical Bayes on vector observations—an extension of Stein's method, *Biometrika* **59** (1972), 335–347.
6. B. Efron and C. Morris, Stein's estimation rule and its competitors—an empirical Bayes approach, *J. Amer. Statist. Assoc.* **68** (1973), 117–130.
7. B. Efron and C. Morris, Combining possibly related estimation problems, (with discussion), *J. Royal Statist. Soc.*, B **35** (1973), 379–421.
8. B. Efron and C. Morris, Data analysis using Stein's estimator and its generalizations, RAND report R-1394–OED.
9. W. James and C. Stein, Estimation with quadratic loss, *in* "Proceedings of the Fourth Berkeley Symposium on Mathematical Statistics and Probability," pp. 361–379, University of California Press, CA, 1961.
10. S. Sclove, C. Morris, and R. Radhakrishnan, Non-optimality of preliminary test estimators for the mean of a multivariate normal distribution, *Ann. Math. Statist.* **43** (1972), 1481–1490.
11. C. Stein, Inadmissability of the usual estimator for the mean of a multivariate normal distribution, *in* "Proceedings of the Third Berkeley Symposium on Mathematical Statistics and Probability," pp. 197–206, University of California Press, CA, 1955.

Algorithms

D. J. KLEITMAN

Mathematics Department, Massachusetts Institute of Technology, Cambridge, Massachusetts 02139

DEDICATED TO S. M. ULAM

The present paper represents an attempt to convey to a group of physicists, engineers and applied mathematicians a taste of the kind of work that has been done recently in the area of combinatorial algorithms.

An algorithm is, according to one definition of it, merely a method for accomplishing something. Thus, unless we qualify the term, the subject of "algorithms" includes in some sense all of mathematics and in fact most of human knowledge.

I will direct my attention to three aspects of the subject: first, the recent attempts to develop a theory capable of determining which problems can be solved by efficient algorithms and which cannot be; second, the recent progress on finding efficient algorithms for simple problems; finally, some aspects of these questions that appear to merit further investigations.

Computational Complexity

The development of computers led people to raise the question: What problems can be solved with them? The first questions of this kind were very theoretical, of the sort, "what problems can be solved by a Türing machine?" Study of statements of this kind led to some surprising results, such as that certain problems might never be solved and others could definitely be solved in a finite length of time. Of the latter kind were results that said that simple statements "in first-order logic" of certain kinds could definitely be proven or disproven in finite time.

Such statements were interesting, but not readily relatable to real computers and real problems. In a given real problem one is willing to spend a "certain" amount of computational time and money, not only a finite amount, but a fixed finite amount. One therefore wants to know whether problems are solvable with that effort rather than in a finite length of time. Formulating the question of the solvability of a given problem in a fixed amount of time is rather difficult, particularly since the fixed time varies from problem to problem. With modern computers one can consider a problem solvable in a practical sense if its solution can be obtained in, say, 10^{10} steps; but this is only a very round number—no fixed number like this lends itself to mathematical analysis. In the last few years real progress has been made on an almost practical analysis of solvability of certain simple problems.

It is apparent that, for finite systems, all problems are solvable in finite time. But for problems that are of any but very small size, methods that involve complete enumeration of cases usually involve numbers of steps that are factorial or exponential in the standard measure of problem size and are impractical. To distinguish from such methods Edmonds introduced the notion of a "good algorithm" as one that operates in a polynomial number of steps in the standard measure of problem size. This notion forms the basis of the following remarks.

We will consider classes of problems each of which can be characterized by a size parameter n. Such a class of problems is said to be polynomially computable (or to "belong to P"), if there is an algorithm that computes a solution of any such problem x in the class in a number of steps bounded by a polynomial function P of the size parameter of x, by $P(n(x))$.

If a problem is a member of a class in P it does not necessarily follow that there is a practical algorithm for computing its solution; the degree or coefficients of the polynomial may be large. But at least there is something in the right direction. On the other hand, if a problem belongs to a class not in P, we know that we must either invoke other special properties of the problem not described by its class membership, or trust to luck, or seek an approximate solution to solve it for sufficiently large n.

There are many well-known problems that are members of P. Sorting numbers, multiplying matrices, multiplying numbers, obtaining determinants of matrices, finding shortest paths or maximal flows or maximal matchings in graphs are all examples of classes in P with the obvious parameters.

Another interesting collection of classes of problems is that known as "*NP*." These are problems for which one can *verify* if a given proposed solution really is one in a polynomial number of steps. A class is in *P* if in worst case each problem in it can be solved in a polynomial number of steps, in *NP* if each problem in it can be solved in a polynomial number of steps, if one stumbles onto a solution. Obviously *NP* includes *P*. There are many problem classes, further, that are in *NP* and not known to be in *P*. Here are some examples.

1. Does a graph on n vertices (specified by vertices and edges that are pairs of vertices) have a Hamiltonian cycle (a cycle that goes through each vertex exactly once)?

2. Does an expression of length n in conjunction form have a solution? Is $[(a \vee \bar{b}) \cdot (c \vee d \vee \bar{a}) \cdot (c \vee b \vee d) \cdot (\bar{b} \vee \bar{c}) \cdot (\bar{d} \vee \bar{a})(a \vee b)]$ satisfiable?

3. Does there exist a k coloring of a given graph on n vertices so that no two neighboring vertices are the same color?

4. Do n linear constraints on n variables have a solution in which each variable is zero–one valued (viz., equals zero or equals one)?

5. Do there exist k vertices that cover all the edges of a graph?

6. Does a graph on n vertices contain a clique (complete subgraph) of at least $n/2$ vertices?

Some of these questions have long been investigated in attempts to find usable "good" algorithms, without success so far.

Another notion that has been introduced in this context is that of *NP*-completeness. A class of problems is said to be *NP*-complete if using it as an "oracle" any *NP* problem can definitely be solved in a polynomial number of steps. If an *NP*-complete problem class is in *P* therefore, *NP* would be contained in *P* and hence identical to it.

The progress made in this general area in the last several years has been in two directions. First, a number of *NP* problem classes including all of our examples above have been shown to be *NP*-complete, by ingenious reduction arguments. Second, a number of problem classes have been shown to require more than an exponential number of steps for their solution.

The first results imply that if a "good" algorithm exists for any of these problem classes, then one exists for all. It is natural to conjecture from this that *NP* is not *P*, and that therefore a "good" algorithm does not exist for any of these problems.

The second kind of result has been primarily of the kind that the general classes of problems that are finitely decidable, such as the truth of a proposition in certain first-order logical theories (e.g., the theory of addition on the integers) are not decidable by "good" algorithms. Thus in a practical sense, they are not attackable by very general machine computation.

The results alluded to here are primarily negative. They suggest the nonexistence of algorithms. Yet they are positive in that they suggest that the concept of polynomial decidability is one that is substantial enough to form the subject matter for nontrivial theory.

They are strong enough to suggest that in order to solve NP-complete class problems, we should perhaps look in other directions than at general algorithms. And of course they emphasize that decidability in the sense of logic does not imply decidability in any practical sense.

The analysis just mentioned has features that point to directions in which further work would be helpful.

First, the concept of "worst case" solvability which is fundamental to the analysis is one whose practical significance, while nonzero, may be surprisingly limited. That is, the worst case of some given algorithm for a problem might rarely if ever occur in the problems encountered in practice. This possibility sounds at first like a quibble; if a method can fail, one would think, it probably does fail on all but a tiny subclass of the problems in the class, and its usefulness is severely limited or limited to this narrow subclass. It is not a quibble, however, as the following case history shows.

One of the most successful algorithms of modern times is the "simplex" algorithm for linear programming suggested by Dantzig in the 1950's. The problem addressed is one of finding minimal point of a linear functional (the "objective function") of n variables within a region bounded by d linear constraints on the variables. The method consists of moving from vertex to neighboring vertex of the "convex polytope" defined by the constraints, always going in a direction that decreases the value of the objective function until one finds the vertex minimizing that function. Going from one vertex to a neighbor is called a "pivot step" and can be performed quite efficiently. The efficiency of the algorithm therefore depends on how many pivots one has to perform to reach the "bottom" of the polytope.

There is some variety within the framework just described, because one can employ several different rules to determine which neighboring vertex to visit when several neighbors of the vertex one has reached are "lower" in the polytope. The most common is to move along on edge of maximal slope magnitude; alternatively one can move to the neighbor that is lowest, or move at random. Now recently, it has been shown that in worst case any of these three approaches can require an exponential number of pivots, so that the simplex algorithm in any of these variations is not a good algorithm (Klee and Minty, 1972, for the maximal slope magnitude; Jeroslow, 1973, for maximal improvement).

In practice, the Dantzig algorithm has been used incredibly often in an enormous number of contexts, with values of n and d in the thousands and, with some improvements, even the hundreds of thousands. General experience is that the number of pivots involved is on the order of $n + d$ and usually less than it. I am unaware of a practical problem class in which a number of pivots that seems to be as much as quadratic in these parameters has even been encountered.

On the other hand in worst case the method requires an exponential number of pivots. N. Zadeh (*Math. Programming*, 1973) has found an artificial class of network flow problems on which it would fail by requiring exponential pivoting.

For the simplex algorithm worst case, this seems to occur on problems whose "practical measure" so far has been zero.

There are those who claim that for network flow problems in which the "good" Ford Fulkerson algorithm can be applied, one can obtain better practical results with efficiently coded versions of the simplex algorithm.

In our scheme of P and NP problem classes, the feasibility or the existence of a point satisfying all the constraints or the solution of a linear program are clearly in NP. Recently someone has claimed that the former problem is NP-complete. I have no idea whether it is or is not, but if linear programming were NP-complete or even required exponential solution time in worst case, one would conclude not that we should stop using the simplex algorithm but that the concepts of worst case behavior and reducibility have deviated hopelessly from practicality.

We conclude from the LP saturation that there is something that we do not understand: why LP appears to be exempt from Murphy's law otherwise so prevalent in science, life and applications of mathematics: if something can go wrong it will.

Until we understand this, our worst case analysis is greatly hindered. In practice an engineer feels he has solved his problem if he can reduce it to a thousand-variable + constraint LP. But a worst case analyst would be constrained to ignore this method of solution.

There are other difficulties with the concepts of good algorithm; P and NP. For one thing the concept of goodness requires good faith in the definition of parameters. By letting $m = 2^n$ we could obtain polynomial behavior in m given exponential behavior in n. We consider this to be a quibble and ignore it. For another thing there are polynomials that are quite impractical (e.g., $10^{10}n^{50}$), and one really should try to distinguish such polynomials from "good" ones.

What directions are suggested by these observations? Among others are the following.

To get around the difficulties with worst case behavior one could seek an analogous analysis that dealt with "average" behavior within some class of problems. This would require defining measures on classes of problems. Some such analysis has occurred for sorting problems where measures are natural (one can assume random ordering of the objects to be sorted) but not much else.

To get around the more specific difficulty presented by the anomalous situation of linear programming, one could repeat the analysis allowing an LP oracle as well as polynomial step operations. This might yield richer results; there may be other practical oracles that might also be used.

To deal with the problem posed by the existence of impractical polynomials one could shift emphasis from polynomially solvable problems to problems solvable in a number of steps that is a polynomial of degree $\leq k$. A difficulty with this is that the definition would likely be machine dependent.

Finally one could take a hint from the fundamentally negative thread of the general theory to focus attention on "heuristic" methods, methods that, for example, provide solutions but not necessarily best ones in rapid time. Such methods exist and have been used extensively (practically all work in physics is of this kind) but has as yet not been the subject of the kind of analysis indicated above.

Some Recent Positive Results

In the last few years there have been a number of new algorithms and tricks developed that provide efficient means of solving a many simple and fundamental problems.

I will do justice to none of these, but will list a few of them and give some idea of the recent progress.

1. *Fast Multiplication of Large Numbers*

The natural way of multiplying two n digit binary numbers requires on the order of n^2 steps. This number can be reduced to something like $cn \log n \log \log n$ steps. There are several ways to do this that are described in detail in Knuth's book.

In one method, one considers the numbers to be polynomials of low degree (e.g., $110111011 - 11 \times X^2 - 001X + 011 =$ polynomial of degree 3 in $X = 2^3$). The product of two of them can be obtained by taking products for $x = 0, 1,...,$ and reconstructing the polynomial product from its values at these integers. Surprisingly all this can be done rapidly.

2. *Fast Multiplication of Matrices*

The obvious way of multiplying two $n \times n$ matrices together requires n^3 multiplications. One can however perform this act in something like $n^{\log_2 7}$ steps or more precisely $7^{\lceil \log_2 n \rceil}$ by use of the following strange fact.

The usual way of multiplying 2×2 matrices requires eight multiplications. There happens to be a way of multiplying them that does not require the entries to commute that involves only seven multiplications. If n is a power of 2 one can break the matrices into $(n/2) \times (n/2)$ blocks, and these into $(n/4) \times (n/4)$ blocks etc. and apply this method at each level of blocks to get the desired product. For $n = 2$ this method requires 18 additions, but for large n values the number of additions involved actually becomes less than the ordinary method requires. This method is due to Strassen. Many people have tried to find analogous 3×3 or 4×4 tricks without success (the optimal way to multiply 3×3's is, I think not known).

3. *Finding Maximal Matching in a Bipartite Graph*

Given a graph on $2n$ vertices such that edges connect only between vertices numbered $\leq n$ and $> n$, (i.e., a bipartite graph) one seeks a maximum sized set of edges that have no vertex in common. It is relatively easy to do this in cn^3 steps in worst case; recently a clever trick has been found to do it in $cn^{5/2}$ steps in worst case. The standard fundamental approach involves alternating path labeling; the tricks involve noticing that one can perform such labelings using three passes over the edges and increase the size of a minimal improvement path by 2. One can do this until this number is $n^{1/2}$, after which one can show that at most $n^{1/2}$ improvements are possible. This result is due to Karp and Hopcroft. The analogous result for ordinary graphs (without the $\leq n$, $> n$ restriction) is uncertain; the best algorithm still requires cn^3 steps in worst case. For this problem average behavior is probably much much better than worst case behavior with suitable algorithms. Definition of averaging in this problem could yield valuable results of this kind.

4. *Testing a Graph for Planarity or for 3 Connectivity*

Tarjan and Tarjan and Hopcroft, respectively, have exploited the depth first variant of the labeling technique to the hilt to obtain linear or essentially linear algorithms (in the number of edges) for these two problems. The labeling technique involves starting at a vertex and "labeling" neighboring vertices connected by certain edges and then labeling from these to neighbors etc. The depth first variant involves labeling to a new vertex when possible and labeling from it immediately looking for another one; if another one is not found one retreats from the given vertex back and labels from the one from which it got its label. The exploitation of this approach is extremely intricate in the case of the planarity algorithm.

5. *Finding the k-th Largest of n Objects*

It is known to take at least $n \log n$ comparisons to sort n objects. Until rather recently it was thought that it took almost as long (in worst case) to pick out the median of the numbers. On the *average* R. W. Floyd has shown that something like $3n/2$ comparisons is enough for large n. By an increasing time sequence of clever tricks, the number of comparisons needed in *worst* case for the median has dwindled to now something like $3n$ in the worst case (Schoehage, Patterson, and Pittinger). I will give an example of an approach that is linear in n but not so good; the best known way is similar but has many improvements and wrinkles.

Divide the n objects into blocks of five; order each block, compare the middle elements, finding the middle of these (by iterating this technique), compare this element with the others, if it is less than the kth omit those below it, otherwise those above it.

One can show that the number of steps $f(n)$ involved here satisfies

$$f(n) \leqslant f(n/5) + (n/5) S_5 + f(7n/10),$$

where the first term counts the effort needed to sort five elements, S_5 is the number of steps necessary to sort five elements and the second term counts the number of steps used in this first sorting. The last term counts the number of steps that may be needed after this elimination which can be seen to eliminate at least three-tenths of the numbers from contention.

6. *Degree Sequences*

Given a set of numbers, one can ask, does there exist a graph having various properties such that these numbers represent the degrees of the vertices? One can also ask, how can we find graphs with the desired properties having these numbers as vertex degrees? Recently algorithms have been produced that answer both questions for the following properties

    (a) Graphs having k-factors (subgraphs all of whose vertices are of degree k) and similar subgraphs (Kundu and, also, Kleitman and Wang).

    (b) $k =$ vertex connected graphs (Kleitman and Wang).

    (c) "Maximally connected graphs" (having $\min(d_1, d_y)$ edge disjoint paths from V_2 to V_j) (Wang).

    (d) Graphs partitionable into k edge disjoint spanning trees (Kleitman and Wang).

The algorithms are fast and are described in the references.

7. *"Distance" Between Sequences*

Given two sequences of digits of lengths m and n, (Uncle) Stan Ulam raised the following question: suppose the distance between two sequences is the minimal number of operations necessary to change one to the other; where an acceptable operation involves one of removing, adding or changing a digit. X and Fischer found a procedure for finding the distance between two sequences, this requiring on the order of $8mn$ steps.

The method is not difficult and may be improvable. (Note: Paterson has improved it to $mn/\log n$ for a fixed alphabet.) It is easy to see that the distance between the sequences is the minimum of

1. the distance between all but the last members of each plus the cost of changing one of these to the other or

2. the cost of removing the last from either one plus the distance between the rest of it and the other. These rules allow us to build up the distance between initial segments of both inductively from the distance between smaller initial segments.

The following result is presented not because it is particularly significant, but because I worked on it recently.

8. *A New Application for Hu–Tucker Algorithm*

Given n words with frequencies $f_1, \ldots, f_n$, suppose one seeks to assign sequences of binary digits to them, such that no assigned sequence is a prefix of any other, and so that the sum $\sum f_i l_i$ is minimized. Here l_i is the length of the ith binary sequence.

A famous algorithm due to Huffman gives an optimal assignment. One simply takes the smallest two f's and replaces them by their sum, assigning 0 and 1 as last digits to the corresponding two words, and iterates the procedure for the new sequence obtained.

A more difficult related question arises if the original words are ordered and the binary sequences ordered lexicographically must retain the same order. T. C. Hu devised the following algorithm for this problem. Replace the two "compatible" f's with smallest sum by their sum left in the position of either one of them, and increment the length of the words associated with them by one. Two f's are compatible when they are separated in the sequence only by f's, each of which is associated with two or more words. (In case of ties always replace the f's that have been formed first.) Iterate the procedure. The lengths of the words obtained in this way uniquely determine a code that can be deduced straightforwardly.

A colleague, M. Fredman, raised a question in the same context as these. Suppose one seeks to minimize not $\sum f_i l_i$ but rather $\max_i f_i t^{l_i}$ for $t \geq 1$. What code works? It can be shown that the Huffman algorithm may be applied here changed only in that one replaces the two smallest f's by t times the larger one. For the T. C. Hu case it is the compatible two f's, the larger of which is smallest that are the ones to replace again by twice the larger. For $t \geq 1$ everything works with smallest replaced by largest at appropriate points. In any case the analogs of the original algorithm work.

An open question for the T. C. Hu algorithm is how to do it with ternary sequences (chosen from 0, 1, 2). For Fredman's problem the exact analog of the T. C. Hu algorithms works precisely for this case, as it does not for the original problem. Other generalized problems may also be attacked by these methods.

There are many more results that could be listed here, and many that deserve listing more than some that do appear. In particular there are nice methods for finding roots of polynomials and for determining if a number is prime.

Additional Comments and Conclusions

In addition to the theoretical developments described and positive results on new or old problems there has been a considerable amount of work on analyzing algorithms and some work aimed at trying to isolate classes of hard problems.

In the former area there has been some interesting work by R. Graham and D. Johnson aimed at understanding the limitations of crude bin packing and scheduling algorithms.

Work in the latter area includes efforts to find classes of Boolean functions of n variables each one of which requires many logical "gates" to construct. The existence of functions requiring exponentially many gates in the number of input variables has been demonstrated, but no classes of examples have been constructed for which it has been verified that that many gates are necessary. It has recently been shown (Meyer) that a machine for determining the truth of statements in certain logical theories requires exponential space or worse. However, new bounds on the number of gates necessary have been put on the class of functions that have the property that, if any three variables are fixed among the eight functions of the other variables obtained, at least five are distinct (Hsieh, Harper, Schnorr, Savage). The bounds are linear in n however, so this class is not excitingly hard to construct at best.

Finally, of course, there is real practical work aimed at solving real problems. Since real problems rarely look like any of those that get studied in detail, the practitioner generally proceeds without benefit of much of the ingenuity contained in the work described above. There is in practice a natural tendency among practitioners to bend problems into those for which solutions are available and to use canned packages where they exist for subroutines performing simple tasks. Surprisingly useful subroutine packages exist for problems that are probably NP complete. These subroutines do not give exact solutions; rather, they give feasible possibilities that are locally optimal within some class of possibility, yet they are used quite often, and quite successfully.

Summary and Conclusions

Looking toward the future, one can expect extensions of theory in some of the directions indicated above, and more ingenious work on new methods for old and new problems.

From a practical standpoint extension of theory to involve heuristic methods, and more comparative analysis of heuristics would be most helpful.

An area suggested by the apparent difficulty of NP problems; study of useful approximate or "heuristic" methods for these problems, and the implication of the polynomial reductions for heuristics of one problem to apply to another. Many NP complete problems are "usually" not so difficult, and are usually amenable to heuristics. Developing useful reductions involving heuristics would be a step in a potentially useful direction.

Developing sensible measures for problems in various classes to permit average case analysis would certainly be valuable.

References

G. Dantzig, Maximization of a linear function of variables subject to linear inequalities, *in* "Activity Analysis of Production and Allocation" (T. C. Koopmans, Ed.), Chap. XXI, Wiley, New York, 1951.

J. Edmonds and E. L. Johnson, Matching: A well solved class of integral linear programs, *in* "Combinatorial Structures and Their Applications, pp. 89–92, Gordon and Breach, New York, 1970.

J. Edmonds and R. M. Karp, Theoretical improvements in algorithmic efficiency for network flow problems, *J. Assoc. Comput. Mach.* **19** (1972), 264–284.

M. J. Fischer and M. S. Paterson, "String-matching and other products," Project MAC Technical Memorandum 41, Massachusetts Institute of Technology, Cambridge, MA, 1974.

R. W. Floyd and R. L. Rivest, "Expected time bounds for selection," Computer Science Dept., Stanford University.

R. L. Graham, Bounds on Multiprocessing anomalies and related packing algorithms, *in* "Proceedings of Spring Joint Computer Conference," Atlantic City, 1972.

J. E. Hopcroft and R. M. Karp, A $n^{5/2}$ algorithm for maximum matchings in bipartite graphs, *Conference Record: Twelfth Annual Symposium on Switching and Automata Dreary* (Oct. 1971), 122–125.

J. E. Hopcroft and R. G. Taijan, Dividing a graph into triconnected components, *SIAM J. Computing* **2** (1973), 135–158.

J. E. Hopcroft and R. E. Tarjan, Efficient planarity testing, *J. Assoc. Comput. Mach.*, to appear.

N. Hsieh and L. Harper, to appear.

T. C. Hu and A. C. Tucker, Optimum binary search trees, *SIAM J. Appl. Math.* **21** (1971), 514–532.

D. A. Huffman, A method for the construction of minimum redundancy codes, *Proc. IRE* **40** (1952), 1098–1101.

D. Johnson, Thesis, Massachusetts Institute of Technology, 1973.

V. L. Klee and G. J. Minty, How good is the simplex algorithm, Boeing, Math. Note No. 643 (1970).

D. J. Kleitman and D. L. Wang, Constructing maximally connected graphs having a given degree sequence, *in* "Proceedings of Seventh Annual Princeton Conference on $IS + S$," pp. 410–413, 1973.

D. Kleitman and D. L. Wang, Algorithms for constructing graphs and digraphs with given valences and factors, *J. Discrete Math.* (1974).

D. E. Knuth, "Seminumerical Algorithms," Addison-Wesley, Reading, MA, 1969.

S. Kundu, The K factor conjecture is true, *Discrete Math.* (1974).

V. Strassen, Gaussian elimination is not optimal, *Numer. Math.* **13** (1969), 354–356.

D. L. Wang and D. J. Kleitman, On the existence of n-connected graphs with prescribed degrees ($n \geqslant 2$), *Networks* **3** (1973), 225–239.

N. Zadeh, A bad network problem for the simplex method and other minimum cost flow algorithms, *Math. Programming* **5** (1973), 255–266.

Whitney Numbers of Geometric Lattices

KENNETH BACLAWSKI*

Department of Mathematics, Harvard University, Cambridge, Massachusetts 02138

DEDICATED TO STAN ULAM

INTRODUCTION

It has been a long-standing conjecture of Rota [14] that there is a homology theory on the category of ordered sets such that the Betti numbers of a geometric lattice are the Whitney numbers of the first kind. The purpose of this paper is to describe such a theory. We will also show that our theory and the usual simplicial theory are related by a spectral sequence.

The theory we develop is a sheaf cohomology on a topological space associated to the ordered set. It is however, also possible to develop the theory in terms of specific simplicial chain complexes. For those who find sheaves unpalatable, we describe these chain complexes in Section 5.

In Section 1, we develop some sheaf theory for the special case in which we will be dealing. In the next section, we examine the cohomology theories of two sheaves: the sheaf of locally constant integer-valued functions and the sheaf $\mathcal{M}$, based on the "valuation ring" of Rota [11]. The former sheaf, of course, gives the ordinary simplicial cohomology theory, while $\mathcal{M}$ gives the same theory in dimensions greater than zero when the ordered set is finite. In Section 3, we define a sheaf $\mathcal{W}$ whose cohomology groups are groups whose ranks are the Whitney numbers when the ordered set (after adjoining a zero) is a geometric lattice. In the next section, we describe a spectral sequence that relates the cohomology theory of $\mathcal{W}$ to the ordinary simplicial cohomology theory. In the last section, we describe a way to relate "transitive" elements of the incidence algebra of the ordered set to certain sheaves on the ordered set. Simplicial chain complexes that give the cohomology of these sheaves are then

---

* This work was done while the author was supported by an NSF Graduate Fellowship.

described, giving an alternative approach to the theories developed in the earlier sections.

The idea of using sheaf theory for studying ordered sets is not new; see, for example, Graves and Molnar [9]. It was their work that inspired this paper. The author wishes to thank Prof. Rota and Prof. Graves for many discussions on this subject.

1. Preliminaries

Let P be an ordered set. An *increasing subset* (or *order-filter*) of P is a subset $U \subseteq P$ such that $x \in U$ and $y \geqslant x$ imply that $y \in U$. One similarly defines *decreasing subset* (or *order-ideal*). The increasing subsets of P are easily seen to be the open sets for a topology on P. In the sequel, we always assume that ordered sets are endowed with this topology. The order-preserving maps from P to another ordered set, Q, are precisely the continuous functions from P to Q. Every point $x \in P$ is contained in a unique smallest open set:

$$V_x = \{y \in P \mid y \geqslant x\}$$

called the *principal filter* of x. See Graves and Molnar [9].

We may also regard P as a category. The objects of this category are the elements of P, and the morphisms are relations of the form $x \leqslant y$, the source of this morphism being x and the target being y. Let $\mathscr{C}$ be the category of abelian groups and homomorphisms. A *sheaf* on P is a covariant functor $\mathscr{F}: P \to \mathscr{C}$. The value of $\mathscr{F}$ at $x \in P$ is a group called the *stalk* of $\mathscr{F}$ at x and will be written $\mathscr{F}(V_x)$. For $x \leqslant y$ in P, the corresponding homomorphism $\mathscr{F}(V_x) \to \mathscr{F}(V_y)$ is called the *restriction* from V_x to V_y. Let $U \subseteq P$ be an open subset, and let U^* be the ordered set obtained by reversing the directions of the inequalities in U (i.e., $x \leqslant y$ in U^* if and only if $x \geqslant y$ in U). Then the restriction of $\mathscr{F}$ to U is an *inverse system* (in the general sense) on U^*. We write $\mathscr{F}(U)$ for the inverse limit $\varprojlim_{x \in U^*} \mathscr{F}(V_x)$. This notation is easily seen to be consistant with the notation $\mathscr{F}(V_x)$ used for the stalks. It is easy to see that $\mathscr{F}$, as a functor on the category of open subsets and inclusions of P, is a sheaf on P in the usual sense of the term.

A sheaf $\mathscr{G}$ on P is said to be *flasque* if, for any open set $U \subseteq P$, the restriction $\mathscr{G}(P) \to \mathscr{G}(U)$ is surjective. Let $\mathscr{F}$ be a sheaf on P. A *flasque resolution* of $\mathscr{F}$ is an exact sequence:

$$0 \to \mathscr{F} \to \mathbf{C}^0 \to \mathbf{C}^1 \to \mathbf{C}^2 \to \cdots$$

of sheaves on P such that $\mathbf{C}^i$ is flasque for all $i \geqslant 0$. Replacing each of these sheaves by the group $\mathbf{C}^i(P)$, defines a complex

$$0 \to \mathbf{C}^0(P) \to \mathbf{C}^1(P) \to \mathbf{C}^2(P) \to \cdots$$

of groups, whose cohomology is called the *cohomology of P with coefficients in the sheaf $\mathscr{F}$* and is denoted $H^i(P, \mathscr{F})$. We write $H^*(P, \mathscr{F})$ for the (graded) direct sum of the $H^i(P, \mathscr{F})$. One can show that the cohomology is independent, up to a natural isomorphism, of the flasque resolution used. See Godement [8, II.4.7.1].

Let $\mathscr{F}$ be a sheaf on P. The *sheaf of discontinuous sections* of $\mathscr{F}$ is the sheaf $[\mathscr{F}]$ defined on each open $U \subseteq P$ by

$$[\mathscr{F}](U) = \prod_{x \in P} \mathscr{F}(V_x),$$

with restrictions given by the obvious projection homomorphisms. Clearly, $[\mathscr{F}]$ is a flasque sheaf, and the canonical morphism $\epsilon: \mathscr{F} \to [\mathscr{F}]$ induced by the restrictions $\mathscr{F}(U) \to \mathscr{F}(V_x)$ is an injective morphism of sheaves. The inductively defined sequence of sheaves:

$$0 \longrightarrow \mathscr{F} \xrightarrow{\epsilon} [\mathscr{F}] \xrightarrow{d_0} [\mathrm{Coker}(\epsilon)] \xrightarrow{d_1} [\mathrm{Coker}(d_0)] \longrightarrow \cdots$$

gives a flasque resolution of $\mathscr{F}$ called the *canonical resolution* of $\mathscr{F}$. Thus every sheaf on P has a flasque resolution, and sheaf cohomology is defined for every sheaf on P.

For the basic properties of sheaves and sheaf cohomology, see Godement [8].

Let P and Q be ordered sets and $f: P \to Q$ an order-preserving map. Let $\mathscr{F}$ be a sheaf on P. The *direct image sheaf* of $\mathscr{F}$ on Q is the sheaf defined by: $(f_*\mathscr{F})(U) = \mathscr{F}(f^{-1}(U))$ for open subsets U of P.

LEMMA 1.1. *Let P be a decreasing subset of Q and $i: P \to Q$ the inclusion. Then, for any sheaf $\mathscr{F}$ on P, there is a natural isomorphism:*

$$H^*(Q, i_*\mathscr{F}) \simeq H^*(P, \mathscr{F}).$$

Proof. Since P is a descending subset of Q, if $x \in Q$ and $V_x \cap P \neq \varnothing$, then $x \in P$. Now $i_*[\mathscr{F}](V_x) = [\mathscr{F}](V_x \cap P) = \prod_{y \in V_x \cap P} \mathscr{F}(V_y \cap P)$, and $[i_*\mathscr{F}](V_x) = \prod_{y \in V_x} i_*\mathscr{F}(V_y) = \prod_{y \in V_x} \mathscr{F}(V_y \cap P)$. Since $V_y \cap P$ is the principal filter of y in P when $y \in P$ and since $V_y \cap P = \varnothing$ when $y \notin P$, these two products coincide. Therefore, $i_*[\mathscr{F}]$ and $[i_*\mathscr{F}]$ are naturally isomorphic. By an inductive procedure, we extend this isomorphism to an isomorphism of the canonical resolution of $i_*\mathscr{F}$ with the direct image of the canonical resolution of $\mathscr{F}$.

2. The Sheaves $\tilde{\mathbf{Z}}$ and $\mathscr{M}$

We define the *sheaf $\tilde{\mathbf{Z}}$ of locally constant integer-valued functions on P* by $\tilde{\mathbf{Z}}(V_x) = \mathbf{Z}$ for all $x \in P$, with the restrictions being the identity maps. More generally, for an arbitrary abelian group G, we define $\tilde{G}$ in a similar fashion.

We regard P as a simplicial complex in the usual fashion: the vertices are the elements of P, and the k-simplices are the totally ordered $k+1$ element subsets of P. We denote the simplicial homology and cohomology of P with coefficients in the abelian group G by $H_*(P, G)$ and $H^*(P, G)$ respectively. As one would expect, the simplicial cohomology and the cohomology with coefficients in $\tilde{G}$ coincide.

Theorem 2.1. *For any ordered set P and abelian group G, there is a natural isomorphism*:

$$H^*(P, \tilde{G}) \cong H^*(P, G).$$

Proof. A more general result is proved by Deheuvels [5, Section 11]. We will give a brief sketch of a proof.

Let $C_*(P, \mathbf{Z})$ be the simplicial chain complex of P. Let $C^*(P, G)$ be the dual of $C_*(P, \mathbf{Z})$, i.e., $\operatorname{Hom}(C_*(P, \mathbf{Z}), G)$. The cohomology of $C^*(P, G)$ is then the simplicial cohomology $H^*(P, G)$. Since $C^*(P, G)$ is functorial in P, we may define a complex of sheaves by $\mathbf{C}^*(P, G)(V_x) = C^*(V_x, G)$ for all $x \in P$. Moreover, there is a natural inclusion of sheaves $\tilde{G} \to \mathbf{C}^0(P, G)$ given by mapping $g \in G(V_x)$ to the element $\chi_g \in C^0(V_x, G)$ defined by $\chi_g(y) = g$ for all $y \in V_x$. Since the simplicial cohomology of a principal filter is trivial, the following is an exact sequence of sheaves on P:

$$0 \to \tilde{G} \to \mathbf{C}^0(P, G) \to \mathbf{C}^1(P, G) \to \cdots.$$

It is easily checked that the sheaves $\mathbf{C}^i(P, G)$ are all flasque. In fact, if we define a sheaf $\mathscr{G}_i$ by $\mathscr{G}_i(V_x) = $ the G-dual of the free abelian group generated by all i-simplices of P with minimal element x, then $\mathbf{C}^i(P, G) \cong [\mathscr{G}_i]$. Therefore, $\mathbf{C}^*(P, G)$ is a flasque resolution of $\tilde{G}$.

Since the group of global sections, $\mathbf{C}^i(P, G)(P)$, is just $C^i(P, G)$, for all i, the result follows.

In [11], Rota introduced an augmented ring $V(L)$ associated to every distributive lattice L, called the *valuation ring* of L. This ring is formed as follows. Let $F(L)$ be the free abelian group generated by L. We give this group a ring structure by defining the product of the basis elements $x, y \in L$ to be $x \wedge y$. Now let J be the ideal of $F(L)$ generated by elements of the form $x + y - x \wedge y - x \vee y$, for $x, y \in L$. The valuation ring of L is then the quotient ring $V(L) = F(L)/J$, adjoining an identity element if $F(L)/J$ does not already have one. We define a homomorphism of rings, $\epsilon : F(L) \to \mathbf{Z}$, by $\epsilon(x) = 1$ for all $x \in L$. Since $J \subseteq \mathrm{Ker}(\epsilon)$, ϵ induces a ring homomorphism $\epsilon : V(L) \to \mathbf{Z}$ which we call the *augmentation* of $V(L)$.

For an ordered set Q, let $L(Q)$ be the lattice of decreasing subsets of Q, and write $M(Q)$ for $V(L(Q))$. The functorial properties of L, V and M should be clear: V is covariant while both L and M are contravariant. Hence we may define a sheaf on P by $\mathscr{M}(V_x) = M(V_x)$, for all $x \in Q$, with the restriction $\mathscr{M}(V_x) \to \mathscr{M}(V_y)$, for $x \leqslant y$ in Q, being induced functorially by the inclusion $V_y \to V_x$. Note that, in general, $\mathscr{M}(Q) \neq M(Q)$. $\mathscr{M}$ is in fact a sheaf of augmented rings, but we do not use this.

An ordered set P is said to be *upper finite* [*lower finite*] if the principal filters V_x [principal ideals $J_x = \{y \in P \mid y \leqslant x\}$] are finite for all $x \in P$.

THEOREM 2.2. *If Q is an upper finite ordered set, then for $i > 0$:*

$$H^i(Q, \mathscr{M}) \cong H^i(Q, \mathbf{Z}).$$

Moreover, if Q is finite and connected, then $H^0(Q, \mathscr{M}) \cong M(Q)$.

Proof. Each stalk $\mathscr{M}(V_x)$ contains an element z_x which corresponds to the minimum element of the lattice $L(V_x)$. We map $\tilde{\mathbf{Z}}$ to $\mathscr{M}$ by sending $1 \in \tilde{\mathbf{Z}}(V_x) = \mathbf{Z}$ to $z_x \in \mathscr{M}(V_x)$ for all $x \in Q$. This map is easily seen to define an injective morphism of sheaves $\tilde{\iota} : \tilde{\mathbf{Z}} \to \mathscr{M}$. We may also map $\mathscr{M}$ to $\tilde{\mathbf{Z}}$ by mapping $\mathscr{M}(V_x) \to \tilde{\mathbf{Z}}(V_x) = \mathbf{Z}$ via the augmentation. This map defines a surjective morphism of sheaves, $\tilde{\epsilon} : \mathscr{M} \to \tilde{\mathbf{Z}}$. Clearly, the composition $\tilde{\epsilon} \circ \tilde{\iota}$ is the identity on $\tilde{\mathbf{Z}}$. Let $\bar{\mathscr{M}}$ be the kernel of $\tilde{\epsilon}$. Then $\bar{\mathscr{M}}$ coincides with the sheaf denoted M in Graves and Molnar [9].

We now prove that $\bar{\mathscr{M}} \cong [\tilde{\mathbf{Z}}]$, and hence that $\bar{\mathscr{M}}$ is flasque. It is a result of Davis [4] that $\bar{\mathscr{M}}(V_x) \cong M(V_x)/(z_x)$ is a free abelian group generated by a set of mutually orthogonal idempotents corresponding bijectively to the set V_x. Moreover, the restriction $\bar{\mathscr{M}}(V_x) \to \bar{\mathscr{M}}(V_y)$ for $x \leqslant y$ is given by the projection homomorphism of the first onto the second as a direct summand. Since for finite sets of abelian groups the direct sum and the direct product coincide, it follows that $\bar{\mathscr{M}} \cong [\tilde{\mathbf{Z}}]$.

Now the morphism $\tilde{\imath} \colon \tilde{\mathbf{Z}} \to \mathscr{M}$ splits the short exact sequence of sheaves:

$$0 \longleftarrow \tilde{\mathbf{Z}} \stackrel{\tilde{\imath}}{\longleftarrow} \mathscr{M} \longleftarrow \bar{\mathscr{M}} \longleftarrow 0.$$

Since flasque sheaves are cohomologically trivial, the long exact sequence of sheaf cohomology gives the result, by Theorem 2.1.

We add that there are results corresponding to those above for cosheaves instead of sheaves. Such objects were studied extensively by Deheuvels [5]. We will just state the results briefly and without proof.

A *cosheaf* on an ordered set P is a contravariant functor $\mathscr{F} \colon P \to \mathscr{C}$. The value of $\mathscr{F}$ at $x \in P$ is called the *stalk* of $\mathscr{F}$ at x, and is denoted $\mathscr{F}(V_x)$. For $x \leqslant y$ in P, the homomorphism $\mathscr{F}(V_y) \to \mathscr{F}(V_x)$ is called the *extension* from V_y to V_x. For an open subset $U \subseteq P$, $\mathscr{F}(U)$ is the direct limit $\varinjlim_{x \in U*} \mathscr{F}(V_x)$, and we call $\mathscr{F}(U)$ the *group of cosections* of $\mathscr{F}$ on U. A cosheaf $\mathscr{F}$ is said to be *flasque* if, for any open set $U \subseteq P$, the extension $\mathscr{F}(U) \to \mathscr{F}(P)$ is injective. A *flasque resolution* of the cosheaf $\mathscr{F}$ is an exact sequence:

$$\cdots \to \mathbf{C}^2 \to \mathbf{C}^1 \to \mathbf{C}^0 \to \mathscr{F} \to 0$$

of cosheaves on P such that $\mathbf{C}^i$ is flasque for all $i \geqslant 0$. As above, we may use flasque resolutions of cosheaves to define the *homology of P with coefficients in a cosheaf* $\mathscr{F}$, denoted $H_i(P, \mathscr{F})$.

We define the cosheaf $\tilde{\mathbf{Z}}'$ by $\tilde{\mathbf{Z}}'(V_x) = \mathbf{Z}$ for all $x \in P$, with the extensions all being the identity map. For an abelian group G, we define $\tilde{G}'$ similarly.

THEOREM 2.3. *For any ordered set P and abelian group G, there is a natural isomorphism*:

$$H_*(P, \tilde{G}') \cong H_*(P, G).$$

For an ordered set Q, we define the *valuation cosheaf* $\mathscr{M}'$ by $\mathscr{M}'(V_x) = \operatorname{Hom}(M(V_x), \mathbf{Z})$, with the obvious extensions.

THEOREM 2.4. *If Q is any ordered set, then for $i > 0$*:

$$H_i(Q, \mathscr{M}') \cong H_i(Q, \mathbf{Z}).$$

Moreover, if Q is connected, then $H^0(Q, \mathscr{M}') \cong \operatorname{Hom}(M(Q), \mathbf{Z})$.

3. The Sheaf $\mathscr{W}$

Let P be a lower finite ordered set. We denote by $\hat{P}$ the ordered set obtained by adjoining a unique minimum element 0 to P (whether or not P already has a minimum element). Then $\hat{P}$ is locally finite so we may speak of the incidence algebra $I(\hat{P})$ of $\hat{P}$ (with coefficients in $\mathbf{Z}$) with zeta function ζ, Möbius function μ and identity element δ. See [6] or [12] for the definitions and terminology.

Let $x \in P$, and let G be an abelian group. We define the sheaf $G(x)$ on P by

$$G(x)(V_y) = \begin{cases} G & \text{if } x = y, \\ 0 & \text{if } x \neq y, \end{cases}$$

with the restrictions, of course, all being zero. We will be examining the cohomology of P with coefficients in the sheaf $G(x)$ in terms of the Möbius function on $\hat{P}$.

LEMMA 3.1. *Let P be a lower finite ordered set, $x \in P$ and G an abelian group. Then*

$$H^0(P, G(x)) \cong \begin{cases} G & \text{if } x \text{ is a minimal element of } P, \\ 0 & \text{if } x \text{ is not minimal in } P, \end{cases}$$

and for $i > 0$

$$H^i(P, G(x)) \cong \tilde{H}^{i-1}((0, x), G).$$

Proof. Let $\eta: G(x) \to [G(x)]$ be the canonical inclusion of sheaves. We then get the short exact sequence of sheaves on P

$$0 \longrightarrow G(x) \xrightarrow{\eta} [G(x)] \longrightarrow \text{Coker}(\eta) \longrightarrow 0.$$

Examining this sequence on stalks, one finds that $\text{Coker}(\eta)$ is the sheaf of locally constant G-valued functions on the open interval $(0, x)$ in P. Hence, by Lemma 1.1 and Theorem 2.1, the cohomology of $\text{Coker}(\eta)$ is just the simplicial cohomology of $(0, x)$ with coefficients in G.

Since $[G(x)]$ is flasque, the long exact sequence of the above short exact sequence of sheaves gives the result for $i > 1$. For $i = 0$ and 1, consider the first part of the long exact sequence:

$$0 \to H^0(P, G(x)) \to H^0(P, [G(x)]) \to H^0(P, \text{Coker}(\eta)) \to H^1(P, G(x)) \to 0.$$

If x is minimal, then $(0, x) = \varnothing$; hence $H^0(P, \text{Coker}(\eta)) = 0$. Thus $H^1(P, G(x)) = 0 = \tilde{H}^0((0, x), G)$, and $H^0(P, G(x)) \cong H^0(P, [G(x)]) \cong G$, in this case. If x is not minimal, then $H^0(P, \text{Coker}(\eta))$ is isomorphic to a direct sum of copies of G, the number of copies being the number of connected components of $(0, x)$. Now the map $G \simeq H^0(P, [G(x)]) \to H^0(P, \text{Coker}(\eta))$ is easily seen to be the inclusion of G as the diagonal of $H^0(P, \text{Coker}(\eta))$. Hence the cokernel of this map is isomorphic to $\tilde{H}^0((0, x), G)$, and the kernel $H^0(P, G(x))$ is zero. This completes the proof.

COROLLARY 3.2. $\chi(\mathbf{Z}(x)) = -\mu(0, x)$.

Proof. By Lemma 3.1, $\chi(\mathbf{Z}(x)) = \sum_{i=0}^{\infty} (-1)^i \text{rank}(H^i(P, \mathbf{Z}(x)) = \text{rank}(H^0(P, \mathbf{Z}(x))) - \sum_{i=0}^{\infty} (-1)^i \text{rank}(\tilde{H}^i((0, x), \mathbf{Z}))$. If x is minimal, the second term vanishes since $(0, x) = \varnothing$; and the first term is 1. For x minimal in P, we have $\mu(0, x) = -1$, so the result holds in this case. If x is not minimal, the first term vanishes, while the second term is one less than the Euler characteristic, $\chi(0, x)$, of the ordered set $(0, x)$ regarded as a simplicial complex. Now Rota [11, Corollary 2 of Theorem 3] has shown that $\chi(0, x) = \mu(0, x) + 1$. Therefore, the result follows also in this case.

Thus we have a cohomological interpretation of the Möbius function. This interpretation is not too far removed from that of Rota [11], the difference being the introduction of sheaf cohomology. Compare also Griffiths [10, Section 16].

COROLLARY 3.3. $H^i(P, G(x)) = 0$ for $i >$ the maximum length of a chain in the ordered set $(0, x)$.

In general, of course, one gets a great deal of garbage in dimensions $0 \leqslant i <$ the maximum length of a chain in $(0, x)$. There is, however, an important case where this does not occur.

A finite ordered set Q is said to be a (*finite*) *geometric lattice* if it satisfies the following conditions (see [3, Chapter 2]):

(a) Q is a lattice (hence has a minimum 0 and a maximum 1);

(b) every $x \in Q$ is a join of atoms;

(c) if $x, y \in Q$ cover $x \wedge y$, then $x \vee y$ covers both x and y.

PROPOSITION 3.4. *Let P be a lower finite ordered set such that $\hat{P}$ is a geometric lattice. Let $x \in P$, and let G be an abelian group. If n is the maximum length of a chain in $(0, x)$, then*

(a) $H^i(P, G(x)) = 0$ for $i < n$,
(b) $H^n(P, G(x))$ is a direct sum of $(-1)^{n+1} \mu(0, x)$ copies of G.

Proof. The result is clearly trivial for $n = 0$. We therefore assume that $n > 0$. The proof of (a) then reduces to proving that $\tilde{H}^i((0, x), G)$ is zero for $i < n - 1$, by Lemma 3.1. Part (b) reduces to showing that $\tilde{H}^{n-1}((0, x), G)$ is a direct sum of $|\mu(0, x)|$ copies of G. By the Universal Coefficient Theorem and Corollaries 3.2 and 3.3, we need only show that $\tilde{H}_i((0, x), \mathbf{Z}) = 0$ for $i < n - 1$, and that $\hat{H}_{n-1}((0, x), \mathbf{Z})$ is a free abelian group. The result is then a consequence of a theorem of Folkman [7, Theorem 4.1].

Let Q be an ordered set with a minimum element 0. A function $r: Q \to \mathbf{Z}$ is a *rank function* for Q if it satisfies:

(a) $r(0) = 0$,
(b) if x covers y in Q, then $r(x) = r(y) + 1$.

It is easily seen that if Q is locally finite, a rank function for Q is unique when it exists. Suppose that Q is locally finite and has a rank function, and that $r^{-1}(m)$ is finite for all $m \in \mathbf{Z}$. We define the kth *Whitney number* (of the first kind) to be

$$w_k = \sum_{r(x)=k} \mu(0, x).$$

In particular, if Q is a geometric lattice, then Q has a rank function; hence the Whitney numbers are defined for Q.

We now define a sheaf $\mathscr{W}$ on P, for an arbitrary ordered set P, by $\mathscr{W}(V_x) = \mathbf{Z}$, for $x \in P$, with the restrictions being the *zero* homomorphism. When P is finite, this sheaf is just the direct sum of the sheaves $\mathbf{Z}(x)$ as x ranges over P. In a similar fashion, we define a sheaf $\mathscr{W}_G$ for any abelian group G.

THEOREM 3.5. *Let P be an ordered set such that $\hat{P}$ is a geometric lattice. Then $H^i(P, \mathscr{W})$ is a free abelian group of rank $(-1)^{i+1} w_{i+1}$ for all $i \geqslant 0$.*

Proof. Since cohomology commutes with finite direct sums, we need only add up the ranks of the groups in Proposition 3.4, being careful to keep them in the right dimension. The result then follows.

Hence the cohomology of $\mathscr{W}$ has the (absolute values of the) Whitney numbers as its set of Betti numbers.

We note that the condition on $\hat{P}$ in Theorem 3.5 cannot be easily removed. Indeed, one can find ordered sets P for which $\hat{P}$ is a finite lattice with a rank function, but for which $(-1)^k w_k$ is negative for some k. In general, one can only assert the following. Let P be an ordered set for which the Whitney numbers are defined. Define the sheaf $\mathscr{W}_i$ on P by

$$\mathscr{W}_i(V_x) = \begin{cases} \mathbf{Z} & \text{if } r(x) = i, \\ 0 & \text{if } r(x) \neq i, \end{cases}$$

for all $x \in P$, with the restrictions being the zero homomorphisms. Then for all $i > 0$ we have $\chi(\mathscr{W}_i) = -w_i$. These sheaves appear in a more natural setting in the spectral sequence to be defined in the next section.

4. A Spectral Sequence

Let P be an ordered set. An *(increasing) filtration* on P is a sequence $\{P^s\}_{s \in \mathbf{Z}}$ of closed subsets of P such that $P^s \subseteq P^{s+1}$ for all s and such that $\bigcap_{s \in \mathbf{Z}} P^s = \varnothing$.

Let Q be any subset of P and $\mathscr{F}$ any sheaf on P. We define a sheaf $\mathscr{F}_Q$ on P by

$$\mathscr{F}_Q(V_x) = \begin{cases} \mathscr{F}(V_x) & \text{if } x \in Q, \\ 0 & \text{if } x \notin Q, \end{cases}$$

with the restrictions being the zero homomorphism unless the entire interval $[x, y]$ is in Q in which case the restriction $\mathscr{F}_Q(V_x) \to \mathscr{F}_Q(V_y)$ is the same as the corresponding restriction $\mathscr{F}(V_x) \to \mathscr{F}(V_y)$ in $\mathscr{F}$. In particular, if P is given a filtration as above, we write $\mathscr{F}^s$ for $\mathscr{F}_{P^s}$ and $\mathscr{F}_s$ for $\mathscr{F}_{P_-P^s}$.

For an open subset U of P, the inclusion $\mathscr{F}_U \to \mathscr{F}$ is a morphism of sheaves. This will not generally hold for an arbitrary subset of P. In particular, if P is given a filtration, we have a short exact sequence of sheaves on P for all $s \in \mathbf{Z}$:

$$0 \to \mathscr{F}_s \to \mathscr{F} \to \mathscr{F}^s \to 0$$

which does not split in general. We also have an inclusion of sheaves $\mathscr{F}_{s+1} \to \mathscr{F}_s$ for all s. Hence we may speak of the sheaf $\mathscr{F}_s/\mathscr{F}_{s+1}$ on P whose support is contained in $P^{s+1} - P^s$.

THEOREM 4.1. *Let P be an ordered set with a filtration. Let $\mathscr{F}$ be any sheaf on P. Then there is an E_1-spectral sequence:*

$$H^{p+q}(P, \mathscr{F}_p/\mathscr{F}_{p+1}) \Rightarrow H^n(P, \mathscr{F}).$$

Proof. This is a pretty standard result in homological algebra. See, for example, Cartan–Eilenberg [2, XV.7]. In the notation developed there, we set $H(p, q) = H^*(P, \mathscr{F}_p/\mathscr{F}_q)$ for all pairs (p, q) such that $-\infty \leqslant p \leqslant q \leqslant +\infty$, where $\mathscr{F}_{-\infty} = \mathscr{F}$ and $\mathscr{F}_\infty = 0$. Axiom (SP5) is a consequence of the fact that cohomology commutes with direct limits. The E_1-differentials are the connecting homomorphisms of the short exact sequences:

$$0 \to \mathscr{F}_{p+1}/\mathscr{F}_{p+2} \to \mathscr{F}_p/\mathscr{F}_{p+2} \to \mathscr{F}_p/\mathscr{F}_{p+1} \to 0.$$

The abutment of the spectral sequence has a filtration given by

$$H^n(P, \mathscr{F})_p = \mathrm{Im}(H^n(P, \mathscr{F}_p) \to H^n(P, \mathscr{F}))$$
$$= \mathrm{Ker}(H^n(P, \mathscr{F}) \to H^n(P, \mathscr{F}^p)).$$

The most interesting special case of Theorem 4.1 is that in which P is an ordered set for which the Whitney numbers are defined and for which $P^s = \{x \in P \mid r(x) < s\}$. Then $P^{s+1} - P^s$ is the set of elements of P of rank s. Since $P^{s+1} - P^s$ is an antichain, i.e., a totally unordered subset of P, the sheaf $\mathscr{F}_s/\mathscr{F}_{s+1}$ has a particularly simple form; namely, it is isomorphic to the direct sum of the sheaves $(\mathscr{F}(V_x))(x)$ as x ranges over the elements of P of rank s. The following result is then an immediate consequence of Lemma 3.1, Theorem 4.1 and the remarks at the end of Section 3.

COROLLARY 4.2. *Let P be an ordered set for which the Whitney numbers are defined. Let $\mathscr{F}$ be any sheaf on P. Then there is a fourth quadrant spectral sequence with*

$$E_1^{p,q} = \begin{cases} \sum_{r(x)=p} \tilde{H}^{p+q-1}((0, x), \mathscr{F}(V_x)) & \text{for } p > 1, \\ \sum_{r(x)=1} \mathscr{F}(V_x) & \text{for } p = 1 \quad \text{and} \quad q = -1, \\ 0 & \text{in all other cases,} \end{cases}$$

such that $E_1^{p,q} \Rightarrow H^n(P, \mathscr{F})$. Moreover, if $\mathscr{F}(V_x) \cong \mathbf{Z}$ for all $x \in P$, then for $p > 0$ we have $\chi(E_1^{p,*}) = w_p$.

Now in the special case of an ordered set P for which $\hat{P}$ is a geometric lattice, we can use Proposition 3.4 to give an explicit computation of the groups in Corollary 4.2.

COROLLARY 4.3. *Let P be an ordered set for which $\hat{P}$ is a geometric lattice. Let $\mathscr{F}$ be a sheaf on P. Then there is a fourth quadrant spectral sequence with*

$$E_1^{p,q} = \begin{cases} \sum_{r(x)=p} \mathscr{F}(V_x)^{\oplus (-1)^p \mu(0,x)} & \text{if } q = -1 \quad \text{and} \quad p \geqslant 1, \\ 0 & \text{in all other cases,} \end{cases}$$

such that $E_1^{p,q} \Rightarrow H^n(P, \mathscr{F})$. Moreover, the spectral sequence degenerates at the E_2-term with $H^{p-1}(P, \mathscr{F}) \cong E_2^{p,-1}$ for $p \geqslant 1$.

The spectral sequence in Corollary 4.3, therefore, has a particularly simple form: $E_2^{p,-1}$ is the pth cohomology group of the complex $E_1^{*,-1}$ at which point the spectral sequence degenerates. The differentials of the complex $E_1^{*,-1}$ are given by the connecting homomorphisms of the short exact sequences in the proof of Theorem 4.1. When these sequences split, e.g., when $\mathscr{F} = \mathscr{W}_G$, these differentials vanish, and we have $E_1^{p,-1} \cong E_2^{p,-1} \cong H^{p-1}(P, \mathscr{F})$. The following result is then immediate.

THEOREM 4.4. *Let P be an ordered set for which $\hat{P}$ is a geometric lattice. Let $\mathscr{F}$ be a sheaf on P all of whose stalks are isomorphic to the same abelian group G. Then there is a structure of a complex on $H^*(P, \mathscr{W}_G)$ such that its cohomology is $H^*(P, \mathscr{F})$.*

5. STANDARD RESOLUTIONS

Let $\mathscr{F}$ be a sheaf on the ordered set P. The *support* of $\mathscr{F}$ is the subset $|\mathscr{F}| = \{x \in P \mid \mathscr{F}(V_x) \neq 0\}$ of P. Suppose also that P is locally finite. An element f of $I(P)$ is said to be *transitive* if, for $x \leqslant y \leqslant z$ in P, it satisfies

$$f(x, y) f(y, z) = f(x, z).$$

Suppose that all the stalks of $\mathscr{F}$ are free abelian groups of ranks 0 or 1. If we fix a choice of isomorphism of each stalk $\mathscr{F}(V_x)$ with **Z**, for $x \in |\mathscr{F}|$, then $\mathscr{F}$ defines a transitive element of $I(P)$ as follows. For $x \leqslant y$ in $|\mathscr{F}|$, $f(x, y)$ is the image of $1 \in \mathscr{F}(V_x)$ under the restriction $\mathscr{F}(V_x) \to \mathscr{F}(V_y)$. For $x \leqslant y$ in P and either $x \notin |\mathscr{F}|$ or $y \notin |\mathscr{F}|$, we define $f(x, y) = 0$. Conversely, every transitive element of $I(P)$ defines such a sheaf on P.

Now, if we allow the choice of isomorphisms of the stalks of $\mathscr{F}$ with $\mathbf{Z}$, on the support of $\mathscr{F}$, to vary, then the element of $I(P)$ defined above may change. Let H be the subgroup of $I(P)$ consisting of invertible diagonal elements, i.e., consisting of the elements $h \in I(P)$ for which $h(x, x) = \pm 1$ for $x \in P$ and for which $h(x, y) = 0$ for $x < y$ in P. Then the conjugacy class of f by elements of H is, nevertheless, uniquely determined by the sheaf $\mathscr{F}$; and, conversely, such a conjugacy class determines the sheaf $\mathscr{F}$ uniquely, up to isomorphism. The sheaf $\mathscr{W}$ corresponds to the identity element δ of $I(P)$, while the sheaf $\tilde{\mathbf{Z}}$ of locally constant functions on P corresponds to the zeta function $\zeta \in I(P)$.

Let $\mathscr{F}$ be a sheaf on P as above with f a corresponding transitive element of $I(P)$. Let $C_*(P, f)$ be a chain complex defined as follows. $C_n(P, f)$ is the subgroup of $C_n(P, \mathbf{Z})$ generated by n-simplices of P whose maximum element lies in $|\mathscr{F}|$. The differential $d_f: C_n(P, f) \to C_{n-1}(P, f)$ is defined by

$$d_f(a_0 < \cdots < a_n) = \sum_{i=0}^{n-1} (-1)^i (a_0 < \cdots < \hat{a}_i < \cdots < a_n)$$
$$+ (-1)^n f(a_{n-1}, a_n)(a_0 < \cdots < a_{n-1}),$$

for each n-simplex $(a_0 < \cdots < a_n) \in C_n(P, f)$. The $\mathbf{Z}$-dual of this complex is denoted $C^*(P, f)$. The differential on $C^n(P, f)$ is then given by

$$d_f(\alpha)(a_0 < \cdots < a_{n+1})$$
$$= \sum_{i=0}^{n} (-1)^i \alpha(a_0 < \cdots < \hat{a}_i < \cdots < a_{n+1})$$
$$+ (-1)^{n+1} f(a_n, a_{n+1}) \alpha(a_0 < \cdots < a_n),$$

for $\alpha \in C^n(P, f)$ and $(a_0 < \cdots < a_{n+1})$ an $(n+1)$-simplex in $C_{n+1}(P, f)$.

PROPOSITION 5.1. *Let P be a locally finite ordered set, and let $\mathscr{F}$ be a sheaf on P all of whose stalks are free abelian groups of rank 0 or 1. Let $f \in I(P)$ be a transitive element which corresponds to $\mathscr{F}$. Then*

$$H^*(P, \mathscr{F}) \cong H^*(C^*(P, f)).$$

Proof. Deheuvels [5, Section 10].

For example, when $\mathscr{F} = \mathscr{W}$, $H^*(P, \mathscr{W})$ is the cohomology of the cochain complex $C^*(P, \delta)$. Explicitly, $C^*(P, \delta)$ is the same as $C^*(P, \mathbf{Z})$ as a group, but the differential in $C^*(P, \delta)$ is

$$d_\delta(\alpha)(a_0 < \cdots < a_{n+1})$$

$$= \sum_{i=0}^{n} (-1)^i \alpha(a_0 < \cdots < \hat{a}_i < \cdots < a_{n+1}),$$

for $\alpha \in C^n(P, \mathbf{Z})$ and for each $(n + 1)$-simplex $(a_0 < \cdots < a_{n+1})$ in $C_{n+1}(P, \mathbf{Z})$.

References

1. K. BACLAWSKI, Automorphisms and derivations of incidence algebras, *Proc. Amer. Math. Soc.* **36** (1972), 351–356.
2. H. CARTAN AND S. EILENBERG, "Homological Algebra," Princeton University Press, Princeton, NJ, 1956.
3. H. CRAPO AND G.-C. ROTA, "On the Foundations of Combinatorial Theory: Combinatorial Geometries" (preliminary edition), M.I.T. Press, Cambridge, MA, 1970.
4. R. DAVIS, Order algebras, *Bull. Amer. Math. Soc.* **76** (1970), 83–87.
5. R. DEHEUVELS, Homologie des ensembles ordonnés et des espaces topologiques, *Bull. Soc. Math. France* **90** (1962), 261–321.
6. P. DOUBILET, G.-C. ROTA, AND R. STANLEY, On the foundations of combinatorial theory (VI): The idea of generating function, *in* "Proceedings of the Sixth Berkeley Symposium on Mathematical Statistics and Probability," Vol. 2, pp. 267–318, Univ. of California Press, Berkeley, CA, 1970–71.
7. J. FOLKMAN, The homology groups of a lattice, *J. Math. Mech.* **15** (1966), 631–636.
8. R. GODEMENT, "Topologie Algébrique et Théorie des Faisceaux," Hermann, Paris, 1958.
9. W. GRAVES AND S. MOLNAR, Incidence algebras as algebras of endomorphisms, *Bull. Amer. Math. Soc.* **79** (1973), 815–820.
10. H. GRIFFITHS, The homology groups of some ordered systems, *Acta Math.* **129** (1972), 195–235.
11. G.-C. ROTA, On the combinatorics of the Euler characteristic, *in* "Studies in Pure Mathematics," Rado Festschrift, pp. 221–233, 1971.
12. G.-C. ROTA, On the foundations of combinatorial theory (I): theory of Möbius functions, *Z. Wahrscheinlichkeitstheorie und Verw. Gebiete* **2** (1964), 340–368.
13. G.-C. ROTA, The valuation ring of a distributive lattice, to appear.
14. G.-C. ROTA, personal communication.

Continued Fraction Expansion of Algebraic Numbers

R. D. RICHTMYER

T-Division, Los Alamos Scientific Laboratory, Los Alamos, New Mexico 87544

DEDICATED TO STAN ULAM

Every irrational x in $[0, 1]$ determines uniquely an infinite sequence $\{a_i\}_1^\infty$ of positive integers by means of its continued fraction expansion

$$x = \cfrac{1}{a_i + \cfrac{1}{a_2 + \cdots}}, \qquad (1)$$

of which a_i are called the *partial denominators*. Conversely, every sequence of positive integers determines uniquely an irrational x in $[0, 1]$ by (1). For a given sequence, let $N_p(K)$ be the number of the a_i among the first K of them that are equal to p, for $p = 1, 2, \ldots$. If the limits

$$\lim_{K \to \infty} \frac{N_p(K)}{K} = f_p \qquad (2)$$

exist, for all p, and if

$$\sum_{p=1}^{\infty} f_p = 1, \qquad (3)$$

then the number x is called *normal*, and f_p may be regarded as the probability that $a_i = p$, for each p.

Khintchine proved in [1] that almost all x in $[0, 1]$ are normal, and, in fact, for almost all x the asymptotic distribution of the partial denominators is given by

$$f_p = \log_2 \frac{(p+1)^2}{p(p+2)}. \qquad (4)$$

Theorems like the one of K. F. Roth [2], which say that algebraic irrationals have the special property of avoiding the rationals as much

117

as possible, in a certain sense, led the author to conjecture in [3] that algebraic irrationals may have a paucity of large partial denominators, so that, for large p, f_p is less than the right member of (4). That is known to be true for quadratic irrationals, which have $f_p = 0$ for large p. (If x is a quadratic irrational, its sequence of partial denominators is periodic from a certain point on.) The conjecture does not contradict Khintchine's theorem, because the algebraic numbers form a set of Lebesgue measure zero.

In [4], the conjecture was studied by numerical tests, in which the continued fraction expansions of several cubic, quartic, and quintic irrationals were obtained to between 700 and 800 terms, using multi-precision integer arithmetic (3900 binary places, $\approx$1174 decimal places) on the Manic I computer. The results did not support the conjecture; instead, the values of $N_p(K)/K$ were quite close to the values of f_p given by (4), and chi-square tests of the hypothesis

$$\text{Prob}\{a_i = p\} = \log_2 \frac{(p+1)^2}{p(p+2)} \tag{5}$$

using our calculated a_i, supported that hypothesis. We were thus led to the alternative conjecture that for any algebraic irrational x of degree $\geqslant 3$ the limits (2) exist and satisfy (4).

The numerical work yielded one very strange result: for $x = 2^{1/3} - 1$, the values of χ^2 obtained were very small, the probability of their being that small, according to tables of the χ^2 distribution, being 3.7% in one test and 0.01% in another, on the hypothesis that the a_i had been chosen at random and independently from the distribution (5). In other words, the a_i seemed to obey the Khintchine law too well to be independent random variables, thereby suggesting some sort of correlation of the a_i in the sequence. A very slight short range correlation is known; namely, if for some i, a_i is smaller than expected, then a_{i+1} tends to be larger, and conversely (see (4)). That correlation is too small, however, to explain the small values of χ^2 observed, but, of course, long-range correlation cannot be excluded, and in fact is present for quadratic irrationals.

In the summer of 1970, further tests were made on the Los Alamos Manic II computer, on which 25,000-decimal integer arithmetic was available, in order to investigate the apparent correlation for $2^{1/3} - 1$, and to make further tests of the hypothesis (5).

The results, which are summarized below, show that the anomalously low value of χ^2 obtained previously for $2^{1/3} - 1$ was accidental, and resulted from having stopped the expansion at 725 terms; other stopping points give quite reasonable values of χ^2, as was predicted by W. A. Beyer (private communication). The results also give a more precise verification of the hypothesis (5) than that reported in [4].

The method of calculation was the same as reported in [4]. A program was written for expanding any cubic irrational x in [0, 1]. In the short time available for this study, most of the calculations were for $x = 2^{1/3} - 1$; here, only that case is reported.

TABLE I

Distribution of Partial Denominators a_i in Five Groups for $x = 2^{1/3} - 1$; Fraction of the First 8000 Denominators in Each Group

Group	Observed fraction	Expected fraction[a]
$p = 1$	0.4185	0.4150
$p = 2$	0.1665	0.1699
$p = 3, 4$	0.1581	0.1520
$5 \leqslant p \leqslant 10$	0.1341	0.1375
$11 \leqslant p < \infty$	0.1227	0.1255

[a] $\sum \log_2 \dfrac{(p+1)^2}{p(p+2)}$, summed over the group.

TABLE II

Distribution of Partial Denominators a_i in Ten Groups for $x = 2^{1/3} - 1$; Fraction of the First 8000 Denominators in Each Group

Group	Observed fraction	Expected fraction[a]	
$p = 1$	0.4185	0.4150	(same as in Table I)
$p = 2$	0.1665	0.1699	
$p = 3$	0.0936	0.0931	
$p = 4$	0.0645	0.0589	
$p = 5, 6$	0.0695	0.0704	
$p = 7$	0.0399	0.0406	
$9 \leqslant p \leqslant 12$	0.0425	0.0451	
$13 \leqslant p \leqslant 19$	0.0359	0.0365	
$20 \leqslant p \leqslant 40$	0.0351	0.0356	
$41 \leqslant p \leqslant \infty$	0.0340	0.0348	

[a] $\sum \log_2 \dfrac{(p+1)^2}{p(p+2)}$, summed over the group.

For the χ^2 tests, the positive integers were divided first into 5 groups, as in Table I, and then into 10 groups, as in Table II. The fractions of the first 8000 partial denominators falling into these groups are given in the Tables, and are seen to agree rather well with the fractions predicted on the hypothesis that Khintchine's law applies for the particular number $x = 2^{1/3} - 1$. The hypothesis was tested statistically by computing

$$\chi^2 = \sum_{r=1}^{R} \frac{(G_r - F_r)^2}{F_r}, \qquad (6)$$

where R is the number of groups, and where G_r and F_r are the observed and expected numbers of the denominators a_i falling in the rth group, namely,

$$G_r = \sum_{(p)} N_p(K), \qquad (7)$$

$$F_r = K \sum_{(p)} \log_2 \frac{(p+1)^2}{p(p+2)}, \qquad (8)$$

the sums being extended over the rth group of integers indicated in Tables I and II, and K being the total number of a_i in the sample ($\leqslant 8000$ in the present calculation).

The probability distribution of χ^2 is given by

$$\text{Prob}\{\chi^2 \leqslant \xi\} = \frac{1}{2^{(R-1)/2}\Gamma((R-1)/2)} \int_0^\xi t^{(R-3)/2} e^{-t/2}\, dt. \qquad (9)$$

A subroutine for evaluating this expression was included in the computer program. For a valid hypothesis, this probability is expected to lie in the range 0.1 to 0.9. If the probability is too close to 1.00, the hypothesis must be rejected, and if it is too close to 0.00, the random variables in question appear to have the right distribution but cannot be independent.

The values of χ^2 obtained and the corresponding probabilities (9) are given in Tables III and IV for the two groupings of the positive integers described in Tables I and II. The values are for the first K of the partial denominators, where K is given in the first column of the Tables. It is seen that the values of χ^2 are all reasonable, except for K near 725, hence it was an unfortunate coincidence that the expansion described in [4] was terminated (by reasons of computer capacity) at $K = 725$.

TABLE III

Chi-Square Test for $x = 2^{1/3} - 1$ Five Groups as in Table I

K	χ^2	Prob.$\{\chi^2 \leqslant$ value in column 2$\}$
100	2.510	0.357
200	1.956	0.256
500	1.388	0.154
700	0.589	0.036
725[a]	0.621[a]	0.037[a]
1000	1.894	0.245
2000	3.081	0.456
3000	3.453	0.515
4000	4.673	0.677
6000	6.784	0.852
8000	3.913	0.582

[a] Result reported in [4].

TABLE IV

Chi-Square Test for $x = 2^{1/3} - 1$ Ten Groups as in Table II

K	χ^2	Prob.$\{\chi^2 \leqslant$ value in column 2$\}$
100	8.135	0.479
200	4.324	0.111
500	1.518	0.003
700	1.320	0.002
725[a]	0.748[a]	0.0001[a]
1000	5.632	0.224
2000	4.270	0.107
3000	5.558	0.217
4000	6.931	0.356
6000	8.477	0.513
8000	6.753	0.337

[a] Result reported in [4].

Owing to the large sample size (8000) this statistical test is rather sharp. If the quantities on the right side of (5) are altered by 3 or 4%, the hypothesis would have to be rejected overwhelmingly.

References

1. A. KHINTCHINE, Metrische Kettenbruch-Probleme, *Compositio Math.* **1** (1935), 361.
2. K. F. ROTH, Rational approximations to algebraic numbers, *Mathematica* **2** (1955), 1 (with corrigendum on p. 168).
3. R. D. RICHTMYER, The evaluation of definite integrals, and a quasi-Monte-Carlo method based on the properties of algebraic numbers, Los Alamos Scientific Laboratory Report LA-1342, 1951.
4. R. D. RICHTMYER, M. DEVANY, AND N. METROPOLIS, Continued fraction expansions of algebraic numbers, *Numer. Math.* **4** (1962), 68.

Random Time Evolution of Infinite Particle Systems

FRANK SPITZER

Department of Mathematics, Cornell University, Ithaca, New York 14850

DEDICATED TO STAN ULAM

The development of this field during the past four years is directly traceable to major developments in equilibrium statistical mechanics during the late 1960's.

In 1968 Dobrushin [2, 3] introduced the notions of an *infinite Gibbs state* (IGS) and of a *Markov random field* (MRF). The latter is a probability measure μ on $\Omega = \{0, 1\}^{\mathbb{Z}_N}$ satisfying

(a) $\mu(C) > 0$ for every finite cylinder set C;

(b) $\mu[\omega(x) = 1 \mid \omega(\cdot) \text{ on } \mathbb{Z}_N \backslash x]$ depends only on $\omega(y)$ for

$$y: |y - x| = 1;$$

(c) The conditional probabilities in (b) are translation invariant.

Here $\mathbb{Z}_N$ is the set of N-dimensional integers and $\omega(x) = 1(0)$ means that the site $x \in \mathbb{Z}_N$ is occupied (vacant).

The notion of an IGS is more general. Finite Gibbs states, on which Gibbs based his theory of equilibrium statistical mechanics, are probability measures on $\{0, 1\}^A$, A being a finite subset of $\mathbb{Z}_N$ determined by a potential V. By using conditional probabilities, Dobrushin defined the notion of IGS on Ω.

Dobrushin showed that every MRF in dimension $N = 1$ is a stationary Markov chain, with values 0, 1, and time parameter in $\mathbb{Z}_1$. Thus the MRF is a natural definition of a multi(time) dimensional Markov process. Secondly it was shown, in increasingly general settings, [1, 6, 22, 23], that an MRF is nothing but an IGS with nearest neighbor potential V. Finally, previous work on the Ising model could be neatly formulated in the present setting, the phenomenon of phase transition (in dimensions $N \geqslant 2$) occurring when there is more than one MRF with given conditional probabilities.

In 1969, Lanford and Ruelle [15] independently defined IGS and moreover proved a variational characterization of these analogous to the classical one on $\{0, 1\}^\Lambda$, Λ finite.

The year 1970 saw the first time-evolutions that have a given MRF or IGS as an equilibrium state. (Actually, for $N = 1$, the first such model was studied by Glauber [5].) Suppose an MRF μ has conditional probabilities

$$p_k = \mu\left[\omega(x) = 1 \,\Big|\, \sum_{y:|y-x|=1} \omega(y) = k\right], \qquad 0 \leqslant k \leqslant 2N. \tag{1}$$

It can be shown that this implies

$$p_k = ar^{k-N}/(1 + ar^{k-N}), \qquad 0 \leqslant k \leqslant 2N, \tag{2}$$

for some $a > 0$, $r > 0$. Dobrushin [4], defined an evolution ω_t, with values in Ω, as follows. Let $\beta_k(\delta_k)$ be given birth (death) rates, i.e., rates for change from 0 to 1 (1 to 0) at an arbitrary site $x \in \mathbb{Z}_N$, when exactly k neighbors of x are occupied. He showed that μ is an equilibrium state for this time-evolution provided the rates satisfy

$$p_k = \beta_k/(\beta_k + \delta_k), \qquad 0 \leqslant k \leqslant 2N, \tag{3}$$

so that one could expect phase transition to manifest itself in the occurrence of more than one equilibrium state. Simultaneously [24] other time-evolutions were proposed in which birth and death are replaced by particle motion with the exclusion of multiple occupancy, which also have MRF's as equilibrium states.

Rigorous existence proofs of such time-evolutions as Markovian Feller semigroups, T_t, were given during 1971–1972 by Dobrushin [4], Harris [7], Holley [9], and Liggett [16].

Here are some major results and open problems concerning Dobrushin's evolutions. When the $\beta_k > 0$ and $\delta_k > 0$, but otherwise arbitrary (i.e., (3) need not hold), he showed [4] that the evolution is ergodic provided the interaction is weak, i.e., the β's and δ's are sufficiently nearly independent of k. (We say that T_t is ergodic if $\mu T_t \Rightarrow \nu$ for every initial measure μ, so that ν is the unique equilibrium state.)

It is a major open problem whether T_t is always ergodic in dimension $N = 1$, when the rates are strictly positive.

When $\beta_0 = 0$ but all other rates are positive, then ergodicity means that the unique equilibrium state ν is concentrated on the atom $\omega \equiv 0$. Harris [8] showed that T_t is ergodic when the β's are sufficiently small compared to the δ's, and non-ergodic if the converse holds. The latter fact is deep and depends on evolutions exhibiting phase transition for which the best results are due to Holley [10–13]. Suppose the MRF μ satisfies (1) and (2) with $r > 1$ (the attractive case). It is known that phase transition occurs only when $a = 1$, and then only for sufficiently large r. Let us assume $N = 2$. Then phase transition occurs if and only if $r > 1 + \sqrt{2}$, a famous result of Onsager. Suppose that (3) holds and $\beta_k \nearrow$, $\delta_k \searrow$ as $k \nearrow$, e.g., $\beta_k = r^{k-2}$, $\delta_k \equiv 1$. Then typical examples of Holley's results are these: T_t is ergodic when $r \leqslant 1 + \sqrt{2}$. When $r > 1 + \sqrt{2}$ let $\mu^+(\mu^-)$ be the initial states with everything occupied (vacant). Then $\mu^+ T_t \Rightarrow \nu^+$, $\mu^- T_t \Rightarrow \nu^-$, where $\nu^+(\nu^-)$ are the high (low) density states for the Ising model, for which the famous Onsager spontaneous magnetization formula reads

$$\nu^+[\omega(x) = 1] = 1 - \nu^-[\omega(x) = 1]$$

$$= \frac{1}{2} + \frac{1}{2}\left[1 - \left(\frac{2r}{r^2-1}\right)^4\right]^{1/8}, \qquad x \in \mathbb{Z}_2. \qquad (4)$$

The reason the theory is much more complete when (3) holds was clarified [21, 25] by showing that (3) holds if and only if T_t has a time reversible equilibrium state which then must be a MRF.

A most surprising irreversible case, where complete results have been obtained by Holley and Liggett [14], is the voter model, with $\beta_k = k$, $\delta_k = 2N - k$, $0 \leqslant k \leqslant 2N$. Here there are only the trivial equilibrium states ($\omega \equiv 0$ or $\omega \equiv 1$) in dimensions $N \leqslant 2$, but a continuum of others when $N \geqslant 3$.

The jump processes T_t, studied primarily by Liggett, take place on a countable set S with irreducible transition function P. When $P(x, y) = P(y, x)$, and the jumping speed is constant, the equilibrium states are completely known. They form a convex set whose extreme points are in a 1 : 1 correspondence with the solutions f of $Pf = f$, $0 \leqslant f \leqslant 1$. In particular, if P is recurrent, or if $S = \mathbb{Z}_N$, $N \geqslant 1$, and P is a random walk transition function, then the equilibrium states are the exchangeable measures on $\{0, 1\}^S$, i.e., convex combinations of Bernoulli product measures μ_α with density α, $0 \leqslant \alpha \leqslant 1$. If μ is ergodic with density α, then in the random walk case $\mu T_t \Rightarrow \mu_\alpha$ [17, 18, 26].

New problems arise in the case when $P(x, y) \neq P(y, x)$. In 1973–1974 Liggett [19, 20] obtained deep partial results toward the following two conjectures:

(a) If P is positive recurrent, with S arbitrary denumerably infinite, and if the evolution starts with infinitely many particles, then we have convergence to total occupancy, i.e.,

$$\lim_{t \to \infty} P[\omega_t(x) = 1] = 1, \quad x \in S.$$

(b) If $S = \mathbb{Z}$, $P(x, y) = 1$ if $y = x + 1$, then the only equilibrium states are: the Bernoulli states μ_α, as above; the states with $\omega(x) \equiv 0$ for $x \leq n$ and $\omega(x) \equiv 1$ for $x > n$, $n \in \mathbb{Z}$; and convex combinations of the above. There is a similar conjecture for the case when $P(x, x + 1) = p$, $P(x, x - 1) = 1 - p$, where $\frac{1}{2} < p < 1$.

References

1. M. B. AVERINTSEV, On a method of describing discrete parameter random fields, *Problemy Peredači Informacii* 6 (1970), 100–109.
2. R. L. DOBRUSHIN, Description of a random field by means of conditional probabilities, *Theor. Probability Appl.* 13 (1968), 197–224.
3. R. L. DOBRUSHIN, Gibbsian random fields for lattice systems with pairwise interactions, *Functional Anal. Appl.* 2 (1968), 292–301.
4. R. L. DOBRUSHIN, Markov processes with a large number of locally interacting components, *Problemy Peredači Informacii* 7 (1971), 70–87.
5. R. GLAUBER, The statistics of the stochastic Ising model, *J. Mathematical Phys.* 4 (1963), 294–307.
6. G. R. GRIMMETT, A theorem about random fields, *Bull. London Math. Soc.* 5 (1973), 81–84.
7. T. HARRIS, Nearest neighbor Markov interaction processes on multidimensional lattices, *Advances in Math.* 9 (1972), 66–89.
8. T. HARRIS, Contact interactions on a lattice, *Ann. Probab.* 2 (1974), 969–988.
9. R. HOLLEY, Markovian interaction processes with finite range interactions, *Ann. Math. Statist.* 43 (1972), 1961–1967.
10. R. HOLLEY, Free energy in a Markovian model of a lattice spin system, *Comm. Math. Phys.* 23 (1971), 87–99.
11. R. HOLLEY, An ergodic theorem for interacting systems with attractive interactions, *Z. Wahrscheinlichkeitstheorie und Verw. Gebiete* 24 (1972), 325–334.
12. R. HOLLEY, Some remarks on the FKG inequality, *Comm. Math. Phys.* 36 (1974), 227–231.
13. R. HOLLEY, Recent results on the stochastic Ising model, *Rocky Mt. Math. J.* 4 (1974), 479–496.
14. R. HOLLEY AND T. LIGGETT, Theorems for weakly interacting infinite systems and the voter model, to appear in *Ann. Probability*.
15. O. LANFORD AND D. RUELLE, Observables at infinity and states with short range correlations in statistical mechanics, *Comm. Math. Phys.* 13 (1969), 194–215.
16. T. LIGGETT, Existence theorems for infinite particle systems, *Trans. Amer. Math. Soc.* 165 (1972), 471–481.

17. T. LIGGETT, A characterization of the invariant measures for an infinite particle system with interactions, *Trans. Amer. Math. Soc.* **179** (1973), 433–453.
18. T. LIGGETT, A characterization of the invariant measures for an infinite particle system with interactions. Part II, *Trans. Amer. Math. Soc.* **198** (1974), 201–214.
19. T. LIGGETT, Convergence to total occupancy in an infinite particle system with interactions *Ann. Probab.* **2** (1974), 989–998.
20. T. LIGGETT, Ergodic theorems for the asymmetric simple exclusion process, *Trans. Amer. Math. Soc.*, to appear.
21. K. LOGAN, Time reversible evolutions in statistical mechanics, to appear in *Illinois Math. J.* (1974).
22. C. PRESTON, "Gibbs States on Countable Sets," Cambridge University Press, 1974.
23. F. SPITZER, Markov random fields and Gibbs ensembles, *Amer. Math. Monthly* **78** (1971), 142–154.
24. F. SPITZER, Interaction of Markov processes, *Advances in Math.* **5** (1970), 246–290.
25. F. SPITZER, "École d'été à St. Flour," Springer Lecture Notes in Math., Springer-Verlag, New York, Berlin, Vol. 390, 1974.
26. F. SPITZER, Recurrent random walk of an infinite particle system, *Trans. Amer. Math. Soc.* **198** (1974), 191–199.

Bifurcations in Reaction–Diffusion Problems

Louis N. Howard

Massachusetts Institute of Technology, Cambridge, Massachusetts 02139

TO STANISLAW M. ULAM ON HIS 65TH BIRTHDAY

I. Bifurcation Theorems

1. Perhaps the simplest example of a bifurcation theorem is the following: We have a family of autonomous nth order systems of differential equations

$$\dot{\mathbf{x}} = \mathbf{F}_\mu(\mathbf{x}) \tag{1}$$

where the right-hand sides $\mathbf{F}$ are continuously differentiable functions of the n dependent variables $\mathbf{x}$ and of the family parameter μ. Each member of the family is assumed to have a critical point at $\mathbf{x} = 0$: $\mathbf{F}_\mu(0) = 0$. The linearization of the system (1) at 0 has the matrix $d\mathbf{F}_\mu(0)$, whose entries are the partial derivatives of the components of $\mathbf{F}_\mu$ with respect to those of $\mathbf{x}$, evaluated at $\mathbf{x} = 0$. We assume that at $\mu = 0$ this matrix has *zero* as a simple eigenvalue. This is easily seen to imply that, for small enough $|\mu|$, $d\mathbf{F}_\mu(0)$ has an eigenvalue $\lambda(\mu)$, also simple and a continuously differentiable function of μ, with $\lambda(\mu) = 0$. Finally, we assume the "transversality condition" $d\lambda/d\mu \neq 0$ at $\mu = 0$. Under these hypotheses, one can show that, in the $n+1$ dimensional $(\mathbf{x}, \mu)$ space, there is another curve of critical points, distinct from the μ-axis, passing through $\mathbf{x} = 0$, $\mu = 0$. Sufficiently close to the origin, all critical points lie on this curve or on the μ-axis.

A "canonical example" for the above is the single first-order equation

$$\dot{x} = \mu x - x^2. \tag{2}$$

Here $\mu = 0$ is the "bifurcation point" at which the two families of critical points, $x = 0$ and $x = \mu$, cross one another. We will refer to the family other than $x = 0$ as the "new" critical points. This example is the typical one, with the new critical points existing on both sides

129

of $\mu = 0$ where the eigenvalue changes sign, but not every case is exactly like this. For instance, if the original system (1) is linear, it is its own linearization and, with the hypotheses made, $\mathbf{x} = 0$ is the *only* critical point for all small μ other than zero. However, at $\mu = 0$, *any* multiple of the eigenvector of $d\mathbf{F}_0(0)$ corresponding to the zero eigenvalue is a critical point: The new family just happens to lie in $\mu = 0$ but is still a curve in $(\mathbf{x}, \mu)$ space, crossing the μ-axis. Another case, rather common in applications, is illustrated by the equation $\dot{x} = \mu x - x^3$. Here the "new" family of critical points is present only for $\mu > 0$ and for each such μ there are *two* critical points. But with the $n + 1$ dimensional perspective obtained by looking at the $(\mathbf{x}, \mu)$-space we recognize that this is not essentially different either. There is again a curve (one-parameter family) of new critical points; here the parabola $\mu = x^2$. It is just that μ itself is not appropriate as the parameter of the family of critical points. In this example, the new critical points all occur on the "supercritical" side—the side of $\mu = 0$ on which $\lambda(\mu) > 0$. $\dot{x} = \mu x + x^3$ illustrates that it is equally possible for them all to be *subcritical*.

Though stated in terms of the differential equation (1), the above result is really only about the roots of the equation $\mathbf{F}_\mu(\mathbf{x}) = 0$. The general structure of this and related bifurcation theorems is that from some hypotheses about the linearized version of the problem, conclusions are drawn about the full nonlinear one, valid in an appropriate neighborhood. This is analogous to the implicit function theorem, and indeed proofs of such bifurcation theorems typically involve transformations that convert the problem into a form in which the implicit function theorem can be applied. (With hypotheses of a less analytic nature than, say, $d\lambda/d\mu \neq 0$, related results can sometimes be obtained by topological methods. These are analogous to the intermediate value theorem rather than the implicit function theorem; more general in some ways, the conclusions are also typically weaker in others, particularly with respect to questions of uniqueness.) Another consequence of the hypotheses in the theorem stated above relates to the stability properties of the new family of critical points. When this is adjoined to the mere existence of the critical points, the bifurcation theorem becomes genuinely a result about differential equations. One can show that sufficiently close to the origin in $(\mathbf{x}, \mu)$-space the linearization of (1) about the new critical point (we assume here that they are not all in $\mu = 0$) also has a simple eigenvalue λ_* that approaches zero as the origin is approached and when the corresponding $\mu \neq 0$, $\lambda(\mu)$

and λ_* have *opposite signs*. Thus, if $\mathbf{x} = 0$ is, say, a *stable* critical point for $\mu > 0$, and if $d\mathbf{F}_0(0)$ has no pure imaginary eigenvalues so that the instability arises solely from the sign change in $\lambda(\mu)$, then the new critical points are unstable when subcritical and stable when supercritical (at least close enough to $\mathbf{x} = 0$, $\mu = 0$). For this reason, such bifurcations are said to exhibit an "exchange of stabilities." If they occur in mathematical models of the temporal evolution of some physical system, one should expect to observe a transition from the state represented by one critical point to that represented by the other when the parameter is moved across the critical value $\mu = 0$.

The bifurcation theorem just discussed was stated in this form by Hopf [1] but must surely have been known much earlier in some form to people working on celestial mechanics such as Lindstedt and Poincaré. Generalizations in various directions exist, for instance, to infinite dimensional cases (see, e.g., [2, 3]). We mention here also a result [4] concerned with dropping the hypothesis of a *simple* eigenvalue. If $d\mathbf{F}_0(0)$ has 0 as a repeated eigenvalue, but is nevertheless of rank $n - 1$ so that there is not more than one independent eigenvector, then it remains true that there is a unique curve of new critical points in a neighborhood of the origin in $(\mathbf{x}, \mu)$-space. For this, a different form of the transversality condition is also needed, namely $d/d\mu \det(d\mathbf{F}_\mu(0)) \neq 0$ at $\mu = 0$, a condition equivalent to $d\lambda/d\mu \neq 0$ when the eigenvalue is simple, but meaningful also when it is not. If the rank is less than $n - 1$ it can be shown that no bifurcation theorem of this kind is possible: the existence or otherwise of new critical points, and when they exist their uniqueness, then depends on higher order terms and cannot be asserted merely on the basis of the properties of the linearization at $\mathbf{x} = 0$. A canonical example for this case is given, in two dimensions, by the system

$$\dot{x} = y$$
$$\dot{y} = \mu x - x^2. \tag{3}$$

2. A deeper kind of bifurcation theorem was given by Hopf in [1]. Again we have the family of systems (1) with $\mathbf{x} = 0$ as a common critical point. But now instead of a zero eigenvalue, $d\mathbf{F}_0(0)$ is assumed to have a conjugate pair of nonzero pure imaginary eigenvalues. These are simple eigenvalues and no other eigenvalues of $d\mathbf{F}_0(0)$ are on the imaginary axis. The transversality condition is $d\lambda_1/d\mu \neq 0$ at $\mu = 0$, where $\lambda_1(\mu) \pm i\lambda_2(\mu)$ is the conjugate pair of eigenvalues of $d\mathbf{F}_\mu(0)$

which reduces to the pure imaginary pair at $\mu = 0$. Hopf then proves that in a neighborhood of the origin in $(\mathbf{x}, \mu)$-space there is a unique one-parameter family of *periodic solutions* to (1), with periods near $2\pi/\lambda_2(0)$. Hopf remarks that this kind of result must have been known also to Poincaré, but the recognition of the relevance of the transversality condition, the general picture, and the proof appear to make this result quite appropriately called (as it often is) the "Hopf Bifurcation Theorem." Hopf also demonstrated that generically, and sufficiently close to $\mathbf{x} = 0$, $\mu = 0$, the periodic solutions exist only supercritically (and are then stable) *or* subcritically (and are then unstable). A canonical example illustrating a Hopf bifurcation is given by the system:

$$\dot{x} = x(\mu - x^2 - y^2) - y$$
$$\dot{y} = y(\mu - x^2 - y^2) + x. \tag{4}$$

For $\mu < 0$ the critical point at the origin is stable, in fact globally so. For $\mu > 0$ the circle $x^2 + y^2 = \mu$ is a stable limit cycle whose domain of attraction consists of everything but the unstable critical point. Infinite-dimensional generalizations of this important result have also been given [5].

3. The result about the occurrence of new critical points when a real eigenvalue crosses zero contrasts with the Hopf theorem in that it at first seems independent of differential equations. However, it can be related to the existence of some nonconstant solutions of (1). This gives a more complete picture of the facts about exchange of stabilities referred to above, and clarifies the role of this set of ideas in the subject of differential equations. The point is well illustrated by the example (2) mentioned before. If, in this equation for $\mu \neq 0$, we set $x = \mu(1 + T)/2$ and $t = 2\tau/\mu$ we obtain

$$dT/d\tau = 1 - T^2. \tag{5}$$

The solutions of (5) which are bounded for all τ are all of the form $T = \tanh(\tau - \tau_0)$, and thus all describe (with differing zero points of τ) the same trajectory joining the critical points $T = \pm 1$ of (5). The change of variable from x to T has merely put the critical points $x = 0$ and $x = \mu$ of (2) into the "standard positions" $T = \pm 1$. Thus we see in this example that the bifurcation at $\mu = 0$ is not only accompanied by a pair of critical points of opposite stabilities for all $\mu \neq 0$, but that there is also a (unique) trajectory going from the unstable critical point to the stable one. That this situation is of some generality is the content of the following theorem:

THEOREM. *Let*

$$\dot{\mathbf{x}} = A\mathbf{x} + \mu A_1 \mathbf{x} + \mathbf{Q}(\mathbf{x}, \mathbf{x}) + \mathbf{R}(\mathbf{x}, \mu) \tag{6}$$

be a family of autonomous systems with $\mathbf{x} = 0$ *as a common critical point. Here the matrix of the linearization at* $\mathbf{x} = 0$ *is* $A + \mu A_1 + O(\mu^2)$, $\mathbf{Q}(\mathbf{x}, \mathbf{x})$ *represents the quadratic terms at* $\mu = 0$ *and* $\mathbf{R} = O(x^3, \mu x^2, \mu^2 x)$ *is everything else. We assume A has 0 as a simple eigenvalue, with corresponding eigenvector* $\mathbf{e}$, *and in addition, that the two matrices of n rows and n + 1 columns,* $[A, A_1\mathbf{e}]$ *and* $[A, \mathbf{Q}(\mathbf{e}, \mathbf{e})]$, *are both of rank n. Finally, we also assume that A has no pure imaginary eigenvalues. One can then show that for all sufficiently small* $|\mu| \neq 0$ *there is (in a neighborhood of* $\mathbf{x} = 0$) *a unique critical point different from* $\mathbf{x} = 0$, *and a unique trajectory joining it to* $\mathbf{x} = 0$.

Remarks. (a) The condition rank $[A, A_1\mathbf{e}] = n$ is equivalent to the transversality condition, and implies that the eigenvalue of the linearization at $\mathbf{x} = 0$ which is zero at $\mu = 0$ is >0 on one (supercritical) side and <0 on the other (subcritical). The connecting trajectory goes from $\mathbf{x} = 0$ to the new critical point on the supercritical side, and the other way on the subcritical side.

(b) The condition rank $[A, \mathbf{Q}(\mathbf{e}, \mathbf{e})] = n$ is a condition of "genuine nonlinearity," which rules out, for instance, the linear case in which all of the new critical points occur at $\mu = 0$ and the question of connecting trajectories does not come up. It implies that for small enough μ the connecting trajectory is in essence like that for the example of Eq. (2). Indeed, for small μ the connecting trajectory is approximately along the direction of $\mathbf{e}$ and is traversed in a manner essentially described by the hyperbolic tangent of an argument that is essentially μt.

(c) That the "characteristic time-scale" of the connecting trajectory should be of order μ^{-1} is already indicated by the fact that one of the eigenvalues of the linearization is of order μ. If the dimension is greater than one, the linearization has other eigenvalues which, with our hypotheses, have real parts of order one; thus for small μ one expects the solutions in general to involve two disparate time-scales. This is not true of the connecting trajectory but is of other solutions starting near it, some of which rapidly approach it and then slowly drift along it. This indicates that for small μ we have here another example of the somewhat vaguely defined class of "singular perturbation problems," problems on which a naive attack by computer is likely to lead to frustration.

(d) The theorem just quoted is essentially the same as results given by Foy [6] in connection with his study of the shock structure problem.

(e) The differential equation $\dot{x} = \mu - x^2$ is very similar to (2), but does not appear as a bifurcation problem. Here there are no critical points for $\mu < 0$, and a pair appears as μ becomes positive; or one may say that a pair of critical points coalesce and disappear as μ decreases through 0. With hypotheses similar to those stated above, one can obtain [7] a general version of this showing that near the point of coalescence there is always a unique trajectory joining the critical points.

(f) If the eigenvalue 0 of A is k-fold but rank $A = n - 1$, there still is, as stated above, a second curve of critical points in the $(\mathbf{x}, \mu)$ space. Furthermore, with the condition of genuine nonlinearity, this curve is transverse to the hyperplane $\mu = 0$, so we may ask whether or not the pair of critical points is joined by a trajectory. If it exists, the connecting trajectory is no longer approximately along the eigenvector $\mathbf{e}$, but is approximately in the k-dimensional generalized eigenspace (null space of A^k). The nature of the transition is approximately described, for k odd, by the canonical example, analogous to (5):

$$y^{(k)} = 1 - y^2. \tag{7}$$

It can be shown that the existence of a connecting trajectory for (6) (for small enough μ) is *equivalent* to the existence of such a trajectory for (7), for k odd. Such a trajectory must, if $k \geqslant 3$, spiral out from one critical point and in toward the other. Thus, the transition is no longer monotonic as with the hyperbolic tangent. For $k = 3$, (7) is known to have a connecting trajectory, and it probably does for all odd k. These results will be presented in [7]. When k is even, the situation is more complex and there may or may not be a connecting trajectory; additional hypotheses are required to decide. The case $k = 2$ will be investigated in some detail in [7].

II. Chemical Waves and Reaction–Diffusion Equations

1. Some chemical reactions proceed in an oscillatory manner: There is an overall reaction that runs down to a final stationary equilibrium state but does so through a sequence of many oscillations whose periods are short compared to the time scale of the overall reaction. Looked at on an intermediate time-scale, the oscillations appear essentially periodic and presumably could be maintained oscillating periodically if new reactants could be continually supplied and the final products removed. Most known oscillatory reactions are biochemical and involve the intervention of enzymes, but two are essentially inorganic. The better studied of these was discovered by Belousov about 15 years ago; it involves the oxidation of malonic acid by potassium bromate, which occurs in the presence of sulfuric acid and a low concentration of cerous ions. With suitable concentrations, the overall time-scale is an hour or two, while the oscillations have a period of about 20 sec. These time scales make this a convenient reaction to study, and, since the oscillations can be made readily visible with a suitable oxidation–reduction indicator, the reaction is also good as a lecture demonstration. In the course of a cycle of oscillation some of the cerous ions are converted to ceric, and then back again; also a small amount of bromide ion appears, decays slowly, drops suddenly to nearly zero, and reappears. The major reactants (like malonic acid) change by only a small fraction during a cycle, but the minor ones like Br^- or Ce^{4+} change a great deal, perhaps by a factor of 100 or more. Under conditions suitable for demonstrations the reaction is a very strong relaxation oscillator, but fairly sinusoidal oscillations can also be obtained. The most extensive investigation of the chemical mechanism appears to be that described in [8].

To keep the oscillations visible in a beaker of the reagent it must be stirred, apparently because different portions of the fluid would otherwise get out of phase with one another. If, however, the fluid occupies a thin layer (1–2 mm) in a flat dish, the variations of phase become organized into rather striking patterns of concentric circles or spirals, meeting along lines of cusps. These patterns, discovered by Zaikin and Zhabotinskii [9], are in fact made up of propagating chemical waves. A typical wavelength is 2 mm and period about 15 sec. The waves radiate outward from the centers. (It is likely that such patterns are not seen in deeper layers merely because then fluid motions produced by thermal convection are almost inevitable.) Good color photographs

Fig. 1. Chemical waves in the Belousov–Zhabotinskii reagent. The small black circles are gas bubbles.

of spirals can be found in [10], and Fig. 1 shows some concentric ring patterns. Notice in particular, the following features.

(a) The rings are essentially circular and are equally spaced within a particular concentric set ("target"), but the spacing differs from one target to another. The same is the case for the period and the speed of propagation.

(b) Two adjacent targets do not overlap, as water waves would, but are separated by a sharply defined "front" on which the wave trains from the two sides meet and seem to annihilate each other. When the two meeting waves have different periods, the front also propagates; the higher frequency region encroaches on its neighbor. What, if anything, is special about the centers of the concentric rings is not clearly understood.

2. A fairly plausible mathematical model is obtained by supposing that the concentrations of the reactants of interest satisfy *reaction–diffusion* equations of the form:

$$\mathbf{c}_t = \mathbf{F}(\mathbf{c}) + K\nabla^2 \mathbf{c}. \tag{8}$$

Here, $\mathbf{c}$ is a vector whose components are the concentrations; $\mathbf{c}_t = \mathbf{F}(\mathbf{c})$ are the equations of the chemical kinetics, after neglect of the long time-scale secular change, and other approximations used so as to concentrate on the oscillation. K is a (positive definite) matrix of diffusion constants.

To investigate the possibility of understanding chemical waves in the Belousov–Zhabotinskii reaction on the basis of such mathematical models, a first question one might ask is whether or not Eq. (8) has wave-like solutions under hypotheses about the chemical kinetics $\mathbf{F}$ which are plausible for such a system; for example, in the spatially homogenous case described by $\mathbf{c}_t = \mathbf{F}(\mathbf{c})$, one might suppose the existence of a stable limit cycle. While the wave structures actually seen are generally circular, a further idealization to plane waves seems appropriate. This means that one looks for solutions of (8) of the form $\mathbf{c} = \mathbf{y}(\sigma t - \boldsymbol{\alpha} \cdot \mathbf{z})$, where $\mathbf{y}$ is a 2π-periodic function of its argument, σ is the angular frequency, $\boldsymbol{\alpha}$ the wavenumber vector, and $\mathbf{z}$ the spatial position vector. Inserting this in (8) one obtains

$$\sigma \mathbf{y}' = \mathbf{F}(\mathbf{y}) + \alpha^2 K \mathbf{y}'', \tag{9}$$

a system of ordinary differential equations whose order is twice that of the chemical kinetics equations. Alternatively, one may set $\mathbf{c} = \mathbf{u}(t - \sigma^{-1}\boldsymbol{\alpha} \cdot \mathbf{z})$, getting instead, with $\beta = \alpha^2/\sigma^2$,

$$\mathbf{u}' = \mathbf{F}(\mathbf{u}) + \beta K \mathbf{u}''. \tag{10}$$

Here, $\mathbf{u}$ is to be a periodic function of its argument; once $\mathbf{u}$ is known and the period determined, σ and α can be obtained from β. One method that can be used to show that (10) has such periodic solutions makes use of the Hopf Bifurcation Theorem. Let us assume about $\mathbf{F}$ that the system $\mathbf{c}_t = \mathbf{F}(\mathbf{c})$ has an unstable critical point (which by a translation may be placed at the origin) at which the matrix $M = d\mathbf{F}(0)$ has a conjugate pair $p \pm iq$ of eigenvalues, with p and q positive. (This is of course, not the same as assuming that $\mathbf{c}_t = \mathbf{F}$ has a limit cycle solution, but is a somewhat similar, and incidentally more easily verified, condition.) The way the Hopf theorem can be used is now easy to see when the diffusion matrix is scalar: $K = kI$, for then the linearization of (10) at the origin (which is also a critical point of (10)) is $\mathbf{x}' = M\mathbf{x} + \beta k\mathbf{x}''$ and if $M\mathbf{e} = (p + iq)\mathbf{e}$ we see that $\mathbf{x} = \mathbf{e}e^{\lambda t}$ is a solution if $\lambda = p + iq + \beta k\lambda^2$. Thus, when $\beta = \beta_c = p/(kq^2)$, $\lambda = iq$; in other words at $\beta = \beta_c$ the linearization of (10) at its critical point has a conjugate pair of pure imaginary eigenvalues. It is easy to check that, as β crosses β_c from above, this pair crosses the imaginary axis transversally, so that Hopf's theorem applies. Thus, in a neighborhood of $(0, 0, \beta_c)$ in $(\mathbf{u}, \mathbf{u}', \beta)$-space, the nonlinear system (10) has a one-parameter family of periodic solutions. Calling the parameter a, which we think of as a measure of the amplitude of the oscillation, and returning to the form (9) we get in this way a family of 2π-periodic solutions $\mathbf{y}_a(\sigma t - \boldsymbol{\alpha} \cdot \mathbf{z})$ with

$$\begin{aligned} \sigma &= \omega(a) \\ \alpha^2 &= \lambda(a). \end{aligned} \tag{11}$$

These relations (11) give a parametric representation of a "dispersion relation"

$$\sigma = H(\alpha^2) \tag{12}$$

for this family of nonlinear plane-waves. (Note that this differs from, say, nonlinear water waves in which there is a two-parameter family; amplitude and wavelength can be specified independently, and frequency is then determined.)

It is also possible to show [11], in a somewhat more complicated way but again using Hopf's theorem, that these results hold even when K is not scalar, provided that it is close enough to a scalar matrix. Reference [11] also shows the existence of plane-wave solutions to (8) when $c_t = F(c)$ has a stable limit cycle solution. These waves are not near the critical point, but near the limit cycle; they are of large amplitude in the sense that the concentrations fluctuate over a wide range, while the waves obtained using Hopf's theorem are of small amplitude. The large-amplitude waves were obtained by a method exploiting the singular perturbation character of the problem for small β, not by a bifurcation approach. Bifurcation methods are often useful, but they are not a panacea.

3. Plane-wave solutions may be expected to give a first approximation to the description of circular or spiral waves more like those observed when the wavelength is short compared to the radius of curvature or other spatial variations. Thus the target patterns of Fig. 1 are presumably rather like plane waves away from the centers and away from the boundaries between different targets. (Quantitative comparison with experiment has not yet been possible, for various reasons, but there is at least some evidence suggesting that these waves do have a one-parameter dispersion relation.) If the plane-wave solutions are regarded as known, one can investigate more general solutions of (8) by the method of "slowly varying waves." Formally, one introduces new time and space scales, $T = \epsilon t$ and $\mathbf{Z} = \epsilon \mathbf{z}$, where ϵ is a small parameter characteristic of the "slowness" of variation, e.g., something like (wavelength/radius) for a large circular wave-pattern. Then look for solutions that have an asymptotic representation for small ϵ of the form $\mathbf{c} = \mathbf{c}^0(\theta(Z, T)/\epsilon, Z, T) + \cdots$, 2π-periodic in the first argument ("phase") θ/ϵ. The plane waves themselves have this form with $\theta = \sigma T - \boldsymbol{\alpha} \cdot \mathbf{Z}$, σ and $|\boldsymbol{\alpha}|$ being given in terms of the amplitude parameter a by (11) and $c^0 = \mathbf{y}_a(\theta/\epsilon)$. For the more general slowly varying waves, one finds $c^0 = \mathbf{y}_A(\theta/\epsilon)$ where the "amplitude" is now a function $A(Z, T)$ of the slow variables, and is related to the phase function θ by (cf. Eq. (11))

$$\theta_T = \omega(A),$$
$$|\nabla_Z \theta|^2 = \lambda(A). \tag{13}$$

If one wishes to eliminate A, paralleling (12), we get the Hamilton–Jacobi equation

$$\theta_T = H(|\nabla \theta|^2) \tag{14}$$

for the phase function θ. This first-order equation can be effectively attacked by the method of characteristics, which can also be applied directly to the system (13). The characteristics turn out to be straight lines along which A and the derivatives of θ are constant and along which the variation of θ is easily computed. Thus, the evolution of initial data which are slowly varying waves can be well described until such time as the projections of the characteristics into the (Z, T)-space intersect. In general, this will happen, and then the slowly varying wave-picture must break down. By analogy with gas dynamics, we say that a "shock" forms. As with inviscid gas dynamics, it is here possible to extend the applicability of the slowly varying wave description by accepting the shocks as discontinuities and from other considerations (e.g., conservation of phase) giving a rule for their propagation. Such shocks are thought (by Kopell and me, at least) to be models for the sharp transition zones between different target patterns.

4. To obtain a more satisfactory understanding of the shocks and to derive the conditions characterizing their propagation, one should return to the full reaction–diffusion equations and, at least in certain idealized cases, show that they do have solutions with a "shock structure," a region of transition between two different plane wave solutions. This can be done for the case of a shock structure joining two plane-wave solutions with nearly equal wavenumbers and frequencies, by bifurcation methods similar to those used by Foy for the weak gas dynamic shock structure problem. The problem reduces to finding a trajectory joining two critical points which can be regarded as having arisen from a bifurcation with a simple eigenvalue crossing zero. In this weak shock case, the thickness of the transition zone is considerably larger than a wavelength (though still small compared to the scale of the slow variation). The prominent interfaces between different target patterns in Fig. 1 are, however, *thinner* than a wavelength. These are transitions between waves whose wavenumbers, though possibly not too different in magnitude, are oppositely directed—they must be regarded as *strong* shocks. They cannot be investigated in the same way as the weak ones. However, by introducing an artificial parameter into the kinetics, it is possible to treat the case of a shock layer between two oppositely directed but otherwise identical plane waves by a bifurcation method. Here again, the problem reduces to finding a trajectory joining two critical points; they have here arisen from a bifurcation point at which 0 is a triple eigenvalue. (It was this

problem which led to some of the investigations of [7]). Some further details about these shock structure problems are given in [12]. A full account is to appear in [13], a paper which, due to the pressures of writing up symposium talks [12, 14, 15], is still in preparation.

REFERENCES

1. E. HOPF, Abzweigung einer periodischen Lösung von einer stationären Lösung eines Differentialsystems, *Ber. Math. Phys. Klasse Sächsischen Akad. Wiss. Leipzig* **94** (1942), 3–22.
2. M. CRANDALL AND P. RABINOWICZ, Bifurcation from simple eigenvalues, *J. Functional Analysis* **8** (1971), 321.
3. D. H. SATTINGER, "Topics in Stability and Bifurcation Theory," Springer Lecture Notes No. 309, Univ. Minnesota, 1972.
4. N. KOPELL AND L. N. HOWARD, Bifurcations under non-generic conditions, *Advances in Math.* **13** (1974), 274–283.
5. D. D. JOSEPH AND D. H. SATTINGER, Bifurcating time periodic solutions and their stability, *Arch. Rational Mech. Anal.* **45** (1972), 79–109.
6. L. R. FOY, Steady state solutions of hyperbolic systems of conservation laws with viscosity terms, *Comm. Pure Appl. Math.* **17** (1964), 177.
7. N. KOPELL AND L. N. HOWARD, Bifurcations and trajectories joining critical points, in preparation.
8. R. J. FIELD, E. KÖRÖS, AND R. M. NOYES, Oscillations in chemical systems. I. Thorough analysis of temporal oscillation in the bromate-cerium-malonic acid system, *J. Amer. Chem. Soc.* **94** (1972), 8649.
9. A. N. ZAIKIN AND A. M. ZHABOTINSKII, Concentration wave propagation in two-dimensional liquid-phase self-oscillating system, *Nature (London)* **225** (1970), 535.
10. A. T. WINFREE, Rotating chemical reactions, *Sci. Amer.* **230**, No. 6 (1974), 82–95.
11. N. KOPELL AND L. N. HOWARD, Plane wave solutions to reaction-diffusion equations, *Studies Appl. Math.* **52** (1973), 291.
12. L. N. HOWARD AND N. KOPELL, Wave trains, shock structures, and transition layers in reaction–diffusion equations, Proceedings of AMS–SIAM Symposium on "Mathematical Aspects of Chemical and Biochemical Problems and Quantum Chemistry," New York, 1974.
13. L. N. HOWARD AND N. KOPELL, Slowly varying waves and shock structures in reaction–diffusion equations, in preparation.
14. N. KOPELL AND L. N. HOWARD, Pattern formation in the belousov reaction, "Lectures on Mathematics in the Life Sciences," Vol. 7, American Mathematical Society, Providence, R.I., 1974.
15. L. N. HOWARD, Bifurcations in reaction–diffusion problems, First Los Alamos Workshop on Mathematics in the Natural Sciences, June, 1974.

Singular Perturbation

J. D. COLE

University of California, Los Angeles, Los Angeles, California 90024

DEDICATED TO STAN ULAM

The main aim of this paper is to provide a general elementary hearistic discussion of the ideas and techniques used in singular perturbation problems. For an attempt at a deeper understanding of basic concepts the reader should consult the work of Lagerstrom and Casten [1] and earlier references cited by them. For a more detailed exposition of the techniques used and many examples of application the reader can consult Refs. [2–4].

Perturbation methods are intimately connected with asymptotic expansion and some details are given below. Generally, in perturbation procedures the asymptotic expansion is carried out with respect to a parameter and has a certain domain of validity, as an asymptotic, with respect to an independent variable. An ultimate aim might be to construct and asymptotic approximation valid over a certain given domain (finite or infinite) of independent variable. The general situation is the following: several asymptotic expansions[1] valid in different domains are needed in order to construct an asymptotic approximation valid over the entire domain. For reasons veiled in antiquity this usual situation is called singular. The situation when a single (limit process) expansion provides an asymptotic approximation valid in the entire domain of interest is called regular. In the following paragraphs we outline the uses and scope of perturbation procedures and describe the notions of asymptotic approximation and limit process expansion. Some simple examples are worked out and some general applications indicated. Finally an illustration is given of a more general type of asymptotic expansion, the multiple scale expansion and it is shown how this expansion can be combined with one of limit process type.

In physical problems perturbation methods are used most often in

---

[1] Here we are speaking of asymptotic approximations of limit process type or equivalent.

143

two ways, (i) to generate approximate solutions to given equations (ii) to generate sets of approximating equations to more exact formulations. The aim is usually not restricted to numerical results but includes the desire for a simplified description. From the mathematical point of view we can think of all physical problems as expressed in dimensionless variables. The various physical constants of the problem will combine to form dimensionless parameters. One of the parameters ϵ (say) may be a small number and we can then consider an approximation valid for small ϵ. ϵ may be, for example, a ratio of characteric lengths, times or energies. We can consider the approximation by considering a family of problems (or solutions) as $\epsilon \downarrow 0$ and then the sense of the approximation becomes asymptotic. That is, the difference between the exact solution and a finite number of approximating terms can be made arbitrarily small as $\epsilon \downarrow 0$. One type of asymptotic expansion which is extremely useful and for which the construction of successive terms follows well defined rules is the limit-process expansion. In its simplest form the limit-process expansion is based on an asymptotic sequence of functions $\mu_k(\epsilon)$ with the property

$$\mu_{k+1}(\epsilon)/\mu_k(\epsilon) \to 0, \quad \text{as} \quad \epsilon \downarrow 0.$$

In a practical problem the $\mu_k(\epsilon)$ are chosen according to certain rules to fit the needs of the particular problem and they measure the orders of magnitude of successive terms in the approximation. Assuming the existence of limit-process expansions for a given problem the basis of the method lies in the following (unproved) theorem: The limit of the (unknown) solution of some exact problem is equal (asymptotically) to the exact solution (known) of the limit of the original problem. For example, from a given sequence $\mu_k(\epsilon)$, a limit process expansion of a function $f(x; \epsilon)$ can be written

$$f(x, \epsilon) = \sum_{k=0}^{K} \mu_k(\epsilon) f_k(x^*) + O(\mu_{K+1}), \tag{1}$$

where $x^* = x^*(x; \epsilon)$ a fixed relationship and the representation (1) is valid in some fixed x^* domain. Valid here means that the asymptotic order of the error is correctly given. Then the $f_k(x^*)$ can be calculated successively as follows:

$$f_0(x^*) = \lim_{\epsilon \downarrow 0, x^* \text{ fixed}} \frac{f(x(x^*; \epsilon); \epsilon)}{\mu_0(\epsilon)},$$

$$f_1(x^*) = \lim_{\epsilon \downarrow 0, x^* \text{ fixed}} \frac{f(x; \epsilon) - \mu_0 f_0(x^*)}{\mu_1(\epsilon)},$$

$$f_2(x^*) = \lim_{\epsilon \downarrow 0, x^* \text{ fixed}} \frac{f(x; \epsilon) - \mu_0(\epsilon) f_0(x^*) - \mu_1(\epsilon) f_1(x^*)}{\mu_2(\epsilon)} \quad \text{etc.}$$

In the situation where we are trying to solve a problem the function $f(x; \epsilon)$ is the (unknown) solution. Then the idea of the theorem above is used and the limits are applied to the original problem to obtain a sequence of problems for $f_0, f_1, \ldots$. This procedure is a more precise version of the rule "equating equal powers of ϵ."

Now the following situation is general. The sequence of problems defining f_0, f_1, etc., is not well posed, for example, certain constants of integration may not be able to be found and the $f_0, f_1, f_2, \ldots$ will not be able to be fully constructed seriatim, even though the original problem for $f(x; \epsilon)$ is well posed. This is an indication that for the given sequence $\mu_k(\epsilon)$ and variable $x^*(x; \epsilon)$ there is a nonuniformity to the approximation. In general this means that more than one expansion (with different x^*, μ_k) must be used to cover the entire x-domain of interest. An outline of the way doing this will now be given.

The question that must be answered is how to choose a suitable coordinate $x^*(x; \epsilon)$ and a suitable asymptotic sequence $\mu_k(\epsilon)$. The general procedure is to consider the problem subject to all possible limits but to choose from all possible limits only those which are distinguished. By distinguished here we mean those limits which assign a definite order of magnitude to $\mu_k(\epsilon)$ with a corresponding $x^*(x, \epsilon)$, rather than those which place $\mu_k(\epsilon)$ in some order class. There are only a finite number of distinguished limits to be considered for a given problem, basically because there are only a finite number of terms in the equations and boundary conditions of a given problem. There are only finite possibilities for dropping out terms and obtaining reduced problems. The usefulness of these distinguished limits depends on the following (unproved) theorem: the expansions corresponding to distinguished limits in adjacent x-regions overlap and match asymptotically. The details of asymptotic matching cannot be discussed here (cf. Ref. [2]) but a simple example will be given. This matching does provide the mechanism for determining unknown constants in the various expansions valid in adjacent regions. If in fact the adjacent regions cover the entire domain of interest then an asymptotic approximation can be constructed from limit process expansions which is valid in the entire region of interest. In this way the original problem is solved in terms of a sequence of simpler problems.

Sometimes, however, the various domains of the limit-process expansions do not cover the entire region of interest. This may happen, for example, if functions which do not have limits occur in the solution (e.g., $\sin t/\epsilon$). Then it still may be possible to construct an asymptotic approximation valid in the entire region by using a more general type of asymptotic expansion in those domains where the limit process expansions do not exist. Under such circumstances the ideas of asymptotic matching can still be used.

The procedure proposed here is certainly not the only way of constructing an asymptotic approximation. However, basing the entire procedure in limits makes it possible often to assign a simple description to the various stages of approximation.

Now, as a simple example, consider the problem of constructing an approximation for the motion of a linear oscillator (spring k, mass m, damping β) initially at rest and subject to an initial impulse, when the mass m is a relatively small value. In terms of a suitable dimensionless y and a time scale t measured in units of β/k the problem for $y(t, \epsilon)$ can be written

$$\epsilon \frac{d^2 y}{dt^2} + \frac{dy}{dt} + y = 0, \qquad 0 \leqslant t < \infty$$

$$y(0) = 0, \qquad \frac{dy}{dt}(0) = \frac{1}{\epsilon}.$$

(2)

Here

$$\epsilon = \frac{mk}{\beta^2} \sim \left(\frac{\text{viscous decay period}}{\text{undamped oscillation period}} \right).$$

We can expect a more rapid rise for smaller ϵ as indicated in Fig. 1. For this problem the limit $\epsilon \downarrow 0$, t fixed and the sequence

$$\mu_k(\epsilon) = \epsilon^k$$

produces a sequence of problems for the terms of $y_k(t)$ of the asymptotic expansion

$$y(t; \epsilon) = y_0(t) + \epsilon y_1(t) + \cdots. \tag{3}$$

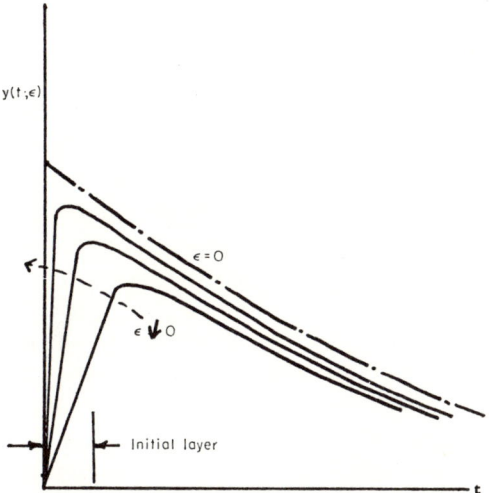

FIG. 1. Response of oscillator as $\epsilon \downarrow 0$.

The first term is found from

$$\frac{dy_0}{dt} + y_0 = 0, \quad y_0 = A_0 e^{-t} \tag{4}$$

and represents the ultimate decay of a massless system. However only one constant of integration appears so that both initial conditions of problem (2) cannot be satisfied. The description is not good for the initial moments and we can only expect the asymptotic expansion (3) to be valid $0 < t_1 \leqslant t \leqslant t_2$ for some $t_{1,2}$.

In order to describe better what happens near $t = 0$ we can use a coordinate $t^* = t/\delta(\epsilon)$, $\delta(\epsilon) \to 0$ so that for fixed t^*, $t \to 0$ as $\epsilon \to 0$. It is sufficient for this example to consider the same sequence $\mu_k = \epsilon^k$. Now the asymptotic expansion has the form

$$y(t; \epsilon) = y_0^*(t^*) + \epsilon y_1^*(t^*) + \cdots \tag{5}$$

and problem (2) reads

$$\frac{\epsilon}{\delta^2} \frac{d^2 y_0}{dt^{*2}} + \cdots + \frac{1}{\delta} \frac{dy_0^*}{dt^*} + \cdots + y_0^* + \cdots = 0$$

$$y_0^*(0) + \epsilon y_1^*(0) + \cdots = 0, \quad \frac{1}{\delta} \frac{dy_0^*}{dt^*} + \frac{\epsilon}{\delta} \frac{dy_1^*}{dt^*}(0) + \cdots = \frac{1}{\epsilon}.$$

Different approximate equations result according as $\delta/\epsilon \to 0$ or $\delta/\epsilon = 1$, but the latter is the distinguished limit and results in the problem

$$\frac{d^2 y_0}{dt^{*2}} + \frac{dy_0}{dt^*} = 0, \qquad t^* = \frac{t}{\epsilon},$$

$$y_0^*(0) = 0, \qquad \frac{dy_0^*}{dt^*} = 1, \qquad (6)$$

with the solution

$$y_0(t^*) = 1 - e^{-t^*} = 1 - e^{-t/\epsilon}. \qquad (7)$$

This approximation gives an accurate picture of the rapid change in the initial moments but cannot be valid for t large because the motion shows no ultimate decay. However according to the ideas of asymptotic matching the ultimate behavior of (5) ($t^* \to \infty$) should correspond to the initial behavior of (3) ($t \to 0+$). Comparing (7) and (4) we see that the unknown constant A_0 must be chosen $A_0 = 1$. The overlapping of the two expansions in the adjacent t regions for this case is essentially guaranteed by the occurence of the common term dy/dt in (4) and (6).

The basic idea used here can be extended in various directions: higher approximations can be constructed and matched, the effects of variable coefficients can be considered (this may demand a different sequence $\nu_k(\epsilon)$ for one of the expansions), and nonlinearities can be considered. In the non-linear case it is possible for even the location of the transition layer to depend on the particular boundary conditions. Further, systems such as

$$\frac{d\mathbf{x}}{dt} = f(\mathbf{x}, \mathbf{y}),$$

$$\epsilon \frac{d\mathbf{y}}{dt} = g(\mathbf{x}, \mathbf{y}),$$

with suitable initial conditions can have rapid layers of transitions (of Ref. [5]). Such systems occur for example in chemical reactions where some reaction rates are much faster than others.

The basic ideas can be extended to boundary value problems and, if the solution exists, rapid layers of transition may occur at the end points or even at interior points. Conceptually, expansions based on limits can also be used for problems described by partial differential equations. An historically important case is the approximate solution of the Navier–Stokes equations for viscous-incompressible flow past a streamlined body according to Prandtl's boundary layer theory [6]. The small parameter is

$\epsilon = 1/(\text{Re})^{1/2}$, Re is the Reynolds' number of flow past object $= UL/\nu$,

U is the velocity, L is the length, ν is the kinematic viscosity. The terms of the expansion valid away from the boundary are given by inviscid potential flow (which would slip past the surface). The flow in the boundary layer is given by the solution to Prandtl's equation and matches to the potential flow. This provides an explanation of skin friction and resistance. For a detailed treatment see Ref. [7].

Thus far in the discussion, the occurence of nonuniformities has been connected with the loss of a boundary or initial condition due to the reduction in order of the differential equation under one of the limits. Such problems are of necessity singular. But there are many other kinds of approximations which lead to nonuniformity. Some of these will now be mentioned briefly.

Thin-domains occur in various problems and the reduction in the number of independent variables associated with trying to exploit thinness inevitably leads to nonuniformity near boundaries. For example steady heat-conduction in a long insulated bar along the x-axis is often treated as one-dimensional. The small parameter is $\epsilon = L/D$ ($L =$ length, $D =$ diameter) and the coordinates held fixed in the limit process are $(x/L, y/D, z/D)$. The distribution of temperature over a (y, z) boundary face cannot be described by the one-dimensional temperature distribution in the interior. A local boundary expansion must be constructed which matches to the solution in the interior. In this heat conduction problem we are concerned with degeneration of a three-dimensional domain to a one-dimensional one. In other problems three dimensional domains are approximated as two-dimensional and boundary layers are again needed. For example elastic shell theory (first approximation equations on a two-dimensional manifold) can be drived from the three-dimensional equations of elasticity in the limit $\epsilon = D/R \to 0$ when D is the thickness of shell, R is the radius of curvature. Another slightly different example occurs in the theory of the electric field induced by a point source of current in a very long thin biological cell. The cell is bounded by a membrane whose electric conductivity is much less than the electric conductivity of the interior of the cell:

$\epsilon \sim$ membrane conductivity/cell interior conductivity.

Representing the cell as an infinite cylinder in a perfectly conducting medium we have a problem for the potential ϕ (dimensionless coordinates)

$$\nabla_\mathbf{x}^2 \phi = -\delta(\mathbf{x} - \boldsymbol{\xi}), \qquad -\infty < x < \infty, \ y, z \text{ in cell},$$

$$\frac{\partial \phi}{\partial n} + \epsilon \phi = 0 \text{ on membrane boundary.}$$

The nonuniformity here occurs as $x \to \pm\infty$. When $\epsilon = 0$ the cell membrane behaves as a perfect insulator and all the current necessarily flows out of the end of the cell. However for $\epsilon > 0$ the current leaks out for all x and there is no current at infinity. In this case different asymptotic expansions must be used near infinity and near the source, but these again match asymptotically. For details see Ref. [8]. Another familiar example involving far fields is that of sonic boom. A supersonic airplane is essentially a thin object moving through the air. The waves produced near the aircraft are basically acoustic waves. If this wave field would be a valid description right down to the ground the shape of the airplane would be heard rather than a boom. The appropriate small parameter for this problem is

$$\epsilon = \text{thickness of aircraft/length of aircraft} = \tau/L.$$

The far field effect in this case is again a cumulative one, but depends on nonlinearity. It is a fact that the speed of a weak shock wave differs by $O(\epsilon)$ from the speed of a sound wave. Thus in the far field the acoustic wave is located in the wrong place and there are also other effects of nonlinearity. The limit process ($\epsilon \downarrow 0$), $\mathbf{x}/L$ fixed generates first order ($O(\epsilon)$) acoustics, second-order acoustics ($O(\epsilon^2)$) etc. This expansion is valid near the airplane and can for instance be used to calculate lift and drag. However within the wave zone as $\mathbf{x} \to \infty$, the second order terms become much larger than the first-order terms. An expansion valid near infinity can be constructed if now $\mathbf{x}^*$ is a point in the wave zone ($\mathbf{x}^*$ fixed means $\mathbf{x} \to \infty$ in the wave zone). When this is done a nonlinear first approximation equation appears for the farfield and shock waves fit in naturally. (cf. Ref. [2]). This far field matches asymptotically to the acoustic nearfield. The farfield description results in the familiar N-wave and sonic boom.

Some further brief remarks will now be made about nonuniformities. The situation of artificial local singularities near noses, corners, under concentrated loads, and other distinguished points appears fairly often in perturbation procedures. The first approximation of some procedure is too crude to give a correct local physical description in the neighborhood of the distinguished point. Sometimes the local singularity can be ignored if the integrated effect is finite. Other times however a better local description is needed. This can be done by considering a local expansion in which the representative point **x** approaches the distinguished point $\mathbf{x}_0$ as $\epsilon \downarrow 0$. That is, a blown-up coordinate $\mathbf{x}^* = (\mathbf{x} - \mathbf{x}_0)/\delta(\epsilon)$ is used. $\delta(\epsilon)$ is chosen so that a distinguished limit results, and asymptotic matching to the expansion valid away from $\mathbf{x}_0$ can be carried out. Another important phenomenon can appear when several dimensionless parameters ϵ, M_1, M_2, M_3,..., etc. enter in the description of a physical problem. There is a tendency toward agglutination of the parameters. An expansion as $\epsilon \downarrow 0$ (M_1, M_2,... fixed) may not be valid for all ranges of the parameters M_j. This will usually become obvious by the occurrence of groups such as $\epsilon/(M_1 - 1)$ etc. In the vicinity of $M_1 = 1$, there is a nonuniformity. This difficulty can be overcome by considering a new limit $\epsilon \downarrow 0$, $M_1 \to 1$ in such a way that $\epsilon/(M_1 - 1)$ is fixed. The natural question of asymptotic matching in the parameter space is one that has not received much attention. These brief remarks conclude our discussion of nonuniformities in limit-process expansions.

I would just like to mention now that the same type of perturbation mathematics as is used in some physical problems appears also in pure numerical analysis. For example, in his study of numerical treatment of a signaling problem for

$$\frac{\partial u}{\partial t} + a \frac{\partial u}{\partial x} = 0. \tag{8}$$

Kreiss [9] was lead to study the properties of the solution of the limit form of the finite-difference approximation

$$\frac{\partial w}{\partial t} + a \frac{\partial w}{\partial x} = \epsilon \text{ (higher derivatives)}. \tag{9}$$

Here ϵ is proportional to the mesh size used in the finite difference approximation. As $\epsilon \downarrow 0$ the solution of the finite difference equations is supposed to approach the solution of (8). Different difference schemes result in different higher derivative terms in (9) and their corresponding boundary layers. A difference scheme in order to be successful must incorporate boundary layers which decay properly.

The last main topic to be discussed is the use of a more general type of asymptotic expansion than limit-process type. A typical example when this is necessary occurs for the same linear spring-mass-damping system as discussed earlier. However, now the initial value problem of an impulse for relatively small damping is considered. In suitable dimensionless units the problem is

$$\frac{d^2y}{dt^2} + 2\epsilon \frac{dy}{dt} + y = 0, \qquad t \geq 0,$$

$$y(0) = 0, \qquad \frac{dy}{dt}(0) = 1,$$

and now $\epsilon = \tfrac{1}{2}\beta/(mk)^{1/2}$. A study of the exact solution for this problem shows that the motion is a slowly damped oscillation. The dominant term is

$$y = e^{-\epsilon t} \sin t + \cdots.$$

Any limit process expansion ($\epsilon \downarrow 0$, t fixed) in effect expands the exponential function, does not show the damping, and is not uniformly valid for large t. It is desirable to keep the coordinate $\tilde{t} = \epsilon t$, a slow time variable. Thus an asymptotic expansion of the form

$$y = \sum_{k=0}^{K} \mu_k(\epsilon) f_k(t, \tilde{t}) + O(\mu_{K+1}) \tag{10}$$

can be tried. The $\mu_k(\epsilon)$ are supposed to give the orders of magnitude of the various terms for some large range of $(t, \tilde{t})$ including the origin. The problems for the $f_k(t, \tilde{t})$ can no longer be obtained from limits, but separate equations for the different orders of magnitude $\mu_k(\epsilon)$ can still be written. Now to enforce uniform validity it is necessary to keep $f_k(t, \tilde{t})$ under control. This is done by considering equations for orders μ_0, μ_1 and preventing fast growth (secular terms) in f_1. For example, for f_0 we find

$$f_0(t, \tilde{t}) = A(\tilde{t}) \cos t + B(\tilde{t}) \sin t$$

$A_0(\tilde{t})$, $B_0(\tilde{t})$ are undetermined from the μ_0 problem but from the restrictions just mentioned on the μ_1 problem some ordinary differential equations for $A(\tilde{t})$, $B(\tilde{t})$ are found. When this process is carried to higher orders provision must also be made for shifts in frequency (or phase) using for example, $t^* = (1 + \Omega_1(\epsilon))t$. In more general problems the choice of suitable fast (t^*) and slow scales $\tilde{t}$ is part of the problem. But basically the approach is the same: the expansion if forced to be "as uniformly valid as possible."

I would like to conclude with a brief discussion of a practical problem in which it was necessary to combine the ideas of limit-process expansions and multiple scale expansions. In normal operation a certain spin-stabilized satellite consists in a box-like stator and a cylindrical rotor. (See Fig. 2.) When the driving torque on the rotor, necessary to overcome bearing friction, failed the satellite was observed to "fall down" into a flat spin about the principal axis of maximum inertia. The process of making the satellite "stand up" and regain, within a residual nutation angle, its original position can be described accurately by a combined

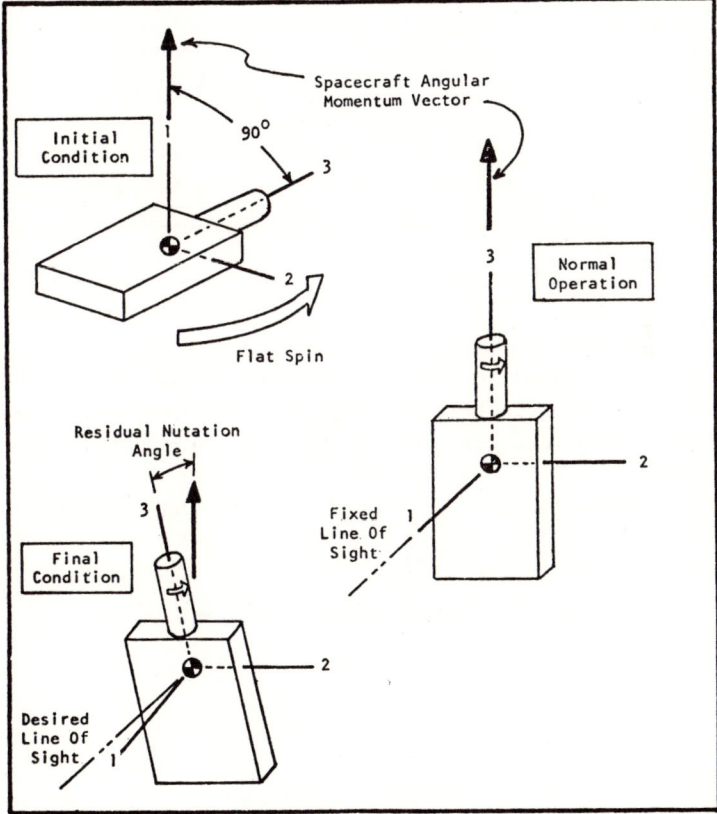

FIG. 2. Illustration of the initial and final conditions and normal operation state.

perturbation method. The perturbation parameter ϵ is proportional to the small constant torque applied to the rotor when it is restarted. The formulation, perturbation expansion, and numerical analysis for this problem are given in the Ph.D. thesis of J. R. Gebman [10]. A study of the (dimensionless) angular momentum $x_1(t; \epsilon)$ about the body-fixed 1-axis shows three stages of motion. (cf. Fig. 3). Regions I, III

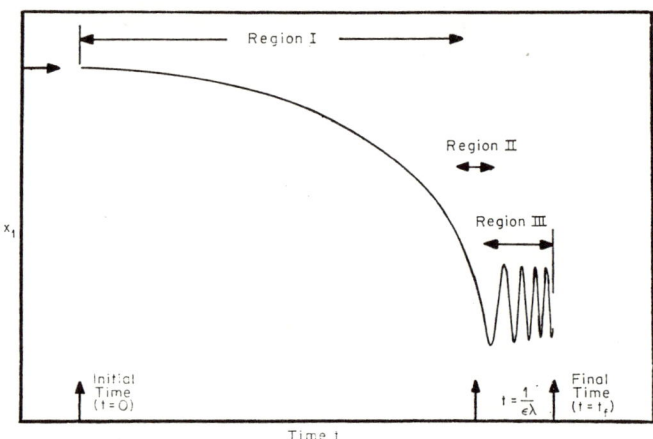

FIG. 3. Illustration of the three perturbation analysis regions in terms of x_1 behavior.

arise from suitable two-time expansions giving varying oscillations while in Region II a rapid transition takes place which is described by a suitable limit-process. There is asymptotic matching from region I to II, and from II to III. The expansions have the following form:

$$x_1^{\mathrm{I}}(t; \epsilon) = (1 - \tilde{t}^2)^{1/2} + \epsilon \lambda \tilde{t}(1 - \tilde{t}^2)^{-1/4} \sin t^* + O(\epsilon^2),$$

$$x_1^{\mathrm{II}}(t; \epsilon) = \epsilon^{1/3}\lambda^{1/3} y(\tau) + \cdots,$$

$$x_1^{\mathrm{III}}(t; \epsilon) = -\epsilon^{1/2}(\text{const.}) \cos\{t^+ + \phi(\tilde{t})\} + O(\epsilon^{3/2}),$$

where

$$\tilde{t} = \epsilon \lambda t, \quad t^* = \left\{\frac{1}{2}(1 - \tilde{t}^2)^{1/2} + \frac{1}{2}\frac{\sin^{-1}\tilde{t}}{\tilde{t}}\right\} t,$$

$$\tau = \lambda^{-2/3}(\tilde{t} - 1); \quad \frac{dt^+}{d\tilde{t}} = \frac{1}{\lambda}[(\tilde{t} - 1)(\tilde{t} - 1 + \lambda^2)]^{1/2},$$

$\lambda = \text{const.} = O(1)$. The transition function $y(\tau)$ satisfies a generalized Airy equation

$$\frac{d^2y}{d\tau^2} + \frac{1}{2}y^3 + \tau y = 0,$$

which must be integrated numerically. Matching however can be carried out from the asymptotic behaviour as $\tau \to \pm\infty$. Similar expansions exist for $X_{2,3}$ so that a complete description can be given. An interesting result comes from the analysis, Gebman's flat spin recovery rule

$$G = \text{const.} = \Omega T N^2,$$

where Ω is the initial flat spin rate, T is the time required for recovery, N is the residual nutation angle.

In conclusion we can say that perturbation methods in their various forms represent an important weapon in the arsenal of the applied scientist. Perhaps the main general point to be learned from our discussion is the importance of a clear understanding of what is actually being done in a given perturbation process.

REFERENCES

1. P. A. LAGESTROM AND R. G. CASTEN, Basic concepts underlying singular perturbation techniques, *SIAM Rev.* 14 (1972), 63–120.
2. J. D. COLE, "Perturbation Methods in Applied Mathematics," Blaisdell, MA, 1968.
3. M. VAN DYKE, "Perturbation Methods in Fluid Mechanics," Academic Press, New York, 1964.
4. A. H. NAYFEH, "Perturbation Methods," Wiley–Interscience, New York, 1973.
5. W. WASOW, "Asymptotic Expansions for Ordinary Differential Equations," Interscience, New York, 1965.
6. L. PRANDTL, "Über Flussigkeiten bei sehr kleiner Reibung," pp. 484–491, Verh. III Internat. Math. Kongr. Heidelberg (1905) Teubner, Leipzig, 1905. English translation in *NACA Tech. Mem.* 452 (1928).
7. P. A. LAGERSTROM, Laminar flow theory, *in* "High Speed Aerodynamics and Jet Propulsion" (F. K. Moore, ed.), Vol. IV. Princeton University Press, Princeton, NJ, 1964.
8. J. D. COLE, A. PESKOFF, AND R. S. EISENBERG, A point source of current in an infinite cylindrical cell, *SIAM J. Appl. Math.*, to appear.
9. H. O. KREISS, Lecture at UCLA, 1970.
10. J. R. GEBMAN, Perturbation analysis of the flat spin recovery of a dual-spin spacecraft, Ph.D. Thesis, School of Engineering and Applied Science, University of California, Los Angeles, 1974.

A Survey of Some Finite Element Methods Proposed for Treating the Dirichlet Problem

JAMES H. BRAMBLE

Department of Mathematics, Cornell University, Ithaca, New York 14850

DEDICATED TO STAN ULAM

1. INTRODUCTION

In this paper I will describe some work of the last few years dealing with the approximation of solutions to elliptic boundary value problems via "finite element" type methods. The main emphasis will be on various approaches to problems with essential boundary conditions. I will describe a number of methods that have been proposed and analyzed and describe briefly their important distinguishing features.

In order to avoid technical difficulties I will consider only the prototype equations

$$-\Delta u + u = f \quad \text{in} \quad \Omega \tag{1.1}$$

or

$$-\Delta u = f \quad \text{in} \quad \Omega, \tag{1.2}$$

where Ω is a bounded domain in R with smooth boundary $\partial\Omega$,

$$\Delta = \sum_{j=1}^{N} \frac{\partial^2}{\partial x_j^2},$$

and f is a given function defined in Ω.

I will first describe the Neumann problem for (1.1) and its Ritz–Galerkin and finite element approximation. Here we will see that there are no major difficulties.

Difficulties arise in treating the Dirichlet problem for (1.2); we choose (1.2) rather than (1.1) for convenience. I will then describe various approaches which have been proposed for dealing with this problem (in a practical sense) and discuss each briefly.

2. The Neumann Problem

It is well known that there exists a unique solution u of (1.1) satisfying

$$\frac{\partial u}{\partial n} = 0 \quad \text{on} \quad \partial\Omega. \tag{2.1}$$

The problem of finding a function u satisfying (1.1) and (2.1) is called the Neumann problem. We shall describe an alternative formulation of this problem as a variational problem and see how this leads naturally to a method of approximation. First we need some notation. Set

$$(v, w) = \int_\Omega vw \, dx \tag{2.2}$$

and

$$(v, w)_1 = \sum_{j=1}^N \left(\frac{\partial v}{\partial x_j}, \frac{\partial w}{\partial x_j}\right) + (v, w).$$

The space L_2 is the usual space of square integrable functions on Ω. Define

$$\|v\|_1 = (v, v)_1^{1/2}$$

and denote by H^1 the completion of $C^\infty(\Omega)$, the infinitely differentiable functions on Ω, with respect to $\|\cdot\|_1$. This is a so-called Sobolev space.

In these terms it is well known that an equivalent formulation of the Neumann problem with $f \in L_2$ is as follows: Find $u \in H^1$ such that

$$(u, \phi)_1 = (f, \phi) \tag{2.3}$$

for all $\phi \in H^1$.

Now let $\{S_h\}$ for $0 < h \leq 1$ be a one-parameter family of finite-dimensional subspaces of H^1. For each fixed h we may now define the Ritz–Galerkin problem as follows: Find $u_h \in S_h$ such that

$$(u_h, \chi)_1 = (f, \chi) \tag{2.4}$$

for all $\chi \in S_h$.

Clearly then from (2.3) and (2.4) we have that

$$(u - u_h, \chi)_1 = 0$$

for all $\chi \in S_h$. Hence we see that u_h is the projection of u onto S_h in H^1.

That is

$$\| u - u_h \|_1 = \inf_{\chi \in S_h} \| u - \chi \|_1.$$

Thus we may view u_h as the best approximation to u in S_h with respect to the H^1-norm.

Now the finite element method in this context is just the Ritz–Galerkin method but with special choices of the subspaces S_h. The following types of subspaces would give rise to what would commonly be called finite element methods (we take for simplicity $N = 2$).

(a) Subdivide the plane into squares with sides of length h. Define on R^2 the set of functions which are polynomials of a given degree on each square and which are continuous on R^2. We take then S_h to be set of functions which are restrictions to Ω of such piecewise polynomial functions (cf. [5]).

(b) For the second example consider instead a uniform subdivision into triangles with largest side h and construct, as above, a class of functions that are polynomials of a given degree on each triangle and continuous on R^2. Let S_h be the set of functions that are restrictions of the above piecewise polynomial functions to Ω (for properties of such subspaces cf. [10, 14]).

These are just two examples which would typically be considered "finite element spaces." Many such spaces have simple "local bases"; i.e., for a given h there are n functions $\phi_1, ..., \phi_n$ that form a basis for the linear space S_h where each ϕ_j has small support (depending on h). Then the solution u_h in (2.4) may be written as

$$u_h = \sum_{j=1}^{n} a_j \phi_j,$$

and the equation (2.4) may be written as

$$\sum_{j=1}^{n} a_j (\phi_j, \phi_i)_1 = (f, \phi_i), \quad i = 1, ..., n. \tag{2.5}$$

Hence (2.5) is just a system of linear algebraic equations for the determination of the coefficients a_j and the matrix of this system has elements $(\phi_j, \phi_i)_1$. In case the ϕ_j's have small support this matrix will be sparse and hence it may be feasible to consider in practice very large systems. This is a feature that has made finite difference methods attractive in practice and is to some extent retained by finite element methods.

We shall now turn our attention to the Dirichlet problem and some finite element methods proposed for its approximate solution.

3. The Dirichlet Problem

It is well known that there exists a unique solution u of (1.2) satisfying

$$u = 0 \quad \text{on} \quad \partial\Omega. \tag{3.1}$$

The problem of finding a function u satisfying (1.2) and (3.1) is called the Dirichlet problem. We want to describe an alternative formulation of this problem.

Denote by $\mathring{H}^1$ the Hilbert space formed by completing $C_0^\infty(\Omega)$ (the infinitely differentiable functions in Ω which vanish in a neighborhood of Ω) with respect to $\|\cdot\|_1$.

These functions may be thought of as vanishing on $\partial\Omega$. $\mathring{H}^1$ includes those functions in H^1 which are continuous on $\bar{\Omega}$ and vanish on $\partial\Omega$. Set $D(v, w) = \sum_{j=1}^N (\partial v/\partial x_j, \partial w/\partial x_j)$. The Dirichlet problem may be formulated, for a given $f \in L_2$, as follows: Find $u \in \mathring{H}^1$ such that

$$D(u, \phi) = (f, \phi)$$

for all $\phi \in \mathring{H}^1$.

We may now formulate the standard Ritz–Galerkin method for the approximation of u.

a). *The Ritz–Galerkin Method*

Let $\{S_h\}$ be as in the Neumann problem but in addition for each h suppose that $S_h \subset \mathring{H}^1$. The Ritz–Galerkin problem is as follows: Find $u_h \in S_h$ such that

$$D(u_h, \chi) = (f, \chi)$$

for all $\chi \in S_h$.

Now it is easily seen that

$$\| u - u_h \|_1 \leqslant C \inf_{\chi \in S_h} \| u - \chi \|_1$$

where C depends only on Ω. The difficulty with this method in practice is that S_h may not be easy to construct so that $\inf_{\chi \in S_h} \| u - \chi \|_1$ is small. In fact if S_h is one of the above mentioned "finite element" subspaces and Ω has a curved boundary then the requirement that $S_h \subset \mathring{H}^1$ is too stringent and functions in S_h will not in general be good approximations to many functions in $\mathring{H}^1$.

The remainder of this paper will describe various approaches to formulating other "finite element" type approximations to u. We start by describing a method proposed independently by Aubin [1] and Babuška [2].

b). *The Aubin–Babuška Penalty Method*

Let $S_h \subset H^1$ and $\gamma > 0$ and $\sigma \geqslant 1$ be two given parameters. The Aubin–Babuška method is as follows: Find $u_h \in S_h$ such that

$$D(u_h, \chi) + \gamma h^{-\sigma} \langle u_h, \chi \rangle = (f, \chi) \qquad (3.2)$$

for all $\chi \in S_h$ where

$$\langle v, w \rangle = \int_{\partial \Omega} vw \, dS.$$

There are no boundary conditions on the functions in S_h, but the form (3.2) corresponds to the boundary value problem

$$-\Delta v = f \quad \text{in} \quad \Omega,$$

$$\frac{h^\sigma}{\gamma} \frac{\partial v}{\partial n} + v = 0 \quad \text{on} \quad \partial \Omega.$$

Thus we are approximating the wrong problem and the penalty which must be paid is that u_h may be much further from u than is its best approximation.

In order to overcome this deficiency King [11] has proposed an extrapolation procedure.

c). *King's Extrapolation*

Let $u_h(\gamma)$ satisfy (3.2) for a given $\gamma > 0$ and take $\sigma = 1$. Depending on the "accuracy" of S_h we may choose any positive distinct values $\gamma_0, ..., \gamma_k$ and find (easily) $a_0, ..., a_k$ such that

$$\tilde{u}_h = \sum_{j=0}^{k} a_j u_h(\gamma_j)$$

is essentially as good as the best approximation to u in S_h.

For the purpose of this exposition we shall say that an approximation

v_h in S_h to a smooth function v is optimal (sometimes called quasi-optimal) with respect to $\|\cdot\|$ if for some $\mu > 0$ whenever

$$\inf_{\chi \in S_h} \| v - \chi \| = \mathcal{O}(h^\mu)$$

as $h \to 0$ it follows that $\| v - v_h \| = \mathcal{O}(h^\mu)$, as $h \to 0$.

In this terminology $u_h(\gamma)$ for a given γ is not in general optimal in H^1 but $\tilde{u}_h$ may be chosen to be. The calculation of $u_h(\gamma)$ for different values of γ should not be much more difficult than the calculation for one value since changing γ involves only a simple change in the matrix of the corresponding linear system and most of the work has been done in setting up the system in the first place. The choice $\sigma = 1$ is the best choice with regard to conditioning of the linear systems which arise from the standard finite elements.

d). The Least Squares Method

This method was proposed and analyzed by Schatz and myself [9]. A simplified analysis was later given by Baker [4].

Let $S_h \subset H^2$ (the space of functions with square integrable partial derivatives up to and including those of order two). The problem may be formulated then as follows: Find $u_h \in S_h$ such that

$$(\Delta u_h, \Delta \chi) + h^{-3} \langle u_h, \chi \rangle = -(f, \Delta \chi)$$

for all $\chi \in S_h$.

This method is optimal in L_2 provided S_h is "rich" enough; e.g., piecewise cubic polynomials would suffice. Possible drawbacks with this method are that the finite elements must be of class C^1 and there could be difficulties resulting from poor conditioning of the linear system arising from the standard finite element bases.

e). Nitsche's Method [12]

For $\gamma \geqslant 0$ define

$$N_\gamma(v, w) = D(v, w) - \left\langle v, \frac{\partial w}{\partial n} \right\rangle - \left\langle \frac{\partial v}{\partial n}, w \right\rangle + \gamma h^{-1} \langle v, w \rangle.$$

Now let $S_h \subset H^1$ and satisfy the additional assumption

$$\left| \frac{\partial \chi}{\partial n} \right| \leqslant C_0 h^{-1/2} D^{1/2}(\chi, \chi) \tag{3.3}$$

for all $\chi \in S_h$, where

$$|v| = \left(\int_{\partial\Omega} v^2 \, dS\right)^{1/2}.$$

The approximation method is as follows: Find $u_h \in S_h$ such that

$$N_\gamma(u_h, \chi) = (f, \chi),$$

for all $\chi \in S_h$. It is easy to see that for $\gamma > C_0^2$ the solution u_h exists and is unique. An interesting feature of this method is that N_γ is not in general positive definite for any γ, but for $\gamma > C_0^2$ it is positive definite when restricted to $S_h \times S_h$. This is, of course, sufficient. Nitsche [12] has shown that u_h is optimal in H^1 and L_2. The subspace S_h must however be constructed so that (3.3) is satisfied.

f). Nitsche's Method with "Nearly Zero Boundary Conditions" [13]

Assume that S_h satisfies the condition (3.3) and in addition that for $\alpha < 1$

$$|\chi| \leqslant \frac{\alpha}{2C_0} h^{1/2} D^{1/2}(\chi, \chi), \tag{3.4}$$

for all $\chi \in S_h$. This means, of course, that χ is "small" on $\partial\Omega$. The approximation method is then as follows: Find $u_h \in S_h$ such that

$$N_0(u_h, \chi) = (f, \chi)$$

for all $\chi \in S_h$. Under the above conditions u_h exists and is unique and has properties similar to those described in (d). The form is in this case simpler ($\gamma = 0$), however additional care must be taken to satisfy (3.4).

g). Least Squares Modified

This method was proposed and analyzed by Nitsche and myself [7] and may be described as follows. Set

$$K_\gamma(v, w) = N_\gamma(v, w) + \gamma h \langle \nabla_S v, \nabla_S w \rangle + h^2(\Delta v, \Delta w),$$

where $\nabla_S v$ is the tangential gradient on $\partial\Omega$. (For $N = 2$, $\nabla_S v = dv/dS$ where S is arc length along $\partial\Omega$.)

We now consider $S_h \subset H^2$ with no further conditions required. The corresponding finite element problem is as follows: Find $u_h \in S_h$ such that

$$K_\gamma(u_h, \chi) = (f, \chi - h^2 \Delta \chi)$$

for all $\chi \in S_h$. It is shown in [7] that there is a $\gamma_0 > 0$ such that for $\gamma \geqslant \gamma_0$, K_γ is positive definite. Hence for $\gamma \geqslant \gamma_0$, u_h exists and is unique. This method has the advantages that it is optimal in H^1 and L_2, does not require that (3.3) be satisfied and the possible conditioning problems of (d) are not present. We pay for these advantages by the added complication of the form and the requirement that $S_h \subset H^2$.

h). *The Babuška Multiplier Method*

The following method was proposed and analyzed by Babuška in [3].

Let $S_h'(\partial\Omega)$ be a space of approximating functions on $\partial\Omega$. The method may be described as follows: Find $u_h \in S_h$ and $u_h' \in S_h'$ such that

$$D(u_h, \chi) - \langle u_h, \chi' \rangle - \langle u_h', \chi \rangle = (f, \chi)$$

for all $\chi \in S_h$ and $\chi' \in S_h'$. There are conditions relating the "sizes" of S_h and S_h' which are precisely described in [3]. It was shown in [3] that u_h (and u_h') exists and is unique and enjoys the property of being optimal in H^1 and L_2. Possible disadvantages of the method are that an additional set of approximating functions are required and the resulting matrix, though nonsingular, is not positive definite.

i). *A Method Using Approximating Polygonal Domains*

The final method which I will describe is one which was proposed and analyzed by Dupont, Thomée and myself [6].

Let $\Omega_h \subset \Omega \subset R^2$ be a domain with a polygonal boundary $\partial\Omega_h$. Now define $S_h \subset H^1(\Omega_h)$; i.e., we shall work only on the domain Ω_h for each h and the forms and integrals below will be taken with respect to this domain of integration. We shall assume that for each $x \in \partial\Omega_h$, $\delta(x)$ is the distance along the normal to $\partial\Omega$ and assumed to be small. At the corners, which are finite in number, the definition is irrelevant.

Now for $\gamma > 0$ and k a given nonnegative integer, set

$$B(v, w) = D(v, w) - \left\langle \frac{\partial v}{\partial n}, w \right\rangle - \left\langle \sum_{j=0}^{k} \frac{1}{j!} \delta^j \left(\frac{\partial}{\partial n}\right)^j v, \frac{\partial w}{\partial n} - \gamma h^{-1} w \right\rangle.$$

We shall assume, for S_h taken here as functions defined on Ω_h, that (3.3) is satisfied. In the case of a domain with a polygonal boundary such subspaces are easy to construct. Now the approximation is given as the solution of the following problem: Find $u_h \in S_h$ such that

$$B(u_h, \chi) = (f, \chi)$$

for all $\chi \in S_h$. This problem is solvable uniquely provided γ is chosen large enough. This method for $k = 0$ is just Nitsche's method (e) relative to Ω_h. The additional terms for $k > 0$ may be thought of as correction terms for having taken Ω_h instead of Ω.

It was shown in [6] that depending on how "good" S_h is, k may be chosen so that u_h is an optimal approximation to u in H^1 and L_2. The principal advantage with this method is that one needs only to work on a polygonal domain. The disadvantage lies in the complicated correction terms near the boundary.

For a summary of results for several of the aforementioned methods cf. [8].

References

1. J. P. Aubin, Approximation des problèmes aux limites non homogènes et régularité de la convergence, *Calcolo* **6** (1969), 117–139.
2. I. Babuška, The finite element method with penalty, *Math. Comp.* **27** (1973), 221–228.
3. I. Babuška, "The Finite Element Method with Lagrangian Multipliers," Tech. Note BN-724, Inst. for Fluid Dynamics and Appl. Math., Univ. of Maryland, College Park, MD, 1972.
4. G. A. Baker, Simplified proofs of error estimates for the least squares method for Dirichlet's problem, *Math. Comp.* **27** (1973), 229–235.
5. G. Birkhoff, M. Schultz, and R. Varga, Piecewise Hermite interpolation in one and two variables with applications to partial differential equations, *Numer. Math.* **11** (1968), 232–256.
6. J. H. Bramble, T. Dupont, and V. Thomée, Projection methods for Dirichlet's problem in approximating polygonal domains with boundary-value corrections, *Math. Comp.* **26** (1972), 869–879.
7. J. H. Bramble and J. A. Nitsche, A generalized Ritz-least-squares method for Dirichlet problems, *SIAM J. Numer. Anal.* **10** (1973), 81–93.
8. J. H. Bramble and J. E. Osborn, Rate of convergence estimates for nonselfadjoint eigenvalue approximations, *Math. Comp.* **27** (1973), 525–549.
9. J. H. Bramble and A. H. Schatz, Rayleigh–Ritz–Galerkin methods for Dirichlet's problem using subspaces without boundary conditions, *Comm. Pure Appl. Math.* **23** (1970), 653–675.
10. J. H. Bramble and M. Zlámal, Triangular elements in the finite element method, *Math. Comp.* **24** (1970), 809–820.
11. J. T. King, New error bounds for the penalty method and extrapolation, *Numer. Math.*, in press.

12. J. Nitsche, Über ein Variationsprinzip zur Lösung von Dirichlet-Problemen bei Verwendung von Teilräumen, die keinen Randbedingungen unterworfen sind, *Abh. Math. Sem. Univ. Hamburg* **36** (1970–71).
13. J. Nitsche, A projection method for Dirichlet-problems using subspaces with nearly zero boundary conditions, preprint.
14. M. Zlámal, On the finite element method, *Numer. Math.* **12** (1968), 394–409.

The Mathematics of Quantum Fields

JAMES GLIMM*

*Courant Institute of Mathematical Sciences, New York University,
New York, New York 10012*

DEDICATED TO STAN ULAM

Quantum field theory is the only satisfactory framework known which unites the principles of quantum mechanics with special relativity (i.e., Lorentz covariance). Its goal is to describe elementary particles. Its primary success is quantum electrodynamics, where it is possible to calculate numbers that are in perfect agreement with extremely accurate experiments. However for the interactions between protons, neutrons, and the many other particles of high energy physics, definitive calculations have not been possible, with the result that one does not know which (if any) interaction between quantum fields is a correct description of nature.

From a mathematical point of view, quantum fields are highly singular, and the elementary operations performed on quantum fields have appeared to lack mathematical meaning. In order to make some progress in understanding the mathematical phenomena involved, the problem has been simplified by taking d, the number of space–time dimensions, to be two or three. To indicate the effect of a change in dimension on the mathematical difficulties of quantum field theory, we note that a nonlinear interaction involves products of functions, which for dimensional reasons are more singular as d increases. As an example, consider the fundamental solution C of $-\Delta + m_0^2$. One can check that

$$C \in L_p, \quad 1 \leqslant p \leqslant \infty, \quad d = 1, \tag{1}$$

$$C \in L_p, \quad \text{all } p < d/d-2, \quad d \geqslant 2, \tag{2}$$

$$C \notin L_p, \quad \text{all } p \geqslant d/d-2, \quad d \geqslant 2. \tag{3}$$

---

* Supported in part by the National Science Foundation, Grant No. NSF-GP-24003.

As d increases, C belongs to L_p for a decreasing range of p. In particular as $x - y \to 0$, the functions

$$C(x-y)^3 \sim |x-y|^{-3}, \quad d = 3, \tag{4}$$

$$C(x-y)^3 \sim |x-y|^{-6}, \quad d = 4, \text{ and} \tag{5}$$

$$C(x-y)^2 \sim |x-y|^{-4}, \quad d = 4 \tag{6}$$

are not locally integrable. Both C^3 and C^2 occur in field theory, and when their integrals diverge, an infinite (mass or charge) renormalization is required.

In this article, we will try to answer in general terms the following two sets of questions:

I. What are quantum fields? Why are they so singular? What kinds of mathematics are used to control them?

II. What progress has been made so far? What problems have not been solved?

The questions of II are easier to answer in nontechnical terms. There are a number of complete and partial theories for $d = 2, 3$. In favorable cases much of the detailed structure can be obtained for the solution, including results of basic interest to physics. The general picture is summarized in the following table. Here $P(\phi)_d$ is a polynomial boson interaction and Y_d is a Yukawa interaction, while d as before, is the number of space–time dimensions.

$P(\phi)_2$: Existence and detailed structure

Y_2: Existence only

ϕ_3^4: Existence

$d = 4$: Only axiomatic results

The detailed structure for $P(\phi)_2$ includes results of basic importance to the physical interpretation of these models. In some cases, further restrictions are imposed on P (e.g., $P = $ even $+$ linear, $P = \phi^4$, or P small). Extensive presentations of the results can be found in [14, 15, 17, 18, 27, 48], and so here we only mention some highlights: the verification of the Wightman and Haag–Ruelle axioms, the existence of particles with discrete, isolated mass spectra, an isometric S matrix, the existence or nonexistence [45] of bound states (depending on the interaction), correlation inequalities (Guerra, Rosen and Simon [48]), bounds on critical exponents [20], asymptotic expansions in the limit of small coupling (Dimock [48]), Borel summability of the perturbation expansion [6] and analyticity of the solution in the coupling constant for small but nonzero coupling. The verification of the Wightman axioms combines a number of results (see [11, 48]). The above results on particles and analyticity are due to Jaffe, Spencer and the author (see [48]).

I am optimistic that the models Y_2 and ϕ_3^4 can be brought up to the level of $P(\phi)_2$. For recent progress in this direction, see [1–3, 7, 13, 30–32, 35, 38, 39, 43, 44, 49, 50].

In four dimensions, new essential difficulties occur, and the only results are in the axiomatic approach, in which one assumes the existence of the quantum fields together with some minimal physical and mathematical properties. Proceeding backwards from the axioms toward the equations of motion, we have the Euclidean axiom scheme of Osterwalder and Schrader [38] and the bounds of [21], but the original problem (existence of interacting quantum fields in four space–time dimensions) is completely open.

Now we turn to the questions mentioned under I. The central mathematical fact underlying quantum field theory is the occurrence of an infinite number of degrees of freedom. Continuum systems in general (e.g., fluids and solids, as well as fields) have an infinite number of degrees of freedom. For linear systems, these degrees of freedom can be decoupled, separated into normal modes, and the time evolution in any given normal mode can be solved independently of the others. Assuming that the dynamics is expressed in terms of a self-adjoint operator, this statement is just a corollary of the spectral theorem. For nonlinear (classical) systems, there is no separation into normal modes, and generally speaking, all degrees of freedom are coupled to each other. In quantum mechanics, the dynamics is again given by a self-adjoint operator H, but if the corresponding classical system is nonlinear, the coupling between the classical degrees of freedom is preserved, in the sense that the eigenfunctions of H can be expected to depend on all the classical degrees of freedom (all coordinate directions in $\mathscr{C}$ below). For particles interacting with a potential force field, we choose

$$H = H_0 + V = \sum (2m_i)^{-1} p_i^2 + V(q) = -\sum (2m_i)^{-1} \frac{\partial^2}{\partial q_i^2} + V(q). \quad (7)$$

H acts on the Hilbert space

$$\mathscr{H} = L_2(\mathscr{C}). \tag{8}$$

Here $\mathscr{C}$ is the "configuration space" of all initial positions (configurations) of the system, $q \in \mathscr{C}$, and $V(q)$ is the potential, a function defined on $\mathscr{C}$. For a continuum system, $\mathscr{C}$ is infinite-dimensional, and H, as given above, is a second-order elliptic operator in an infinite number of variables. We summarize this discussion as follows: Quantum field theory involves analysis over infinite dimensional spaces. Here infinite-dimensional stands in contrast to the conventional case of analysis over a finite dimensional space, $\mathscr{C} = R^N$, or a finite dimensional manifold, $\mathscr{C} = \mathscr{M}_N$.

From this point of view, it is clear that quantum field theory involves operators and Hilbert space theory. Both the theory of single self-adjoint operators and the theory of operator algebras (C^* and W^* algebras) have played important roles. Because H is a second-order elliptic operator, probability methods are useful, including function space integrals (e.g., (8)), the Feynman–Kac formula and stochastic processes. We remark that the case $d = 1$ has no space variable, and $P(\phi)_1$ reduces to the quantum mechanics of a single anharmonic oscillator. The stochastic variable $\phi = \phi(t)$ depends on time alone; for a linear theory $(P(\phi) = \phi^2)$, this stochastic process is an Ohrenstein–Uhlenbeck process. In the general case, $d \geqslant 2$, $\phi = \phi(x, t)$ is a stochastic field depending on $(x, t) \in R^d$.

Also important for quantum field theory are analytic functions of several complex variables (in studying scattering amplitudes) and group representations. The most common groups are the inhomogeneous Lorentz group, as the symmetry group of space–time, and $SU(2)$ or $SU(3)$, which act as symmetries of the vector components of the field ϕ, in case $\phi = \phi(x, t) \in R^2$ is vector valued. Using such symmetry considerations alone, Gell-Mann achieved a remarkable classification ("the eightfold way") of the strongly interacting particles. It follows that the equations for strongly interacting particles should be $SU(2)$ or $SU(3)$ invariant, as well as Lorentz invariant. In summary, we see that quantum field theory draws on some old and well-developed mathematical theories. It also provides a qualitatively new class of examples, which are in no sense contained in existing general theories (of stochastic fields or of partial differential operators in infinitely many variables).

It remains to answer the question: What are quantum fields, and why are they so singular? To simplify the exposition we assume the existence of a Schrödinger representation (in technical terms, we assume cyclicity of the time zero fields) so that the Hilbert space $\mathcal{H}$ of quantum mechanical states is given by (8). The initial position of a classical field is a function, defined on R^{d-1}. Thus $\mathscr{C}$ is some space of functions defined on R^{d-1}. We want $\mathscr{C}$ to be sufficiently large to contain the solutions we are seeking and we want $\mathscr{C}$ to have a nice integration theory defined on it. A convenient choice is $\mathscr{C} = \mathscr{S}'(R^{d-1})$, the Schwartz space of tempered distributions, at most polynomially increasing at infinity. The choice of a measure dq on $\mathscr{C}$, implicit in the definition $\mathcal{H} = L_2(\mathscr{C})$, is a very deep question, which we postpone. In case the quantum field ϕ is vector valued, the generalized functions

$$q \in \mathscr{C} = \mathscr{S}'(R^{d-1})$$

are also taken to be vector valued.

A quantum field is a linear operator on $\mathcal{H}$, defined as follows. For any test function $f \in \mathscr{S}(R^{d-1})$, $q \to \langle q, f \rangle$ is a linear coordinate function defined on $\mathscr{C}$. The quantum field $\phi(f)$ is multiplication by this linear coordinate function. Thus if $F = F(q) \in \mathcal{H}$, then

$$(\phi(f)F)(q) = \langle q, f \rangle F(q). \tag{9}$$

In the identification between self-adjoint operators and observables, $\phi(f)$ measures the field strength or intensity in the coordinate direction f. In the case of a finite number of degrees of freedom, $\mathscr{C} = R^N$, the position operators correspond to $\phi(f)$. A position operator on $L_2(\mathscr{C}) = L_2(R^N)$ is a multiplication operator $\theta(q) \to q_i\theta(q)$ which acts as multiplication by a linear coordinate function q_i on $\mathscr{C}$. q_i measures the expected value of the position of the particles in the coordinate direction q_i. Symbolically, we write

$$\phi(f) = \int \phi(x) f(x)\, dx, \tag{10}$$

so that $\phi(x)$ is an operator valued distribution, equal to multiplication by the (densely defined) coordinate function $q \to \langle q, \delta_x \rangle = q(x)$, and

$$(\phi(x)F)(q) = q(x) F(q). \tag{11}$$

It turns out that for measures dq of interest on $\mathscr{C}$, even in the case of free fields, $\phi(x)$ as defined above, is defined only on a set of measure zero in $\mathscr{C}$, unless $d = 1$. In the Hilbert space $\mathscr{H}$, this statement translates into the fact that $\phi(x)$ is a densely defined bilinear form, but, as an operator, it is defined only on the zero vector. Thus (10) is meaningful as an identity between bilinear forms but not as an identity between operators. A more detailed analysis of these points follows from the support properties of dq given in [4, 5, 11, 26, 37, 40–42] and the contribution of Collela and Lanford in [48].

It is now easy to see why the field $\phi(x)$ is singular. The interaction potential V is a function on $\mathscr{C}$. The standard and simplest choice is that V is a polynomial of degree at least three (so that the classical field equations are nonlinear). Because of Lorentz invariance, V contains no factors $\phi(x)\phi(y)$ with $x \neq y$. For this reason and because of translation invariance

$$V = \int P(\phi(x))\,dx. \tag{12}$$

The local singularities resulting from $\phi^2(x)$, $\phi^3(x)$,... in (12) are removed easily (and rigorously) by a device known as Wick ordering. In essence, the definition is as follows:

$$:\phi^2(x): = \phi^2(x) - \int \phi^2(x)\,dq,$$

$$:\phi^3(x): = \phi^3(x) - 3\phi(x)\int \phi^2(x)\,dq.$$

The higher Wick powers are defined similarly, as Hermite polynomials in $\phi(x)$. It turns out that the left side is defined even when the individual terms on the right are not, as is the case for $d \geq 2$. In fact for the free field measure dq,

$$\int \phi^2(x)\,dq = C(0),$$

which is finite for $d = 1$ only, by (1)–(3). Thus we modify (12) with the definition

$$V = \int :P(\phi(x)):\,dx. \tag{12'}$$

Having removed the singularities in V, we find that for $d \geq 3$, they recur in the higher powers of V, in the perturbation solution of the equations. The singularities occur as (divergent) integrals of quantities such as (4)–(6). They are removed by mass ($d = 3, 4$) and charge ($d = 4$) renormalization. Since these renormalizations are accomplished by an infinite change in the coefficients of P in (12'), they may raise doubts on two grounds. First, can the subtraction $\infty - \infty = $ finite be made mathematically rigorous? The answer is yes for the Y_2 and ϕ_3^4 interactions, which have infinite mass and vacuum energy renormalizations. The correct mathematical treatment involves first inserting a mollifier

$$\phi(x) \to \phi_\kappa(x) = \int \phi(y) j_\kappa(x - y) \, dy$$

to remove local singularities, then subtracting finite, but κ-dependent quantities, and finally taking the limit $\kappa \to \infty$, $j_\kappa \to \delta$ in the differences (see [15]). Second, is it physically correct to modify the equations of motion, or the potential V, even by a finite amount? Again the answer is yes. The coefficients in P are not directly observed quantities. Thus they can and must be restricted by the condition that the mass and charge, as defined by the solution of the $P(\phi)$ interaction, coincide with some (experimentally given) values. Thus on physical grounds, P can and must be replaced by some new polynomial P_{ren}. The coefficients of P_{ren} are determined in some complicated fashion by the mass and charge of the particles that the field is describing. This physically correct choice of the renormalized interaction polynomial P_{ren} coincides, on the level of formal perturbation theory, with the previous requirement of subtracting divergent quantities from powers of V. Similar arguments for $d = 2$ lead to a renormalized polynomial P_{ren} which still has finite coefficients. However the constant term in P_{ren} may be nonzero, which because of the infinite range of integration in (12'), leads to an infinite constant in V. Again the mathematics can be handled with complete rigor, and the infinite subtractions are defined by a limit process, as the limit of the difference of finite quantities.

The final topic that we consider is the choice of the measure dq. dq depends in a sensitive fashion on P. Distinct choices of P should lead to mutually singular measures dq. Again the origin of this phenomena is the infinite number of degrees of freedom, as reflected in the fact that $\mathscr{C}$ is infinite-dimensional and does not possess a distinguished measure class, such as Lebesgue measure. dq must be chosen so that $\mathscr{H} = L_2(\mathscr{C}, dq)$ contains the ground state Ω of H. Using the similarity transform

$$H \to \Omega H \Omega^{-1} = H'; \quad dq \to \Omega^2 \, dq = dq', \tag{13}$$

we suppose that $\Omega = 1$. This choice of dq may be regarded as a renormalization. It is inessential when there is no change in measure class (as in the case of a finite number of degrees of freedom), but it is essential in typical limit processes with an infinite number of degrees of freedom.

We use the Feynman–Kac formula to define $\mathscr{H}$ (and presumably the measure dq). The use of this formula in field theory can be traced back to Feynman and Schwinger and was further considered by Symanzik [46, 47] as part of his covariant Euclidean program for the construction of quantum fields. See also [17] and the references cited there for mathematically rigorous but noncovariant applications of this formula and [36, 22] for the first applications which are simultaneously rigorous and covariant. In order to concentrate on the formal ideas, we consider first the case $\mathscr{C} = R^N$ of N degrees of freedom, with $H = -\frac{1}{2}\Delta + V$, and we add a constant to V, so that

$$0 = \inf \text{spectrum } H.$$

This change in V and H is the vacuum energy renormalization. Since $H \geqslant 0$, we may make an analytic continuation $t \to it = s$ in the solution

$$U(t) = e^{-itH}$$

of the Schrödinger equation and obtain thereby a solution e^{-sH} of the heat equation.

For $V = 0$, the heat equation is solved explicitly in terms of its fundamental solution,

$$u(q, s) = (e^{-sH}u)(q) = \int F(q - q', s)\, u(q', 0)\, dq',$$

$$F(q - q', s) = (2\pi s)^{-N/2} e^{-|q-q'|^2/2s}.$$

F also defines the density for the transition probability for a Wiener path $\omega = \omega(s)$ with values in $\mathscr{C} = R^N$ to pass from $\omega(0) = q'$ to $\omega(s) = q$ in time s. Thus by definition of Wiener measure, u has a Wiener integral representation

$$u(q, s) = \int_{\mathscr{W}(q,s)} u(\omega(0), 0)\, d\omega_0 \tag{14}$$

where $d\omega_0$ is a conditional Wiener measure and

$$\mathscr{W}(q, s) = \{\omega : \omega(s) = q\}.$$

is a set of Wiener paths. In this notation, the Feynman–Kac formula states that

$$v(q, s) = \int_{\mathscr{W}(q,s)} \exp\left\{-\int_0^s V(\omega(\sigma))\, d\sigma\right\} v(\omega(0), 0)\, d\omega_0 \tag{15}$$

is a solution of the heat equation

$$v_t = -Hv, \quad H = -\tfrac{1}{2}\Delta + V.$$

On a formal level, (15) is a corollary of (14) and the product formula

$$e^{-(A+B)} = \lim_{n\to\infty} (e^{-A/n} e^{-B/n})^n.$$

We use (13) to transform (15) and obtain

$$\langle u(q), e^{-sH'} v(q)\rangle = \int u(\omega(0))^{-} v(\omega(s))\, d\omega, \tag{16}$$

where

$$\langle u(q), v(q)\rangle = \int u(q)^{-} v(q) \Omega^2\, dq \text{ and}$$

$$d\omega = \lim_{T\to\infty} \frac{\exp\{-\int_{-T}^{T} V(\omega(\sigma))\, d\sigma\}\, d\omega_0}{\int \exp\{-\int_{-T}^{T} V(\omega(\sigma))\, d\sigma\}\, d\omega_0}.$$

To extend these ideas to field theory, we choose the path space

$$q(s) \in \mathscr{C} = \mathscr{S}'(R^{d-1})$$

as before. The Hilbert space $\mathscr{H}$ and Hamiltonian H are now defined by the formulas

$$\langle u, e^{-sH} v\rangle = \int u(q)^{-} v(q(\cdot + s))\, dq$$

where u is a function of $q(x, t)$, $t \leq 0$, and v is a function of $q(x, t)$, $t \geq 0$. Moreover, we let dq_0 be the Gaussian measure on $\mathscr{W}$ with covariance C and define dq as

$$dq = \lim_{\Lambda\to\infty} \frac{\exp\{-\int_\Lambda :P(\phi(x)):\, dx\}\, dq_0}{\int \exp\{-\int_\Lambda :P(\phi(x)):\, dx\}\, dq_0}. \tag{17}$$

Thus the mathematical problems in quantum field theory can be reduced to the existence and properties of the limit (17). For P small, or for Dirichlet data in C with $P =$ even $+$ linear, the existence and certain properties of this limit in the case $d = 2$ are due to Glimm–Jaffe–Spencer, Nelson, Guerra–Rosen–Simon, and Fröhlich (see [9, 10, 48]). An earlier and more general construction of $\mathscr{H}$ and H uses C^*-algebra states for the functional analysis, while estimates on function space integrals occur only in technical intermediate steps [16].

References

1. L. Accardi, On the noncommutative Markov property, *Functional Anal. Appl.* (in Russian), to appear.
2. D. Brydges, Boundedness below for Fermion model theories, Part I, preprint.
3. D. Brydges and P. Federbush, A semi-euclidean approach to boson fermion model theories, *J. Math. Phys.* 15 (1974), 730–732.
4. J. Cannon, Continuous sample paths in quantum field theory, *Comm. Math. Phys.* 35 (1974), 215–233.
5. R. L. Dobrushin and R. A. Minlos, Construction of a one dimensional quantum field via a continuous Markov field, *Functional Anal. Appl.* 7 (1973), 324–325 (English translation).
6. J.-P. Eckmann, J. Magnen, and R. Seneor, Decay properties and Borel summability for the Schwinger functions in $P(\phi)_2$ theories, *Comm. Math. Phys.*, to appear.
7. J. Feldman, The $\lambda \phi_3^4$ field theory in a finite value, *Comm. Math. Phys.* 37 (1974), 93–120.
8. J. Feldman, On the absence of bound states in the ϕ^4 quantum field model without symmetry breaking, *Canad. J. Phys.* 52 (1974), 1583–1587.
9. J. Fröhlich, Schwinger functions and their generating functions, I, *Helv. Phys. Acta* 47 (1974), 265–306.
10. J. Fröhlich, Schwinger functions and their generating functionals, II, preprint.
11. J. Fröhlich, Verification of axioms for Euclidean and relativistic fields and Haag's theorem in a class of $P(\phi)_2$-models, *Ann. Inst. H. Poincaré* 21.
12. J. Fröhlich, The reconstruction of quantum fields from Euclidean Green's functions at arbitrary temperatures in models of a self-interacting Bose field in two space-time dimensions, preprint.
13. J. Fröhlich and K. Osterwalder, Is there a Euclidean Fermi field? Preprint.
14. J. Glimm, Models for quantum field theory, *in* "Local Quantum Theory" (R. Jost, Ed.), Proceedings of the International School of Physics "Enrico Fermi," Course 45, Academic Press, New York, 1969.
15. J. Glimm, Quantum field theory models, *in* "Proceedings International Congress of Mathematicians, Nice 1970, Gauthier-Villars, Paris, 1971.
16. J. Glimm and A. Jaffe, The $\lambda(\phi^4)_2$ quantum field theory without cutoffs. III. The physical vacuum, *Acta Math.* 125 (1970), 203–261.
17. J. Glimm and A. Jaffe, Quantum field models, *in* "Statistical Mechanics and Quantum Field Theory" (C. deWitt and R. Stora, Eds.), Gordon and Breach, New York, 1971.

18. J. GLIMM AND A. JAFFE, Boson quantum field models, *in* "Mathematics of Contemporary Physics" (R. Streater, Ed.), Academic Press, New York, 1972.
19. J. GLIMM AND A. JAFFE, Critical point dominance in quantum field models, *Ann. Inst. H. Poincaré* **21** (1974), 27–41.
20. J. GLIMM AND A. JAFFE, ϕ^4 quantum field model in the single phase region: Differentiability of the mass and bounds on critical exponents, *Phys. Rev.* **D10** (1974), 536–539.
21. J. GLIMM AND A. JAFFE, A remark on the existence of $\phi_4{}^4$, *Phys. Rev. Lett.* **33** (1974), 440–442.
22. F. GUERRA, Uniqueness of the vacuum energy and Van Hove phenomena in the infinite volume limit for two dimensional self-coupled Bose fields, *Phys. Rev. Lett.* **28** (1972), 1213.
23. F. GUERRA AND P. RUGGIERO, A new interpretation of the Euclidean-Markov field in the framework of physical Minkowski space-time, preprint.
24. F. GUERRA, L. ROSEN, AND B. SIMON, The pressure is independent of the boundary conditions, *Bull. Amer. Math. Soc.* **80** (1974), 1205–1209.
25. F. GUERRA, L. ROSEN, AND B. SIMON, Correlation inequalities and the mass gap in $P(\phi)_2$, III. Preprint.
26. G. HEGERFELD, Probability measures on distribution spaces and quantum field theoretical models, preprint.
27. A. JAFFE, Constructing the $\lambda(\phi^4)_2$ theory, *in* "Local Quantum Theory" (R. Jost, Ed.), Proceedings of the International School of Physics "Enrico Fermi," Course 45, Academic Press, New York, 1969.
28. J. KLAUDER, Field structure through model studies: Aspects of nonrenormalizable theories, *Acta Phys. Austriaca* Suppl. XI (1973), 341–387.
29. A. LENARD AND C. NEWMAN, Infinite volume asymptotics in $P(\phi)_2$ field theory and some applications, *Comm. Math. Phys.* **39** (1974), 243–250.
30. O. MCBRYAN, Vector currents in the Yukawa$_2$ quantum field theory, preprint.
31. O. MCBRYAN, The N-th order renormalized pull through formula for the Yukawa$_2$ quantum field theory, preprint.
32. O. MCBRYAN, Higher order estimates for the Yukawa$_2$ quantum field theory, *Comm. Math. Phys.*, to appear.
33. O. MCBRYAN, Local generators for the Lorentz group in the $P(\phi)_2$ model, *Nuovo Cimento* **18A** (1973), 654–662.
34. O. MCBRYAN, Self adjointness of relatively bounded quadratic forms and operators, *J. Funct. Anal.*, to appear.
35. O. MCBRYAN AND Y. PARK, Lorentz covariance of the Yukawa$_2$ quantum field theory, *J. Math. Phys.* **16** (1975) 104–110.
36. E. NELSON, Quantum fields and Markov fields, *in* "Proceedings of Summer Institute of Partial Differential Equations, Berkeley, 1971," Amer. Math. Soc., Providence, RI, 1973.
37. C. NEWMAN, The construction of two-dimensional Markoff fields with an application to quantum field theory, *J. Functional Analysis* **14** (1973), 44–61.
38. K. OSTERWALDER AND R. SCHRADER, Axioms for Euclidean Green's functions, II, to appear.
39. Y. PARK, Lattice approximation of the $(\lambda \phi^4 - \mu \phi)_3$ field theory in a finite volume, preprint.
40. M. REED AND L. ROSEN, Support properties of the free measure for Boson fields, *Comm. Math. Phys.* **36** (1974), 123–132.

41. J. ROSEN, "Logarithmic Sobolev Inequalities and Supercontractivity for Anharmonic Oscillators," Thesis, Princeton University.
42. R. SCHRADER, On the Euclidean version of Haag's theorem in $P(\phi)_2$ theories, *Comm. Math. Phys.* **36** (1974), 133–136.
43. R. SCHRADER AND D. UHLENBROCK, Markov structures on Clifford algebras, preprint.
44. E. SEILER, Schwinger functions for the Yukawa model in two dimensions with space-time cutoff, preprint.
45. T. SPENCER, The absence of even bound states in ϕ_2^4, *Comm. Math. Phys.* **39** (1974), 77–79.
46. K. SYMANZIK, A modified model of Euclidean quantum field theory, N.Y.U. preprint, 1964.
47. K. SYMANZIK, Euclidean quantum field theory, *in* "Local Quantum Theory" (R. Jost, Ed.), Proceedings of the International School of Physics "Enrico Fermi," Course 45, Academic Press, New York, 1969.
48. G. VELO AND A. WIGHTMAN (Eds.), "Constructive Quantum Field Theory," Lecture Notes in Physics 25, Springer-Verlag, Berlin, 1973.
49. J. FELDMAN AND K. OSTERWALDER, The Wightman axioms and mass gap for φ_3^4, *in* "Proceedings of a Symposium on Mathematical Problems in Theoretical Physics," Lecture Notes in Physics, Springer–Verlag, Berlin, 1975.
50. J. MAGNEN AND R. SENEOR, The infinite volume limit of the $\lambda\varphi_3^4$ model. Preprint.

Renormalization Group Methods*

KENNETH G. WILSON

Laboratory of Nuclear Studies, Cornell University, Ithaca, New York 14850

TO STANISLAW ULAM ON THE OCCASION OF HIS 65TH BIRTHDAY

> An especially intractable breed of problems in physics involves those with very many or an infinite number of degrees of freedom and in addition involve "renormalization." Renormalization is explained as the existence of very many length or energy scales of importance in the physics of the problem. The renormalization group approach is a way of reducing the complexity of these problems to the point where numerical-methods can be used to solve them. The Kondo problem (dilute magnetic alloys) is used as an illustration.

1. INTRODUCTION

When one surveys the use of numerical methods and computers in theoretical physics, one finds that there are four main classes of problems. First, there are problems involving one degree of freedom: an ordinary differential equation or a one-dimensional integral. These can usually be solved rapidly to high precision on a computer that fits into one's pocket. The second class of problems are those involving several degrees of freedom: a partial differential equation in three variables or a fourfold integration, for example. Solving these problems can challenge even a CDC-7600. The third class of problems are those involving very many degrees of freedom. The only purely numerical approach for these problems is the Monte Carlo technique, which one uses for example to compute many-fold integrals or properties of simple liquids. Monte Carlo methods are considerably less accurate and reliable than the more controlled methods available for simpler problems. Accordingly, purely

---

* Supported in part by the National Science Foundation.

numerical methods remain less important than the various formal methods, such as Feynman graph expansions, for the many degree of freedom case. (There can be a lot of numerical work involved in calculating Feynman graphs, but this will not be discussed here.)

There is a fourth class of problems which until very recently lacked *any* convincing numerical approach. This fourth class is a subclass of problems involving a large or infinite number of degrees of freedom. The special feature of this suclass is the problem of "renormalization." Originally, renormalization was the procedure for removing the divergences of quantum electrodynamics and was applied to the Feynman graph expansion. The difficulties of renormalization prevent one from formulating even a Monte Carlo method for quantum electrodynamics. Similar difficulties show up in a number of problems scattered throughout physics (and chemistry, too). These problems include: turbulence (a problem in classical hydrodynamics), critical phenomena (statistical mechanics), dilute magnetic alloys, known as the Kondo problem (solid state physics), the molecular bond for large molecules (chemistry), in addition to all of quantum field theory.

In this paper the problem of renormalization will be shown to be the problem of *many length* or *energy scales*. This problem will be explained and illustrated in the first part of this paper.

The second part of the paper is an introduction to a new approach, called the "renormalization group," which can overcome the problems of renormalization. At the present time precise numerical calculations exist for the Kondo problem [1] (previously unsolved) and the two-dimensional Ising model of a critical point [2] (previously exactly solved by Onsager). Rough numerical calculations exist for a modification of the three-dimensional Ising model, but these latter have been superseded by an expansion technique about four dimensions which will not be discussed here (see Ref. [3]). In this paper the renormalization group approach will be illustrated using a watered-down form of the Kondo problem; a description of the Kondo problem will also be given as well as some of the results of the calculation.

No numerical calculations using the renormalization group exist at present for quantum field theory, nor do they exist for turbulence or the molecular bond. Formidable obstacles still remain in all these subjects. But, as will be seen below, these problems are broadly similar to the Kondo problem. In consequence, one can expect that further development of the renormalization group approach will solve these problems also, although a long time will probably be required.

2. Multiple Scales

Consider a glass of water at room temperature. There will be hydrodynamic fluctuations of the water, i.e., surface waves, say with a wavelength of order 1 cm. In addition, there are random motions of the individual water molecules leading to density fluctuations with wavelengths of order 1 Å. Not much happens in between. This is not yet a problem involving renormalization: it has only two length scales, easily separated.

Next consider the same water brought to the critical values of temperature and pressure (647 K, 217 atm). At this "critical point" the density of water and steam become equal. As a result, very close to the critical point one finds large-scale density fluctuations, i.e., large bubbles of steam interspersed among large water drops. These density fluctuations occur at all wavelengths from 1 Å up to a maximum wavelength ("correlation length") ξ. At the critical point, ξ is infinite. The fluctuations on the scale of several thousand angstroms cause strong light scattering ("critical opalescence").

The large-scale fluctuations near the critical point in no way diminish the fluctuations on an atomic scale. In consequence, an adequate description of water near the critical point requires that one consider fluctuations at all wavelength scales from 1 Å to ξ. This is an example of *many wavelength scales*.

Simple statistical methods for studying fluids (theoretically) emphasize length scales near 1 Å. The same is true of Monte Carlo methods. By setting up hydrodynamics one can treat very long wavelength fluctuations also. But all these methods break down when one has a large range of wavelengths of equal importance. In the simplest situation (for critical phenomena the simplest situation is four dimensions, or close to four dimensions, rather than three or two dimensions) the importance of many wavelengths shows up through the integral $\int_0^\kappa dk/k$ where k is the wavenumber and $\kappa = 1$ Å^{-1}. All wavenumber scales below 1 contribute equally to this integral. For example, the range $1/200 < k < 1/100$ is as important as the range $1/20 < k < 1/10$. In consequence, the integral diverges. (The divergence occurs for $\xi \to \infty$; for finite ξ the lower limit on the integral is ξ^{-1}, not 0, and the integral is finite.) The problem of eliminating logarithmically divergent integrals was the basic renormalization problem in quantum electrodynamics. Such divergences are often a symptom of a problem involving multiple scales.

As an example of a problem with multiple energy scales, consider the conduction band of a metal. A conduction band has a Fermi surface in momentum space at momentum k_F. At low temperatures the conduction-band states with momentum $k < k_F$ are filled with electrons; the states with $k > k_F$ are unoccupied. To be precise this is the situation for the conduction band in its ground state. Consider the excited states of the band. One can take an electron from deep inside the Fermi surface (for example, with $k = k_F/2$) and put it in a state well outside the Fermi surface. Such a state has an energy ~1 eV above the ground state. This is the largest energy scale for a typical conduction band. We can illustrate these two states by an energy level diagram; see Fig. 1.

FIG. 1. Ground and excited-state energies for conduction band on scale of 1 eV.

One can also excite an electron near the Fermi surface to a state just outside the Fermi surface. This leads to a smaller energy change, say 0.1 eV. This electron is distinct from an electron deep inside the Fermi surface, so now one has four possible states when both electrons are considered. The four states are illustrated in Fig. 2. One now has two

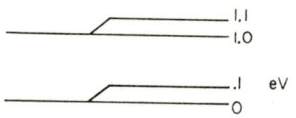

FIG. 2. Ground and excited states for conduction band including scales of 1 eV and 0.1 eV.

states separated by 1 eV each of which is split into two states with separation 0.1 eV. The diagram looks similar to the energy level splitting caused by the fine structure in hydrogen. One now has two energy scales: 1 and 0.1 eV.

Since the Fermi surface is infinitely sharp, one can produce further energy scales, say 0.01, 0.001 eV etc., by exciting electrons closer and closer to the Fermi surface (Fig. 3).

To picture the conduction band as having separate discrete energy scales, as in Fig. 3, is a gross oversimplification. Actually the possible energy scales form a continuum: all energies less than ~1 eV are possible.

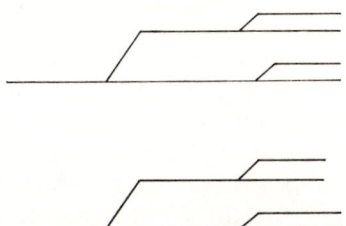

FIG. 3. Conduction-band energy levels: three energy scales.

For many band-problems these multiple energy scales are not a difficulty. The reason for this is that in many cases the interactions between electrons in the band can be ignored. As long as one has to discuss only single electron states the difficulties of the multiple energy scales are not so severe. The energy level structure of Fig. 3 is relevant only when multiple electron states have to be considered.

The Kondo problem [4, 5] is the problem of the behavior of dilute magnetic alloys, such as 0.01 %-iron impurities in copper. The particular question of interest is the zero temperature limit of the susceptibility, specific heat, resistivity etc. due to the impurity. The interaction of the impurity with the conduction band of the host metal cannot be treated in a one-electron approximation due to spin–flip scattering of conduction-band electrons by the impurity. Two successive electrons cannot undergo independent spin–flip scatterings by the same impurity because the first electron changes the spin of the impurity (see Fig. 4).

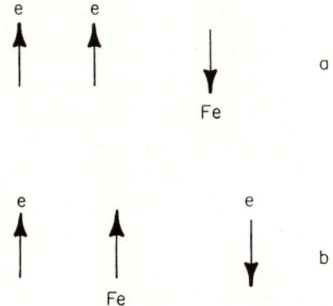

FIG. 4. Successive scatterings of two spin-up electrons by an iron impurity. The first scattering is spin–flip, the second scattering cannot flip spins. (a), Configuration before any scattering; (b), configuration after the first scattering.

There was great interest in the Kondo problem over the last ten years, but the problem turned out to be depressingly difficult to solve, the multiple energy scales being the bottleneck. Anderson, Yuval, and Hamann [6, 5] gave a qualitative solution of the zero temperature behavior by setting up an analogy between the Kondo problem and a one-dimensional Coulomb gas (another not-so-easy problem). Now one can calculate properties of the simplest model (one impurity coupled to s-wave electrons) to a few percent accuracy by renormalization group methods [1].

In turbulence, for example in the atmosphere, energy is generated at very long wavelengths by an external disturbance and is dissipated into heat by viscosity effects at very short wavelengths. Turbulent fluctuations develop at all intermediate wavelengths; these multiple length scales are a principal reason for the slow progress of theories of turbulence. In quantum field theory one has multiple energy scales; for example, in quantum electrodynamics one has single electron states with any energy from 0.5 MeV to ∞, and multiple particle states must be considered just as in the conduction-band case. The divergences arise due to the lack of a cutoff on the high energy scales. In the molecular problem one has multiple energy scales, from the inner s-shell energies of the atoms ($\sim$100 eV to keV) to the small van der Waals forces between distant atoms.

3. The Renormalization Group: Technique

The problem of renormalization has two parts. There is a technical problem, namely, how to solve problems involving renormalization. Until recently, renormalization was applied only to a Feynman graph expansion except for a few exactly soluble models [7]. Unfortunately, an expansion makes sense only when the expansion parameter is small. Thus one searches for alternative methods, and with the availability of large computers one turns to purely numerical methods. The renormalization group approach is a numerical approach [8]: it will be described in detail shortly.

The second part of the problem is determining the qualitative physics of systems requiring renormalization. This turns out to be quite different from naive expectations, and as a result a special language has developed to describe the results one finds: fixed points, universality etc. This language will be explained in Section 4. The peculiar results of renormalization are a direct consequence of the existence of multiple length or energy scales, and therefore the same set of results occurs in many different kinds of problems so long as they involve renormalization.

To illustrate the technical aspects of the renormalization group approach, a simplified version of the Kondo problem will be discussed. Namely, we shall invent a quantum system that has the energy level structure illustrated in Figs. 1–3. Then the solution of this problem will be described. Nothing beyond elementary quantum mechanics will be involved.

First consider the space of states of the system. One starts on an energy scale of 1 eV with two states, a ground state and an excited state. One might set up Pauli matrices σ_{0x}, σ_{0y}, σ_{0z} to be the operators on this space. Next one brings in the energy scale 0.1 eV; on this scale each state splits into two, giving four states total. A new set of Pauli operators σ_{1x}, σ_{1y}, σ_{1z} can describe the splitting. One splits each state into two again on the scale 0.01 eV, introducing operators σ_{2x}, σ_{2y}, and σ_{2z}.

Suppose one has 11 energy scales *in toto*: 1, 0.1, 0.01 eV etc., down to 10^{-10} eV. Then there are 11 sets of Pauli matrices and 2,048 states after the final splitting.

Secondly, consider the Hamiltonian of the system. The Hamiltonian must have 11 separate terms corresponding to the 11 energy scales:

$$\mathcal{H} = \mathbf{H}_0 + 0.1\mathbf{H}_1 + 0.01\mathbf{H}_2 + \cdots + 10^{-10}\mathbf{H}_{11}. \tag{1}$$

The operator $\mathbf{H}_0$ gives the unperturbed energies of the original two states (Fig. 1). In other words, $\mathbf{H}_0$ depends only on $\vec{\sigma}_0$; $\mathbf{H}_0$ can be represented by a 2×2 matrix. For example, $\mathbf{H}_0$ might be

$$\mathbf{H}_0 = \begin{vmatrix} 1 & 2 \\ 2 & -1 \end{vmatrix}. \tag{2}$$

The operator $\mathbf{H}_1$ determines the first set of energy level splittings (Fig. 2); $\mathbf{H}_1$ depends on both $\vec{\sigma}_0$ and $\vec{\sigma}_1$ and can be represented by a 4×4 matrix. See, e.g., Eq. (6) below. Similarly, $\mathbf{H}_2$ is represented by an 8×8 matrix, etc.; $\mathbf{H}_{11}$ is represented by a $2,048 \times 2,048$ matrix.

The full Hamiltonian $\mathcal{H}$, like $\mathbf{H}_{11}$, is represented by a $2,048 \times 2,048$ matrix. Assuming this matrix has been constructed, one's first temptation (especially if one is at Los Alamos) might be to feed it to the 7600 and go fishing.

This is not the most economical procedure. A $2,048 \times 2,048$ matrix is rather large; diagonalizing it would require several hours of computing time. To make matters worse, high precision is required: the energy eigenvalues of $\mathscr{H}$ are of order 1 but the energy level splittings are as small as 10^{-10}. To see these splittings one should have an accuracy at 10^{-12} or better.

The structure of the energy levels suggests an alternative approach: perturbation theory.

To illustrate what one can accomplish with perturbation theory, consider a much simpler model, namely one with only the energy scales 1 and 10^{-10}, say

$$\mathscr{H} = \mathbf{H}_0 + 10^{-10} \mathbf{H}_1 . \tag{3}$$

The first step is to diagonalize $\mathbf{H}_0$. Suppose that this calculation is performed with a roundoff error $\sim 0.1\%$. Then $\mathbf{H}_0$ in diagonal form is

$$\mathbf{H}_0 = \begin{vmatrix} -2.236 & 0 \\ 0 & +2.236 \end{vmatrix}. \tag{4}$$

Next one adds $\mathbf{H}_1$; to do this one must first rewrite $\mathbf{H}_0$ as a 4×4 matrix:

$$\mathbf{H}_0 = \begin{vmatrix} -2.236 & 0 & 0 & 0 \\ 0 & -2.236 & 0 & 0 \\ 0 & 0 & 2.236 & 0 \\ 0 & 0 & 0 & 2.236 \end{vmatrix}. \tag{5}$$

(As indicated by Fig. 2, each eigenvalue of $\mathbf{H}_0$ occurs twice.) Then one adds $10^{-10} \mathbf{H}_1$ to $\mathbf{H}_0$; suppose

$$10^{-10} \mathbf{H}_1 = 10^{-10} \times \begin{vmatrix} 1.2 & 0.5 & 0.6 & 0.8 \\ 0.5 & 1.4 & 0.4 & 0.3 \\ 0.6 & 0.4 & 2.2 & 0.1 \\ 0.8 & 0.3 & 0.1 & 0.9 \end{vmatrix}. \tag{6}$$

To add $10^{-10} \mathbf{H}_1$ to $\mathbf{H}_0$ one must use the same representation for both, so one has to compute matrix elements of $\mathbf{H}_1$ with respect to eigenstates of $\mathbf{H}_0$. Since these eigenstates are only computed to 0.1% accuracy, the matrix elements of $\mathbf{H}_1$ are only determined to 0.1% accuracy also. Let Eq. (6) be the result of calculating these matrix elements.

The next step in the (degenerate) perturbation theory calculation is to neglect the excited states of $\mathbf{H}_0$. The result of neglecting these states is an effective Hamiltonian

$$\mathscr{H}_{\text{eff}} = 10^{-10} \begin{vmatrix} 1.2 & 0.5 \\ 0.5 & 1.4 \end{vmatrix} \quad (7)$$

(apart from the constant -2.236). In diagonal form $\mathscr{H}_{\text{eff}}$ is

$$\mathscr{H}_{\text{eff}} = 10^{-10} \begin{vmatrix} 0.790 & 0 \\ 0 & 1.810 \end{vmatrix}. \quad (8)$$

The lowest two eigenvalues of $\mathscr{H}$ are therefore

$$E_1 = -2.236 + 0.790 \times 10^{-10}, \quad (9)$$

$$E_2 = -2.236 + 1.810 \times 10^{-10}. \quad (10)$$

Note the following:

(1) The roundoff error in E_1 or E_2 separately is about 10^{-3}, but the energy splitting $E_2 - E_1$ is known to much greater accuracy:

$$E_2 - E_1 = 1.020 \times 10^{-10}, \quad (11)$$

with an error of about 10^{-13}.

(2) The perturbation calculation of $\mathscr{H}$ involves the diagonalization of two separate 2×2 matrices [Eqs. (2) and (7)] each to an accuracy $\sim 0.1\%$. In contrast, the brute force diagonalization of $\mathscr{H}$ involves diagonalization of a 4×4 matrix to an accuracy of $10^{-11}\%$.

(3) The error due to using perturbation theory is of order 10^{-20}, much smaller than the round-off error.

Return to the full Hamiltonian [Eq. (1)]. A lowest-order degenerate perturbation treatment consists of the following steps:

(1) Diagonalize $\mathbf{H}_0$ and keep only the ground state.

(2) Construct the ground state matrix elements of $0.1\,\mathbf{H}_1$, producing a 2×2 matrix which one diagonalizes.

(3) Now one treats $0.01\,\mathbf{H}_2$ as a perturbation on $\mathbf{H}_0 + 0.1\,\mathbf{H}_1$. This means one keeps only the ground state of $\mathbf{H}_0 + 0.1\,\mathbf{H}_1$, i.e., the lowest energy eigenstate from step 2. One constructs the ground state matrix elements of $0.01\,\mathbf{H}_2$, thus generating a third 2×2 matrix.

Continuing in this way the entire Hamiltonian $\mathcal{H}$ is solved by constructing and diagonalizing 11 2 × 2 matrices, starting with $\mathbf{H}_0$ and ending with $\mathbf{H}_{11}$. The perturbation calculation is represented schematically in Fig. 5.

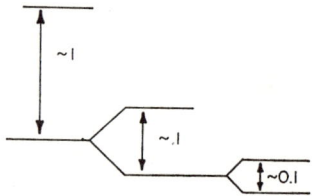

FIG. 5. Energy levels calculated in perturbation theory: for each energy scale the lowest two states are calculated, all higher states being neglected.

The calculation just described yields only the ground state and first excited-state energies of $\mathcal{H}$. Other excited-state energies can be obtained by repeating the calculation except that an excited state is kept instead of the ground state at some stages of the calculation.

The accuracy of the degenerate perturbation calculation is not very good. At each stage the perturbation is about 10% of the unperturbed level spacing, which means one makes about a 10% error at each stage, and this error can accumulate over the 11 stages. This error is easily reduced. Suppose that one starts by diagonalizing $\mathbf{H}_0 + 0.1\ \mathbf{H}_1$ (a 4 × 4 matrix) exactly except for roundoff error. Then instead of keeping only the ground state of $\mathbf{H}_0 + 0.1\ \mathbf{H}_1$, one keeps both the ground state and first excited state, neglecting only the excited states of $\mathbf{H}_0$. Then one adds $0.01\ \mathbf{H}_2$, leading to a new 4 × 4 matrix which again one diagonalizes exactly. Continuing this procedure, one has ten 4 × 4 matrices to diagonalize. This scheme is illustrated in Fig. 6. The error in this procedure can easily be seen to be the ratio of the perturbation energy to the neglected unperturbed energies (relative to the ground state energy), i.e., 1%. Now even the accumulated error of ten iterations cannot be more than 10%. If further accuracy is required one can keep four states after each diagonalization and diagonalizes 8 × 8 matrices at each step, starting with $\mathbf{H}_0 + 0.1\ \mathbf{H}_1 + 0.01\ \mathbf{H}_2$. This gives 0.1% accuracy at each step [9].

The full 2,048 × 2,048 matrix $\mathcal{H}$ has to be diagonalized only if one needs an accuracy better than 10^{-8}%.

The lessons of the above calculation are the following. There is one

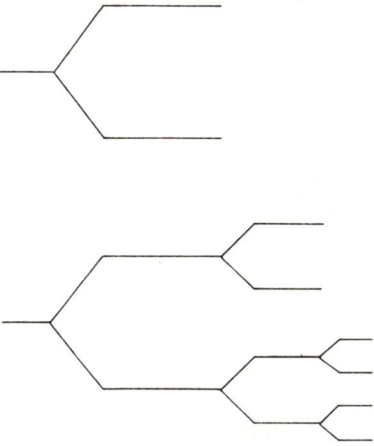

FIG. 6. Energy levels calculated in improved perturbation theory: four energy levels are calculated at each stage.

degree of freedom (one set of Pauli operators) for each energy scale in $\mathcal{H}$. Considering the whole of $\mathcal{H}$ at once, one has 11 degrees of freedom resulting in 2,048 states and an unpleasant matrix to diagonalize. The degenerate perturbation treatment allows one to break up the calculation into 11 parts, each of which involves only *one* energy scale, and hence only *one* degree of freedom. The result was 11 separate 2×2 matrices to diagonalize. For greater accuracy one can diagonalize 8×8 matrices at each step, which corresponds to considering three neighboring energy scales at each step. These 8×8 matrices need only be diagonalized to 0.1% accuracy and one will still have 10^{-12} absolute accuracy on the energy splittings of order 10^{-10} in $\mathcal{H}$.

The artificial model described above illustrates the basic idea of the renormalization group approach. This idea is that whenever one has a problem with many length scales, or many energy scales etc., one breaks up the calculation into separate stages, with one stage for each length or energy scale. This break-up is to be defined so that each stage involves only a few length or energy scales (for example, the 8×8 matrix diagonalization in the above model, involving three energy scales from the original 11). As a result each stage of the calculation involves fewer degrees of freedom than appear in the full problem. This means each stage is more amenable to numerical techniques than the full problem.

In the model, the separate energy scales were clearly identified, and it was straightforward to treat them one at a time. In real life, the situation is more complicated. Instead of separate discrete scales one has a continuum, for example, a continuum of energy scales in a conduction band. It is a nontrivial task to break down the problem to a few energy scales in a workable form. Accomplishing this is as much an art as a science and will not be discussed further here. Several procedures exist for separating length scales in critical phenomena; see Refs. [2] and [3]. The procedure for handling the Kondo problem has been described in Ref. [1]. The classic renormalization group approach of Gell-Mann and Low accomplishes this breakdown for weakly coupled quantum field theories, but no procedure exists yet for strongly coupled field theories. No procedures exist yet for turbulence or the molecular band.

In the model, the reduction in degrees of freedom was from 11 total to three or less per stage. In real life one often can think of techniques that reduce 10^{23} degrees of freedom, say, to 60 or so. For example, one can think of generalizations of the Niemeijer–Van Leeuwen technique to the three-dimensional Ising model which involve a $4 \times 4 \times 4$ block of spins at each stage. Unfortunately, this does not seem to be much of an improvement in practice: a block of 64 spins has 2^{64} configurations: a sum over all these configurations would require 30,000 years of computer time. The hope is that by being clever enough the number of degrees of freedom per stage can be further reduced.[1] Alternatively, one may be able to develop Monte Carlo methods to handle 64 or more degrees of freedom. So far I have been reluctant to try Monte Carlo techniques, preferring to locate and study special cases such as the Kondo problem where Monte Carlo methods are not needed.

4. The Renormalization Group: Consequences

When one solves a problem by renormalization group techniques, one often finds surprising results. This is due to the iterative character of the renormalization group. In the model of the previous section, one had ~ 10 matrices to diagonalize; in the actual Kondo calculations, as discussed below, there were as many as 80 stages, each stage involving a matrix diagonalization. The results to be explained below are natural and normal for any iterative calculation, and hence they recur in many renormalization group calculations. They are surprising in the context

---

[1] See, e.g. [12].

of problems such as the Kondo problem only because physicists are not used to thinking of these problems in terms of multiple energy scales and the role that each scale plays in the physics of the problem.

The example that will be used in this section is the Kondo problem. The Hamiltonian actually used in the Kondo calculation had the following form

$$\mathcal{H} = -J\mathbf{f}_0^+ \vec{\sigma} \mathbf{f}_0 \cdot \vec{\tau} + \sum_{n=0}^{\infty} 2^{-n/2}(\mathbf{f}_n^+ \mathbf{f}_{n+1} + \mathbf{f}_{n+1}^+ \mathbf{f}_n). \quad (12)$$

It is not necessary to understand this Hamiltonian in complete detail; for further information see Ref. [1]. The operators f_n^+ create electrons in states with energies of order $2^{-n/2}$. The parameter J is the strength of the magnetic coupling of a single impurity to the conduction band electrons. The successive energy scales differ by $\sqrt{2}$ in this Hamiltonian, instead of the factor 10 of the previous model. (The factor $\sqrt{2}$ is only an approximation: the exact Kondo model is obtained by replacing $\sqrt{2}$ by $\sqrt{\Lambda}$ and taking the limit $\Lambda \to 1$. But very good accuracy (much better than 1%) appears to result from choosing $\Lambda = 2$; see Ref. [1].

The solution of $\mathcal{H}$ was obtained by the same procedure outlined for the previous model. An iteration consisted of adding one more term from the sum over n to the Hamiltonian, diagonalizing, and keeping only the lowest eigenstates. However, since the ratio of energy scales was only $\sqrt{2}$ per stage, it was necessary to keep of the order 600 states after each iteration instead of 4 or 8. (Calculations were performed keeping either 526 or 1620 states; they differed by a few percent only.)

TABLE I

First Two Excited-State Energies of Kondo Hamiltonian
(Ground State Energy E_0 Normalized to 0)
After N Iterations, Apart from Scale Factor S

N	44	46	48	45
E_0	0	0	0	0
E_1	0.6532	0.6545	0.6550	0.0021
E_2	1.301	1.306	1.308	1.292
S	$\sqrt{2}^{-44}$	$\sqrt{2}^{-46}$	$\sqrt{2}^{-48}$	$\sqrt{2}^{-45}$

Perhaps the most important term in the lexicon of the renormalization groups is "fixed point." A fixed point comes about in the Kondo calculation as follows. After 44 iterations of a typical calculation, the first few energy levels were those of Table I. The energy levels after 46 iterations are also shown. They are almost identical except for a change in scale of a factor of 2. There is even less change between 46 and 48 iterations. In the limit of an infinite number of iterations there is no difference at all except for the change in scale. This limiting set of energy levels is called a fixed point.

A fixed point is a special solution of the iterative calculation. To see this let $\mathbf{H}_N$ be the Hamiltonian after N steps, rescaled so that the low lying energies are of order 1 instead of $\sqrt{2}^{-N}$. Then the iteration formula is

$$\mathbf{H}_{N+1} = \mathbf{f}_N{}^+\mathbf{f}_{N+1} + \mathbf{f}_{N+1}^+\mathbf{f}_N + \sqrt{2}\,\mathbf{H}_N \tag{13}$$

Suppose $\mathbf{H}_{N+1}$ has the same set of energy levels as $\mathbf{H}_N$; this is a fixed point. Then it is evident that $\mathbf{H}_{N+2}$ will have the same set of energy levels as $\mathbf{H}_{N+1}$ because the formula generating $\mathbf{H}_{N+2}$ from $\mathbf{H}_{N+1}$ has the same form as the formula generating $\mathbf{H}_{N+1}$ from $\mathbf{H}_N$.

There is no guarantee that a given renormalization group calculation will lead to a fixed point, and in the Kondo case what one obtains is not precisely a fixed point. In the Kondo case it is only $\mathbf{H}_{N+2}$ that has the same energy levels as $\mathbf{H}_N$; $\mathbf{H}_{N+1}$ has a quite different set of energies (see Table I). It is more correct to call this a "limit cycle." If one studies $\mathbf{H}_N$ only for even N it looks like a fixed point.

One of the results of the Kondo calculation was a determination of the ratio of the specific heat C to the susceptibility χ of the impurity in the limit of zero temperature. The results obtained were:

$$J = 0: \lim_{T \to 0} \left(\frac{C}{T\chi}\right) = 0, \tag{14}$$

$$J < 0: \lim_{T \to 0} \left(\frac{C}{T\chi}\right) = \frac{2\pi^2}{3}(1 \pm 0.03)k^2, \tag{15}$$

the latter result valid for small negative J. There is a discontinuity in this ratio at $J = 0$; then for small negative J it is a constant independent of J. The discontinuity is a typical outcome of a renormalization group calculation.

From a conventional (non-renormalization group) approach the discontinuity is unexpected. For small enough J one would think that the Hamiltonian $\mathcal{H}$ of Eq. (12) could be solved in a perturbation expansion in J. This means that one should have

$$\frac{C}{T\chi} = a_0(T) + Ja_1(T) + J^2a_2(T) + \cdots. \tag{16}$$

There is no hint of a discontinuity at $J = 0$ in this formula.

When $a_0(T)$, $a_1(T)$ etc. are calculated [4] it is found that $a_4(T)$ behaves as $\ln T$ for $T \to 0$. In consequence the expansion becomes infinite for $T = 0$. Unfortunately, $a_5(T)$ behaves as $\ln^2 T$, $a_6(T)$ as $\ln^3 T$, etc., so that one cannot neglect higher orders in J when T is very small. Thus a knowledge of the coefficients $a_4(T)$ etc. does indicate something peculiar will occur at $T = 0$ but leaves one with no insight as to what will happen.

The discontinuity at $J = 0$ develops naturally in the renormalization group calculations. Consider two calculations, one for $J = 0$ and one for small negative J, say $J = -0.001$. In the first stage of the calculation, the energy levels for $J = 0$ and $J = -0.001$ differ only by about 0.1% due to the small value of J. The energy levels are analytic functions of J; there are no discontinuities in J at $J = 0$. In the second stage of the calculation, with a smaller energy scale, there are somewhat larger differences between the $J = 0$ energy levels and the $J = -0.001$ energy levels. This might be a 0.2% difference (I am exaggerating the actual increase). The energy levels are still analytic in J.

As one goes through the third, fourth, and higher stages the difference between the $J = 0$ and $J = -0.001$ energy levels increases further, to to 0.4, 0.8% etc., until after sufficiently many iterations (say 10) there is a 100% difference. Each stage of the calculation is completely analytic, but, no matter how small the initial value of J is, the energy levels after N stages can be 100% different from the energy levels for $J = 0$ if N is large enough.

For $N \to \infty$, what happens is that both the $J = 0$ and $J = -0.001$ calculations lead to fixed points, but they are *different* fixed points. Denote these fixed points by $\mathbf{H}_A^*$ ($J = 0$) and $\mathbf{H}_B^*$ ($J = -0.001$). Both $\mathbf{H}_A^*$ and $\mathbf{H}_B^*$ reproduce themselves when subsituted in Eq. (13) (to be precise both require two iterations to reproduce themselves), but $\mathbf{H}_A^*$ is a completely different list of energies than $\mathbf{H}_B^*$. Any negative value of J no matter how small results in $\mathbf{H}_B^*$.

Thus it is natural that the physics for $J = 0$ is qualitatively different than for $J < 0$.

The discontinuity at $J = 0$ occurs only at zero temperature. It is easily seen why this is so. When T is greater than zero there is a minimum energy scale, namely kT. To determine quantities like the susceptibility and specific heat at temperature T, one needs to know the energy level structure for energies of order kT above the ground state. Roughly speaking this structure is determined when one has reached the iteration N for which $\sqrt{2}^{-N}$ is of order kT. If $kT > 0$, then N is finite and the energy levels are still analytic in J. Only when $T \to 0$ is one forced to go to the limit $N \to \infty$ for which the discontinuity occurs.

The constancy of the ratio $(C/\chi T)$ for small J is also easily understood. The same fixed point is reached for $N \to \infty$ for any small value of J; thus one would expect zero temperature properties of the system to be independent of J. (This ratio is, correctly speaking, not quite a zero temperature quantity and does vary with J when J is sufficiently large. This is a technical detail, one of many, that cannot be discussed further here. See Ref. [1]. This independence of J is a particular example of "universality," namely the independence of properties of a system described by a fixed point on parameters in the initial Hamiltonian. Universality plays an important role in the present understanding of critical phenomena [3].

5. Outlook

The examples discussed in the previous sections show the importance of the renormalization group both for solving problems requiring renormalization and for revising one's expectations about what the solution will look like. Nevertheless, the renormalization group continues to be less important than one might expect. It is at present an approach of last resort, to be used only when all other approaches have been tried and discarded. The reason for this is that it is rather difficult to formulate renormalization group methods for new problems; in fact, the renormalization group approach generally seems as hopeless as any other approach until someone succeeds in solving the problem by the renormalization group approach. Where the renormalization group approach has been successful a lot of ingenuity has been required: one cannot write a renormalization group cookbook. (In contrast, Feynman diagram techniques can be reduced to simple strict rules.) Even if one succeeds in formulating the renormalization group approach for a particular problem, one is likely to have to carry out a complicated computer calculation, which makes most theoretical physicists cringe. Especially in the case of strong interactions of elementary particles, most theorists hope to solve the problem without turning to modern renormalization group methods. It will probably require several years of stagnation in elementary particle theory before theorists will accept the inevitability of the renormalization group approach despite its difficulties.

Postscript on terminology: The "group" in "renormalization group" refers to the iteration which is always part of a renormalization group approach. An example is Eq. (13), which can be thought of as defining a transformation T on a space of Hamiltonians: $\mathbf{H}_{N+1} = T[\mathbf{H}_N]$. Iterating this transformation defines a simple semi-group.

Added notes. (1) A short introduction to the renormalization group approach is given in Ref. [10]. (2) There is a more formal "multiple time scale" technique discussed, e.g., in Ref. [11]. The multiple time scale method is another example of a renormalization group approach.

References

1. K. Wilson, *in* "Nobel Symposia—Medicine and Natural Sciences" (B. and S. Lundqvist, Eds.), Vol. 24, p. 68, Academic Press, New York, 1973; K. Wilson, Cargèse Lecture notes (*Revs. Mod. Phys.*, to be published).
2. Th. Niemeijer and J. M. J. Van Leeuwen, *Phys. Rev. Lett.* 31 (1973), 1411; *Physica* 71 (1974), 17.
3. K. Wilson and J. B. Kogut, *Physics Reports* 12C (1974), 75.
4. For a review, see J. Kondo, *Solid State Physics* 23 (1969), 184.
5. P. W. Anderson, Comments on Solid State Phys. 5 (1973), 73.
6. P. W. Anderson, G. Yuval, and D. Hamann, *Phys. Rev. B* 1 (1970), 4464.
7. For example, the Thirring model: see K. Johnson, *Nuovo Cimento* 20 (1961), 773.
8. Originally the renormalization group approach was a particular technique for summing a subclass of Feynman graphs: see, for example, N. N. Bogoliubov and D. V. Shirkov, "Introduction to the Theory of Quantized Fields," Chap. 8, Interscience, New York, 1969. The numerical procedure that will be discussed here is much more powerful than the earlier analytic approach.
9. An alternative procedure to obtain higher accuracy is to use degenerate perturbation theory to higher orders. This procedure has been discussed in rigorous detail for a similar model: see K. Wilson, *Phys. Rev. D* 2 (1970), 1438. Higher order perturbation theory is less useful when the ratio of energy scales is of order 1 (see the following section).
10. K. Wilson, *in* "Magnetism and Magnetic Materials–1972" (C. D. Graham, Jr., and J. J. Rhyne, Eds.), A.I.P. Conference Proceedings No. 10, p. 843, American Institute of Physics, New York, 1973.
11. B. B. Varga and S. Ø. Aks, *J. Mathematical Phys.* 15 (1974), 149.
12. L. Kadanoff and A. Houghton, *Phys. Rev. B* 11 (1975), 377.

Remarks on Turbulence Theory

ROBERT H. KRAICHNAN

Dublin, New Hampshire 03444

Presented at the First Los Alamos Workshop on Mathematics in the Natural Sciences, June 1974

TO STANISLAW ULAM ON HIS 65TH BIRTHDAY

> Some recent work in turbulence theory is reviewed, with emphasis on methods related to renormalized perturbation theory. The topics discussed include: constants of motion and equilibrium statistical mechanics of the Euler equations; cascade phenomena in turbulence and intermittency in the small scales; renormalized perturbation expansions and renormalization of viscosity; the difficulty of distinguishing convection from distortion effects in turbulence dynamics, the need for Lagrangian description, and related difficulties in the quantum field theory of many-body systems; the direct-interaction approximation for turbulence; what can presently be computed from turbulence theory.

1. INTRODUCTION

The word turbulence has been used to label many different phenomena, most of which have, in some degree, the common characteristics of complexity and disorder. We shall not attempt in this paper an overall survey of even hydrodynamic turbulence. The discussion will be limited to incompressible, Navier–Stokes turbulence with cyclic boundary conditions, and the point of view will be to examine this turbulence as a specimen problem in classical, statistical field theory. The chosen turbulence phenomena serve to exhibit deep-lying problems in a simple physical setting. The hard problems in turbulence theory have to do with essential nonlinearity (there is no hint of the nonlinear solutions in linearized approximations), with strong departure from absolute statistical equilibrium, and with the existence (at high Reynolds numbers) of excited degrees of freedom with widely differing length and time scales.

The need for a statistical description of turbulence arises both from the complexity of individual flow realizations and from the strong instability of realizations to small perturbations in initial or boundary conditions. This makes it natural to examine ensembles of realizations rather than individuals, in the hope that aptly chosen ensembles will be insensitive to the details of perturbations and will exhibit relatively simple behavior.

The statistical properties revealed by experiments on laboratory and geophysical flows involve an interplay of randomness and order whose efficient mathematical description is difficult. In a wide variety of turbulent flows, univariate distributions of the velocity field at given points in the flow are close to normal. But the multivariate distributions of the velocity at several space–time points is essentially different from normal. The deviations from normality are strongest for small spatial separations at large Reynolds numbers. They appear to be associated with the concentration of vorticity into intense, thin ribbons or related structures.

A peculiar feature of turbulence as a statistical–mechanical problem arises from the non-Hamiltonian form of the dynamical equation. The kinetic energy, which is conserved if viscosity is zero, is a diagonal quadratic form in the amplitudes of spatial Fourier components. Thus, although the dynamical equations give essential nonlinear interaction of Fourier modes, there is no reflection of this interaction in the form of the energy. A Liouville theorem holds in the phase space of the Fourier amplitudes, with the immediate consequence that the familiar canonical distribution of Hamiltonian equilibrium statistical mechanics is an equilibrium distribution here also. But since the energy is a diagonal quadratic form, the whole equilibrium statistical mechanics in the present case is almost trivial compared to that of usual interacting Hamiltonian systems. The essence of turbulence as a statistical–mechanical problem lies in the necessity to consider strong departures from absolute statistical equilibrium. The difficulties which then arise are much the same for Hamiltonian and non-Hamiltonian systems.

The nonequilibrium states which constitute actual turbulence are ones in which hydrodynamic kinetic energy is dissipated into heat through the action of viscosity. A crucially important feature is the peculiar dependence of dissipation rate upon the viscosity parameter. As viscosity goes to zero, the rate of energy dissipation approaches, or at least appears to approach, a nonzero limit that is independent of viscosity. This experimentally observed behavior appears to be associated with a thinning and intensification of the vortex ribbons that, on the average, compensates the lessened dissipation that otherwise would accompany the reduction of viscosity.

2. The Dynamical Equations

If the velocity field in the cyclic box is expanded in the Fourier series

$$\tilde{\mathbf{u}}(\mathbf{x}) = \sum_{\mathbf{k}} \mathbf{u}(\mathbf{k}) \exp(i\mathbf{k} \cdot \mathbf{x}), \tag{1}$$

the incompressible Navier–Stokes equation can be written as

$$(\partial/\partial t + \nu k^2) u_i(\mathbf{k}) = -ik_m P_{ij}(\mathbf{k}) \sum_{\mathbf{p}+\mathbf{q}=\mathbf{k}} u_j(\mathbf{p}) u_m(\mathbf{q}). \tag{2}$$

Here ν is the so-called kinematic viscosity (viscosity/density), the summation is over all allowed wavevectors, and $P_{ij}(\mathbf{k}) = \delta_{ij} - k_i k_j/k^2$ is the solenoidal projection operator which includes the pressure force and maintains the incompressibility, or transverseness, condition

$$\mathbf{k} \cdot \mathbf{u}(\mathbf{k}) = 0. \tag{3}$$

The energy per unit mass is

$$\hat{E} = \sum_{\mathbf{k}} |\mathbf{u}(\mathbf{k})|^2/2. \tag{4}$$

It is formally conserved under (2) if $\nu = 0$.

Some essential features of this dynamical system are more clearly displayed by introducing as variables the real and imaginary parts of the Fourier amplitudes that are linearly independent under (3) and under the reality condition $\mathbf{u}(-\mathbf{k}) = \mathbf{u}^*(\mathbf{k})$ [1]. Let these variables be arranged in a single linear sequence denoted by y_α, where α takes integer values. Then (2) can be written as

$$(d/dt + \nu_\alpha) y_\alpha = \sum_{\beta\gamma} A_{\alpha\beta\gamma} y_\beta y_\gamma, \tag{5}$$

while

$$E = \sum_\alpha y_\alpha^2. \tag{6}$$

The conservation property of (2) then implies the identity

$$A_{\alpha\beta\gamma} + A_{\beta\gamma\alpha} + A_{\gamma\alpha\beta} \equiv 0. \tag{7}$$

We make the convention $A_{\alpha\beta\gamma} = A_{\alpha\gamma\beta}$ (only the symmetrical part can contribute to (5)). Moreover, we exclude excitation at $\mathbf{k} = 0$ (uniform convection of the flow), a condition preserved by (2). With $\mathbf{u}\,(\mathbf{k} = 0)$ removed from the equations, $A_{\alpha\beta\gamma}$ vanishes if any two indices are equal.

In addition to energy, there are an infinite number of other inviscid constants of motion. In three-dimensional flow, the circulation (integrated tangential velocity component) around each closed curve that moves with the fluid is a constant. In two-dimensional flow, the vorticity of each fluid element is an inviscid constant. In particular, there are additional quadratic constants: in three dimensions, the dot product of velocity and vorticity, integrated over the whole flow; in two dimensions the squared vorticity, similarly integrated. These inviscid constants are called helicity and enstrophy, respectively. In the Fourier representation, the enstrophy per unit volume is

$$\hat{\Omega} = \sum_{\mathbf{k}} k^2 |\mathbf{u}(\mathbf{k})|^2/2 = \sum_{\alpha} k_{\alpha}^2 y_{\alpha}^2, \tag{8}$$

where $\mathbf{k}_{\alpha}$ is the wavevector for mode y_{α}. Enstrophy conservation is expressed by the additional identity

$$k_{\alpha}^2 A_{\alpha\beta\gamma} + k_{\beta}^2 A_{\beta\gamma\alpha} + k_{\gamma}^2 A_{\gamma\alpha\beta} \equiv 0 \quad \text{(two dimensions)}. \tag{9}$$

The helicity per unit volume is

$$i \sum_{k} \epsilon_{imj} k_m u_i(\mathbf{k})\, u_j^*(\mathbf{k}), \tag{10}$$

where ϵ_{imj} is the alternating tensor. Its conservation is again associated with an identity like (7) and (9), but to write the identity explicitly we would have to elaborate the notation of the y_{α}.

The quadratic constants energy, enstrophy, and helicity are distinguished from the host of other inviscid constants by the fact that they are conserved in detail by the interaction of each triad of Fourier modes. This detailed conservation is assured by the identities (7) and (9). A consequence is that the quadratic constants survive if the system is truncated by removing from the dynamical equations every term involving a wavevector whose magnitude exceeds an arbitrary cutoff k_{max}. In contrast, the circulation and fluid-element-vorticity constants, which have a simple meaning only in x space, do not recognizably survive such truncation.

3. Absolute Equilibrium Distributions

If $\nu = 0$, (5) together with the fact that $A_{\alpha\beta\gamma}$ vanishes if $\beta = \alpha$ or $\gamma = \alpha$, yields a detailed form of Liouville's theorem:

$$\partial \dot{y}_\alpha / \partial y_\alpha = 0. \tag{11}$$

Because of (11), many aspects of the equilibrium statistical mechanics of Hamiltonian systems carry over to the present non-Hamiltonian dynamics. Thus, if $K(y)$ is any constant of motion under (5), with $\nu = 0$, then $N \exp(-K)$, where N is a normalization factor, is a stable absolute equilibrium distribution, provided only that $\exp(-K)$ is integrable over y space [2]. This distribution is stable under all perturbations of the A's, or couplings to external systems, which do not destroy the conservation of K. The stability property implies an infinity of fluctuation–dissipation relations in equilibrium [2] of which the simplest is

$$\langle y_\beta(t)(\partial K/\partial y_\alpha)_{t'}\rangle = g_{\beta\alpha}(t, t') \qquad (t \geqslant t') \tag{12}$$

Here $\langle\ \rangle$ denotes distribution average and $g_{\beta\alpha}(t, t') = \langle \delta y_\beta(t)/\delta f_\alpha(t')\rangle$ is the mean response of y_β to an infinitesimal forcing term f_α added to the right-hand side of (5).

At $t = t'$, (12) reduces to the generalized equipartition law

$$\langle y_\beta\, \partial K/\partial y_\alpha \rangle = \delta_{\alpha\beta}. \tag{13}$$

If K is additive over the modes, as are the quadratic constants of Section 2, then (13) exhibits the ultraviolet catastrophe that characterizes all classical fields: $\langle K \rangle$ is infinite in equilibrium unless the system is truncated at a finite k_{max}. Consider the consequences of (13) under this truncation. First take $K = \beta L^d \hat{E}$, where β is an inverse temperature parameter, L is the side of the cyclic box, and $d = 2$ or 3 is the dimensionality. The resulting distribution is trivially simple (if attention is restricted to single-time averages) and gives equipartition of kinetic energy among the Fourier modes. Nontrivial distributions result if K

is taken more generally as a linear combination of energy and enstrophy ($d = 2$) or of energy and helicity ($d = 3$).

For $d = 2$, the general equilibrium distribution of this form yields [3]

$$\langle y_\alpha^2 \rangle \propto (\xi k_\alpha^2 + \beta)^{-1}, \qquad (14)$$

where ξ is a thermodynamic potential for enstrophy. Equation (14) gives a variety of behaviors, depending on the relative values of β and ξ. The cases $\xi = 0$ and $\beta = 0$ give energy and enstrophy equipartition, respectively. The most interesting distributions are those with negative temperatures ($\beta < 0$) in which there can be enormously enhanced excitation of the modes with the smallest k [4]. The equilibrium (14) is formally identical with that of a finite gas of free bosons if the identification is made of flow kinetic energy with boson particle number and of enstrophy with boson kinetic energy. In contrast to the two-dimensional boson gas, the Navier–Stokes system includes nonlinear interactions of the modes which, although they do not affect the form of the constants of motion, give a mechanism of relaxation toward equilibrium.

For $d = 3$, the equilibria with nonzero helicity give [5]

$$\langle y_\alpha^2 \rangle \propto \beta/(\beta^2 - \xi k_\alpha^2). \qquad (15)$$

There are no negative-temperature equilibria in this case. However, the corresponding equilibria for hydromagnetic flows show highly interesting effects, including strong enhancement of magnetic energy above equipartition with kinetic energy when the magnetic helicity (integrated dot product of magnetic field and vector potential) is nonzero [6].

It should be stressed that the interaction coefficients $A_{\alpha\beta\gamma}$ do not enter at all in the constants of motion we have used to form absolute equilibria and, hence, they do not enter any single-time averages over the equilibrium distributions. Every modification of the A's which preserves (7), (9), and (11) yields the same equilibria; the A's could even be functions of time. As we shall point out later, this is in marked contrast to the nonequilibrium distributions which correspond to actual turbulence; they depend essentially on the detailed form of the A's. Even in the present absolute equilibrium case, however, many-time averages, such as those in (12), do depend on the A's.

4. Nonequilibrium: Cascade Phenomena

The observed states of high Reynolds-number turbulence are very far from the absolute equilibrium ensembles. At wavenumbers between those where energy is fed into the turbulence and the higher wavenumbers where the energy is dissipated by viscosity, the energy spectrum of three-dimensional turbulence without helicity is observed to be close to the form

$$E(k) \doteq C\epsilon^{2/3}k^{-5/3} \tag{16}$$

predicted in 1941 by Kolmogorov and others [7]. Here ϵ is the rate of energy dissipation per unit mass and $E(k)$ is normalized by

$$\int_0^\infty E(k)\,dk = \langle \hat{E} \rangle. \tag{17}$$

Since the mode density is $\propto k^2$, (16) implies

$$\langle y_\alpha^2 \rangle \propto k_\alpha^{-11/3} \tag{18}$$

in contrast to the equipartition behavior in absolute equilibrium.

Equation (16) represents a proposed form for the so-called inertial range of wavenumbers, in which energy is transferred from small to large wavenumbers by the process of vortex stretching. The chaotic nature of turbulence tends to separate any two fluid elements initially near each other; consequently there is a tendency to stretch initial vorticity distributions into ever-lengthening and thinning vortex ribbons, until viscosity stops the thinning. By Kelvin's circulation theorem, the circulation about a curve encircling the ribbon cross-section remains constant so that, as the cross section decreases under stretching, the fluid in the vortex ribbon must spin harder. The combination of spin-up and thinning means an increase of enstrophy and, in the Fourier representation, a transfer of energy from lower to higher wavenumbers.

Interactions between wavenumbers that differ by as much as two octaves contribute strongly to the inertial-range energy transfer. This shows how strong is the disequilibrium represented by (18). A comparable disequilibrium in a gas of interacting particles would require that the temperature change by its own order of magnitude in a distance comparable to the range of the interaction potential. The analogy here is between Fourier modes of the turbulence and particles of the gas.

Equation (16) follows from similarity-scaling arguments which Kolmogorov originally stated in x space. If $1/r$ is in the inertial range of wavenumbers, Kolmogorov's 1941 similarity hypothesis says that the n-variate distributions of the velocity differences $\tilde{\mathbf{u}}(\mathbf{x} + \mathbf{r}, t) - \tilde{\mathbf{u}}(\mathbf{x}, t)$ are universal isotropic functions solely of the vectors r and of the mean rate of dissipation ϵ. Dimensional analysis then leads immediately to the following forms for moments, where the B_n are universal constants:

$$\langle |\tilde{\mathbf{u}}(\mathbf{x} + \mathbf{r}, t) - \tilde{\mathbf{u}}(\mathbf{x}, t)|^n \rangle = B_n (\epsilon r)^{n/3}. \tag{19}$$

On the assumption that the spectrum in the dissipation range of wavenumbers falls off rapidly enough with k (19), with $n = 2$, leads by Fourier transformation to (16), and C is determined by B_2.

The physics behind Kolmogorov's similarity hypothesis is rooted in the idea of localness of energetic interaction in scale size or wavenumber. Consider a vortex structure of given scale size. It seems plausible that excitation on much larger scales should act principally to convect the structure, without significantly stretching it. On the other hand, the shears associated with excitation on much smaller scales than the given one should tend to cancel out over the extent of the vortex structure. This suggests that distortions of the given vortex structure, and, therefore, transfer of the energy of the given structure to higher wavenumbers, should be due principally to interactions with other structures of similar scale.

This consideration leads to the idea of the energy cascade as a kind of diffusion process in k space. Kolmogorov's 1941 hypothesis implies that in this process all detailed statistical information about the source of energy in the large spatial scales is lost: the only macroscale parameter which controls the cascade is ϵ, the rate of cascade. This parameter enters because the cascade is conservative.

An elementary consistency check on the localness argument comes easily from (16). It is readily found by Fourier transformation that when (16) holds the space-averaged shear (velocity gradient) over a domain of size r comes from wavenumbers $k \sim 1/r$.

According to (2), the total dissipation rate ϵ from all wavenumbers must satisfy

$$\epsilon/\nu = 2 \int_0^\infty k^2 E(k)\, dk. \tag{20}$$

Thus the inertial range spectrum (16) must cut off at a wavenumber of the order of $k_d = (\epsilon/\nu^3)^{1/4}$. As a corollary to the inertial-range hypothesis, Kolmogorov further proposed in 1941 that velocity-difference statistics for difference vectors $\mathbf{r}$ satisfying $r \lesssim k_d^{-1}$ should be universal functions of the difference vectors, ϵ, and ν. In particular, the spectrum in the dissipation range of wavenumbers should satisfy

$$E(k) = C\epsilon^{2/3}k^{-5/3}f(k/k_d), \qquad (21)$$

where $f(0) = 1$ to yield (16) in the inertial range $k \ll k_d$ and

$$C\int_0^\infty x^{1/3}f(x)\,dx = 1$$

to yield (20).

Measurements of high Reynolds-number flows support the similarity law (21) well and are consistent with universality of C over a wide range of ϵ. However, the experimental results on higher statistics are not universal in the sense of Kolmogorov's 1941 hypotheses. Instead measured values of the normalized moments

$$\langle |\tilde{\mathbf{u}}(\mathbf{x}+\mathbf{r},t) - \tilde{\mathbf{u}}(\mathbf{x},t)|^n \rangle / \langle |\tilde{\mathbf{u}}(\mathbf{x}+\mathbf{r},t) - \tilde{\mathbf{u}}(\mathbf{x},t)|^2 \rangle^{n/2}$$

increase dramatically with both n and $1/r$ at high Reynolds numbers, indicating strong scale-dependent intermittency at small scales. Moreover, at fixed ϵ and r, the intermittency at a scale r in the inertial range appears to increase with L_0/r, where L_0 is a length scale of the largest eddies in the flow. The data are strongly suggestive, but they are not conclusive because it is uncertain whether the Reynolds numbers attained are large enough to produce a truly asymptotic regime.

Already in 1962, when the data on small-scale intermittency were just beginning to appear, Kolmogorov [8] and Oboukhov [9] suggested a modification of the 1941 theory in which small-scale statistics depended on L_0 as well as on ϵ and ν. The basic physical idea behind the modified theory, as developed by these authors and others later, is that the cascade of excitation from large to small scales effectively takes place in finite logarithmic steps in wavenumber or scale size. The number of steps required for excitation to cascade from L_0 to r is then measured by $\ln(L_0/r)$. Now it is assumed that each cascade step consists of a breakdown process of turbulent eddies of given scale into daughters of smaller scale and that the breakdown is stochastic in nature and statistically independent of previous breakdowns in an appropriate sense. The result is a cumulative increase in intermittency at each cascade step.

According to one class of models of this kind [10], the velocity-difference statistics approach log-normal-like forms for small separations and (19) is replaced by

$$\langle |\tilde{\mathbf{u}}(\mathbf{x}+\mathbf{r},t)-\tilde{\mathbf{u}}(\mathbf{x},t)|^n\rangle = B_n(\epsilon r)^{n/3}(r/L_0)^{\mu n(3-n)/2}, \qquad (22)$$

while (16) is changed to

$$E(k) = C\epsilon^{2/3}k^{-5/3}(kL_0)^{-\mu}, \qquad (23)$$

with μ a positive parameter. Equation (22) is reasonably consistent with measurements at several values of n if $\mu \sim 0.05$ is taken. With this value of μ, (23) and (16) are not experimentally distinguishable. An overall analysis of ideas connected with Kolmogorov's 1962 theory and its developments has recently been attempted by the present author [11], and the reader is referred there for further details.

In the context of the physics of vortex stretching, systematic increase of intermittency along the cascade chain means that the vorticity becomes concentrated in an increasingly sparse collection of intense ribbons. An implication of $\mu > 0$ in (22) is that as $\nu \to 0$ with fixed macroscale excitation the dissipation approaches a finite limit but takes place principally in a fraction of the fluid volume that goes to zero in the limit [11].

The exponent in (16) is completely fixed by dimensional analysis. The Navier–Stokes dynamics enter only to the extent of the conservation property, which makes ϵ a meaningful parameter, and of some qualitative support which (2) gives to the idea of localness of energetic interaction. On the contrary, the value of μ in (23), if it is nonzero, must depend on the detailed structure of the coefficients $A_{\alpha\beta\gamma}$ in (5) if the general idea of stepwise cascade, which led to (22) and (23), is valid. To see this consider the effects of varying the A's. Suppose that $A_{\alpha\beta\gamma}$ is replaced in (5) by

$$A'_{\alpha\beta\gamma} = f(k_\alpha, k_\beta, k_\gamma) A_{\alpha\beta\gamma}, \qquad (24)$$

where f is a symmetrical function of its three wavenumber arguments. Then the conservation identity (7) and the corresponding identity for helicity conservation are unaffected. Suppose that f depends on the ratios of the three wavenumber arguments and varies smoothly from the value one when the ratio of any two wavenumbers is large to a value greater than one when all the wavenumbers are nearly the same. This alteration of (5) decreases the effective step size of the cascade by emphasizing the contribution of local interactions in wavenumber. If the build-up of intermittency depends on the effective number of logarithmic steps, then this build-up, and the value of μ, should be affected by the alteration.

The build up of intermittency in the inertial-range cascade is highly interesting from the point of view of fundamental statistical mechanics. Suppose that systematically increasing intermittency really does occur asymptotically as $\nu \to 0$. Consider the evolution of a statistical ensemble whose initial distribution is multivariate-normal, with all energy concentrated in wavenumbers the order of k_0. This energy will cascade to larger k with the eventual production of very strong departure from normal statistics because of the intermittency at small scales, if ν is very small. Now instead of small ν, take $\nu = 0$ and truncate the system at a high wavenumber $k_{\max}$, as in Section 2. It is plausible that the cascade and intermittency increase should proceed as in the finite ν case until the wavenumbers $k \sim k_{\max}$ are appreciably excited (we may take $k_{\max} = k_d$, the viscous cut-off wavenumber for the finite ν case). After wavenumbers $k \sim k_{\max}$ are strongly excited, it is plausible that the system should evolve toward the absolute equilibrium ensembles of Section 3, which again are normal. The evolution toward absolute equilibrium is in fact supported by recent computer experiments [12]. We have then the paradoxical situation of a normally distributed final state obtained from a normally distributed initial state through a highly nonnormal transient stage. The transient intermittency means that, in effect, the system chooses preferred paths in phase space for its evolution toward eventual equilibrium rather than taking all *a priori* possible paths with equal probability. The behavior is perhaps somewhat analogous to the intermittent spatial distribution of water flowing down a mountainside, with the formation of canyons where the flow is preferred. The question arises of whether intermittency build-up during approach to equilibrium is a general characteristic of systems where the dynamical equations couple many degrees of freedom with different characteristic scales.

Both the 1941 and the 1962 Kolmogorov theories imply a mathematically interesting behavior of the inviscid Navier–Stokes equation (Euler equation). This is that smooth initial conditions can lead to the formation of lines or surfaces of infinite shear in finite times of evolution. It is easiest to see this by assuming the 1941 theory, associated with (16), and estimating the time required for the energy cascade to extend itself from k_0 to the viscous cut-off wavenumber $k_d = (\epsilon/\nu^3)^{1/4}$. According to that theory, the characteristic time for breakdown of structures of wavenumber k into structures of wavenumber the order of, say, $2k$ can depend only on ϵ and k and by dimensional analysis must therefore be of order $(\epsilon k^2)^{-1/3}$. This corresponds physically to the eddy-circulation time $(v_k k)^{-1}$ where v_k is the rms velocity associated with excitation in wavenumbers $\sim k$. The times required for successive doublings of the highest excited wavenumber then go like $2^{-2n/3}$, where n is the number of doublings, so that the time required to reach k_d approaches a finite limit as $k_d \to \infty$ ($\nu \to 0$). The 1962 theory involves only minor modifications.

To the author's knowledge, no solution of the Euler equation exhibiting such singularity formation has yet been demonstrated. It may be that such solutions, assuming they exist, are intrinsically complicated.

5. Eddy Viscosity and Renormalization

The nonequilibrium Navier–Stokes system presents severe difficulties for systematic theoretical treatment. A number of things can be done formally, but they seem always to lead to convergence problems and to overwhelming complications of analysis if carried far in a systematic way. Moreover, we shall point out that peculiar features of the turbulence problem make it hard to set up a formalism which is physically appropriate.

Hopf [13] has formulated the Liouville equation governing the evolution of the single-time, multivariate probability density P for the Navier–Stokes system. It follows readily from (5) that this equation can be written as

$$\partial P/\partial t + \mathscr{L}P = 0,$$
$$\mathscr{L} = \sum_\alpha (\partial/\partial y_\alpha)(-\nu_\alpha y_\alpha + \sum_{\beta\gamma} A_{\alpha\beta\gamma} y_\beta y_\gamma). \tag{25}$$

A related equation for the many-time probability density has also been described [14]. A formal solution for P in terms of the exponentiated operator $\mathscr{L}$ comes immediately from (25), but so far it seems not to have led to useful results.

The formal procedure that seems to have been most successful to date involves the use of renormalized perturbation series which generalize those used in quantum field theory. This approach has grave deficiencies, which we shall discuss, but it does have the advantage of describing in a natural way a physical phenomenon noticed early in the study of turbulence: namely, that the small scales of turbulence react on the large scales like a dynamical or eddy viscosity that augments the molecular viscosity ν.

Consider the problem of following the evolution of the Navier–Stokes system when the initial statistical ensemble is Gaussian. The quadratic energy constant, whose value can only be decreased by the viscous damping term, assures that this problem is well posed provided the system is truncated at a wavenumber $k_{\max}$. It is to be expected on physical grounds that this truncation should have negligible effects if $k_{\max}$ is taken large compared to whatever turns out to be the characteristic viscous cut-off wavenumber.

Let an ordering parameter λ (eventually to be set equal to one) be inserted as a factor in the right-hand side of (5). Then the initial value problem can be formally solved by expanding all moments and other statistical functions in powers of λ, thereby treating the nonlinear interaction as a perturbation on the linear viscous decay of the velocity field. If the initial time is $t = 0$, the linear equations have the solution

$$y_\alpha^{(0)}(t) = y_\alpha(0) \exp(-\nu_\alpha t) \qquad (26)$$

and the Green's function matrix

$$G_{\alpha\beta}^{(0)}(t, t') = \delta_{\alpha\beta} \exp[-\nu_\alpha(t - t')]. \qquad (27)$$

If the solution $y_\alpha(t)$ of the full equation is expanded in powers of λ, the coefficient of λ^n is homogeneous of degree $n + 1$ in the $y^{(0)}$ and homogeneous of degree n in the $G^{(0)}$, and it involves an n-fold time integration. Each coefficient consists of a number of terms, which rises rapidly with n according to the branching properties of (5) under iteration.

The λ expansion of any moment of the y distribution can be formed by multiplying out the λ expansions of the factors of the moment and then averaging over the assumed Gaussian distribution of the $y_\alpha(0)$. According to a well-known rule, moments of a Gaussian distribution are reducible to sums of products of covariances by taking all possible pairings of the factors (odd-order moments vanish). This reduction, together with the branching topology of the λ expansions of the y

factors, leads to a diagram representation of the λ expansion of the moment [15].

Two quantities of fundamental importance are the covariance

$$Y_{\alpha\beta}(t, t') = \langle y_\alpha(t) y_\beta(t') \rangle$$

and the infinitesimal response matrix

$$G_{\beta\alpha}(t, t') = \langle \delta y_\beta(t)/\delta f_\alpha(t') \rangle,$$

where $f_\alpha(t)$ is an infinitesimal driving term added to the right-hand side of (5). The perturbation expansions of these quantities contain only even powers of λ. In the expansion of $Y_{\alpha\beta}(t, t')$, the coefficient of λ^{2n} is homogeneous of degree $n + 1$ in the unperturbed covariances $Y^{(0)}$ and degree $2n$ in the $G^{(0)}$, while these degrees become n and $2n + 1$, respectively, in the expansion of $G_{\alpha\beta}(t, t')$.

The line-renormalized expansions for Y and G are reworkings of the perturbation expansions such that the exact Y and G functions, rather than the unperturbed ones, appear in the coefficients of the various terms and, in the language of field theory, all diagrams with self-energy parts are dropped. The renormalized expansions are most easily constructed as follows. First, write the equations of motion for G and Y obtained from (5) (with λ inserted):

$$(\partial/\partial t + \nu_\alpha) G_{\alpha\mu}(t, t') = 2\lambda \sum_{\beta\gamma} A_{\alpha\beta\gamma} \langle y_\beta(t) \, \delta y_\gamma(t)/\delta f_\mu(t') \rangle, \tag{28}$$

$$(\partial/\partial t + \nu_\alpha) Y_{\alpha\mu}(t, t') = \lambda \sum_{\beta\gamma} A_{\alpha\beta\gamma} \langle y_\beta(t) y_\gamma(t) y_\mu(t') \rangle, \tag{29}$$

where the symmetry of $A_{\alpha\beta\gamma}$ is used in (28). Second, write out the perturbation expansions of the right-hand sides, reducing all averages to expressions in $G^{(0)}$ and $Y^{(0)}$. Third, discard all diagrams with self-energy parts in these expansions. Fourth, in the surviving diagrams replace every $G^{(0)}$ and $Y^{(0)}$ by a G and Y with the same indices and time arguments.

The renormalized expansions thus generated can be formally justified in several ways: by straightforward partial summation of the original expansions, by considering effects of variations of the A's [16], or by functional techniques [17]. The result is that (28) and (29) are converted into a pair of coupled infinite-series, integro-differential equations for G and Y.

The final form taken by the G equation is

$$(\partial/\partial t + \nu_\alpha) G_{\alpha\mu}(t, t') + \sum_\delta \int_{t'}^{t} \eta_{\alpha\delta}(t, s) G_{\delta\mu}(s, t') \, ds = 0, \tag{30}$$

where

$$\eta_{\alpha\delta}(t, s) = 4\lambda^2 \sum_{\beta\gamma\tau\sigma} A_{\alpha\beta\gamma} A_{\tau\sigma\delta} G_{\gamma\tau}(t, s) Y_{\beta\sigma}(t, s) + \text{higher terms}. \tag{31}$$

The higher terms in (31) involve more G and Y factors and integrations over intermediate times.

Equations (30) and (31) have the simplest explicit form when the turbulence is isotropic. In this case, taking the box size L very large, we define covariance and response scalars by

$$(L/2\pi)^3 \langle u_i(\mathbf{k}, t) u_j^*(\mathbf{k}, t') \rangle = \tfrac{1}{2} P_{ij}(\mathbf{k}) U(k, t, t'), \tag{32}$$

$$\langle \delta u_i(\mathbf{k}, t)/\delta f_j(\mathbf{k}, t') \rangle = P_{ij}(\mathbf{k}) G(k, t, t'), \tag{33}$$

where $f_i(\mathbf{k})$ is an infinitesimal forcing term added to the right-hand side of (2). Then, with the use of $(2\pi/L)^3 \sum_\mathbf{k} \to \int d^3k$, (30) and (31) may be written as

$$(\partial/\partial t + \nu k^2) G(k, t, t') + \int_{t'}^{t} \eta(k, t, s) G(k, s, t') \, ds = 0, \tag{34}$$

$$\eta(k, t, s) = \pi \lambda^2 k \iint_\Delta pq \, dp \, dq \, b_{kpq} G(p, t, s) U(q, t, s) + \text{higher terms}. \tag{35}$$

In (35), $\iint_\Delta$ denotes integration over all parts of the p, q plane such that k, p, and q can form the legs of a triangle, and b_{kpq} is given by

$$b_{kpq} = (p/k)(xy + z^3), \tag{36}$$

where x, y, and z are the interior-angle cosines opposite k, p, and q, respectively. The corresponding integro-differential equation for U is

$$(\partial/\partial t + \nu k^2) U(k, t, t') = S(k, t, t'), \tag{37}$$

$$S(k, t, t') = \pi \lambda^2 k \iint_\Delta pq \, dp \, dq \, b_{kpq} \left[\int_0^{t'} G(k, t', s) U(p, t, s) U(q, t, s) \, ds \right.$$

$$\left. - \int_0^{t} U(k, t', s) G(p, t, s) U(q, t, s) \, ds \right] + \text{higher terms}. \tag{38}$$

The η term in (34) represents a dynamical damping with memory. Its resemblance to viscous damping shows most clearly for the inviscid truncated system in the energy-equipartition absolute equilibrium. Then (12) yields

$$U(k, t, t') = U(k, t, t) \, G(k, t, t') \qquad (t \geqslant t'), \qquad (39)$$

$G(k, t, t') = G(k, t - t')$, a function only of time difference, and $U(k, t, t) = U$, a value independent of k. [We note that $E(k, t)$, the energy spectrum previously introduced, is related to $U(k, t, t')$ by $E(k, t) = 2\pi k^2 U(k, t, t)$.] Then (34) and (35) can be written

$$\partial G(k, t)/\partial t + \int_0^t \eta(k, t - s) \, G(k, s) \, ds = 0, \qquad (40)$$

$$\eta(k, t) = \pi \lambda^2 k U \iint_\Delta pq \, dp \, dq \, b_{kpq} G(p, t) \, G(q, t) + \text{higher terms}. \qquad (41)$$

These equations are opaque because of the complicated coefficient b_{kpq}. But it can be shown that the solution of (40), (41) is qualitatively similar to that of the simplified set

$$\partial G(k, t)/\partial t + k^2 \int_0^t \eta(t - s) \, G(k, s) \, ds = 0, \qquad (42)$$

$$\eta(t) = cU \int_0^{k_{\max}} [G(p, t)]^2 p^2 \, dp + \text{higher terms}, \qquad (43)$$

where c is a numerical factor and we have finally set $\lambda = 1$.

The solution of (42) and (43) has the following properties. For small t most of the contribution to $\eta(t)$ comes from $p \sim k_{\max}$. For large t, $\eta(t)$ falls off like $t^{-3/2}$ and the dominant contributions come from $p/k_{\max} \sim (t/t_*)^{-1/2}$, where $t_* = (Uk_{\max}^5)^{-1/2}$. For $k \ll k_{\max}$, the decay time of $G(k, t)$ is $\gg t_*$ and the solution of (42) is asymptotically

$$G(k, t) = \exp[-\nu' k^2 t], \qquad \nu' = \int_0^\infty \eta(s) \, ds. \qquad (44)$$

These properties are verified in the following sense. They follow directly if only the explicitly shown term in (43) is retained. They are then consistent, term-by-term, with the higher-order terms in the renormalized series.

Thus, subject to questions about convergence properties of the series, the truncated Euler system in thermal equilibrium exhibits a dynamical damping of low-wavenumber disturbances just like the viscous damping of the Navier–Stokes system at zero temperature. If $k_{\max}$ is taken as some kind of intermolecular spacing scale or mean free path, then the truncated Euler system constitutes a nontrivial model of a molecular liquid, with the equilibrium excitation corresponding to normal molecular thermal energy. The $t^{-3/2}$ tail on $\eta(t)$ then corroborates similar tails on transport coefficients deduced from kinetic theory considerations [18].

If the analysis is repeated for two dimensions, (42) again emerges while (43) is replaced by

$$\eta(t) = cU \int_0^{k_{\max}} [G(p, t)]^2 p\, dp + \text{higher terms.} \qquad (45)$$

Again c is a numerical factor and U is now related to $E(k)$ by $E(k) = \pi k U$. In this case, the tail on $\eta(t)$ goes like t^{-1} and there exists no effective viscosity coefficient for low k, again corroborating kinetic-theory results. The divergence of two-dimensional transport coefficients has a physical interpretation in terms of the dynamics of two-dimensional turbulence. It is well known [4] that, because of enstrophy conservation, the transport of kinetic energy in two-dimensional turbulence is toward lower instead of toward higher wavenumbers. This suggests that the whole idea of a transport coefficient based on the small-scale excitation is inapplicable. Moreover, if the truncated Euler system is again viewed as a model of a molecular fluid, the existence of the general equilibrium distribution (14), of which energy equipartition is only a limiting case, suggests that the entire equilibrium statistical mechanics of the two-dimensional fluid may exhibit deep-lying anomalies.

6. Convection Invariance and Lagrangian Coordinates

Let us now apply (34)–(38) to isotropic Navier–Stokes turbulence with finite ν, taking physically appropriate initial spectra in which the excitation per mode falls off rapidly with wavenumber, in contrast to the equipartition equilibrium spectrum. There are essential differences from the Euler system in absolute equilibrium. First, the fluctuation–dissipation relation (39) no longer holds, even approximately, for most k, so (34) and (37) must be solved together as a pair. Further, a basic distinction must be made between the dynamical damping function $\eta(k, t, s)$ which appears in the Green's function equation and the effective dynamical damping which describes the erosion of the energy of large scale motions by small-scale motions. The latter is the eddy viscosity usually talked about in turbulence theory. The two kinds of damping can be very different out of absolute equilibrium. If k is small compared to all appreciably excited wavenumbers of the turbulence, then $\eta(k, t, s)$ measures both amplitude decay and energy decay, like an ordinary viscosity. But if substantial energy lies in wavenumbers below k, then $\eta(k, t, s)$ may have little relation to energy-transfer dynamics. This is because $G(k, t, t')$ is an averaged Green's function; its decay arises not only from decay of amplitudes in individual realizations but also from phase decorrelation.

Consider the effect at wavenumber k of excitation at much smaller wavenumbers. Energy transfer in wavenumber involves distortion in x space; it depends essentially on the intensity of shear (velocity gradient) in the low-wavenumber excitation. On the other hand, phase decorrelation of Fourier amplitudes at wavenumber k depends essentially on the magnitude of the velocity itself of the low-wavenumber excitation. To see this most clearly, take the limit case where the affecting excitation is at zero wavenumber and consists of a spatially uniform velocity field which is isotropically and Gaussianly distributed over ensemble. This field convects all structures without distortion in every realization and therefore has no effect at all on energy transfer. In other words, energy transfer is invariant to random Galilean transformation of a statistically homogeneous turbulence. But if $\mathbf{v}$ is the convecting velocity, it induces in all Fourier amplitudes a phrase variation of the form $\exp(i\mathbf{v} \cdot \mathbf{k}t)$. If this is averaged over Gaussian v, the resulting decay of mean amplitude is $\exp(-\frac{1}{2}v_0^2 k^2 t^2)$, where v_0 is the rms value of each vector component of v.

The effects of random Galilean transformation appear with qualitative correctness in each order of the renormalized expansion (35) for $\eta(k, t, s)$. But the situation is different with energy transfer, which is given by the time-diagonal function $S(k, t, t)$. Each order of the renormalized expansion (38) for $S(k, t, t)$ describes correctly the fact that $\mathbf{v}$ transfers no energy because it does not distort; this is expressed in the explicitly shown order by cancellation of the two integrals in square brackets when p or $q = 0$. However there is a more subtle, but still essentially important, violation of random Galilean invariance. The Green's functions on the right-hand side of (38) all have decay times that are affected by $\mathbf{v}$. The result is that, in each order of the renormalized expansion, the magnitude of the terms is typically depressed by introducing **v**. The invariance of energy transfer is recovered only if the whole series is summed. In contrast, the primitive perturbation in λ exhibits proper invariance in each order. The rearrangement of this series into the renormalized series groups together partial sets of terms of different orders in the primitive expansion, and this is mathematically how the trouble arises.

The physical significance of the failure of random Galilean invariance in the renormalized series is that any truncation of these series, or any approximant based on a truncation, gives a spurious effect of convection on energy transfer at any wavenumbers such that a lot of kinetic energy lies at lower wavenumbers. In particular, such truncations cannot correctly describe the inertial-range dynamics. Instead of (16), or (23) with positive μ, they give an inertial range of the form [19]

$$E(k) = C(\epsilon v_0)^{1/2} k^{-3/2}, \qquad (46)$$

where v_0 is now the rms component of the total turbulent velocity.

In addition to the line renormalization we have described, it is also possible to carry out a vertex renormalization for the turbulence equations [16, 17]. It is quite complicated because of the statistical disequilibrium. In addition to the distinct functions G and U, three distinct vertex functions must be carried. The vertex-renormalized expansions also violate random Galilean invariance in each order and lead to (46) if truncated [19]. In fact, the vertex renormalization has so far shown no qualitative advantage over line renormalization to compensate for its enormously increased complexity.

The troubles with random Galilean invariance lie deeper than the particular techniques used for renormalization. They originate in the underlying representation of the dynamical system by an Eulerian velocity field, measured relative to fixed coordinates. The natural separation of convection and distortion effects calls for a Lagrangian coordinate system that moves with the fluid and thereby transforms away the effects of uniform convection. But there are two severe problems in using a Lagrangian representation. First, the pressure and viscous-damping terms of the Navier–Stokes equation display highly intractable kinds of nonlinearity in Lagrangian coordinates. Second, and more important in principle, Lagrangian coordinates label fluid elements according to their Cartesian coordinates at some initial instant. After a substantial time of evolution of a complicated turbulent motion, this labeling has such a mixed-up relation to current positions of the fluid elements that it becomes inappropriate. What is needed is some way of updating the reference time for the Lagrangian coordinate system.

A possible resolution of both difficulties may lie in adopting as the fundamental field variable a generalized velocity field $\mathbf{u}(\mathbf{x}, t \mid t')$ which has both Eulerian and Lagrangian characteristics [20]. This field is defined as the velocity measured at time t' in that fluid element which passes through $\mathbf{x}$ at time t. For $t = t'$ it is the ordinary Eulerian field $\mathbf{u}(\mathbf{x}, t)$, while for $t = 0$ it is the usually defined Lagrangian velocity. The evolution of $\mathbf{u}(\mathbf{x}, t \mid t')$ is determined by differential equations involving derivatives with respect to t as well as t', and these equations present only the same kinds of nonlinearity already present in the Navier–Stokes equation. The degree of nonlinearity is reduced over that of the customary Lagrangian equations by enlarging the dimensionality of the function space.

The generalized velocity-field provides a tool, in principle at least, for reworking the renormalized series so that random Galilean invariance is preserved in each order. So far, this has been carried out explicitly only for the lowest-order terms, which are written out in (35) and (38). The technique used [20] is first to carry out the renormalized expansion for the response functions and covariance of the full generalized velocity-field. Then integrals over past history, which appear in (34)–(38), are altered so as to go back in time along particle trajectories rather than at fixed points in space. At each order this gives rise to additional higher-order terms and amounts to a re-ordering of the series, but one which has no simple expression in terms of familiar diagram representations. The lowest-order approximation retains all essential invariance properties and yields (16) [20]. In higher orders, the term-by-term re-working is unacceptably clumsy, and it is to be hoped that some powerful functional technique will be discovered to replace it.

The failure of standard renormalized perturbation theory to preserve random Galilean invariance, which means physically an inability of the theory to handle problems where there is a wide range of scales excited, is not unique to turbulence theory. A closely analogous difficulty arises in quantum field theory. This shows up most strikingly in the classical limit $h \to 0$ of the field theory of many-body systems. In this limit the wavelengths of wavefunctions become vanishingly small compared to spatial scales of the potentials, and the consequence is that scattering cross-sections vanish in every order of renormalized perturbation theory. We shall illustrate this effect, which seems not much appreciated, by means of the simplest example: a quantum-mechanical particle in a random potential.

Consider the motion of a classical particle in a statistically homogeneous, Gaussianly distributed random potential whose rms value is V_0 and whose wavenumber spectrum peaks smoothly around a wavenumber k_0. The typical force is then $F_0 = V_0 k_0$. Let the mean velocity u of the particle be such that the mean kinetic energy is large compared to V_0. The particle will then undergo a diffusion process in momentum space. On time scales large compared to the characteristic time $\tau_0 = (k_0 u)^{-1}$ for penetration of a potential hump, the diffusion of the momentum probability distribution $P(\xi)$ will be well described by

$$\partial P(\xi)/\partial t = \eta \partial^2 P(\xi)/\partial \xi^2, \qquad \eta \sim \tau_0 F_0^2, \tag{47}$$

where the form of the diffusion coefficient η can be obtained by elementary physics and follows already from dimensional analysis. Here ξ ($\ll$ mu) is a momentum component normal to u.

If the same problem is treated quantum-mechanically, the Schrödinger equation is

$$\hbar \partial \psi/\partial t = [-iv(\mathbf{x}) + i(\hbar^2/2m) \nabla^2]\psi, \tag{48}$$

where $v(\mathbf{x})$ is the potential field and m is the particle mass. The renormalized perturbation expansions for this problem [16] are closely analogous to those for the convection of a passive scalar field by a random velocity field. They resemble those for the turbulence problem but have fewer terms because (48) is linear in the field ψ. In the Fourier representation, the lowest-order terms in the line-renormalized equations for the Green's function $G_k(t)$ and the covariance

$$\Psi_k(t, t') = \langle \psi_k(t) \psi_k^*(t') \rangle$$

are

$$[\partial/\partial t + i(\hbar/2m)k^2] G_k(t) = -\hbar^{-1} \int_0^t ds \int d\mathbf{k}' V_{k-k'} G_{k'}(t-s) G_k(s), \tag{49}$$

$$\hbar^2 \, d\Psi_k(t, t)/dt = 2\,\text{Re} \int_0^t ds \int d\mathbf{k}' V_{k-k'} [G_k^*(t-s)\Psi_{k'}(t, s) - G_{k'}(t-s)\Psi_k(s, t)]. \tag{50}$$

Here V_k is the Fourier transform with respect to $\mathbf{x} - \mathbf{x}'$ of $\langle v(\mathbf{x}) v(\mathbf{x}') \rangle$ and (50) is specialized to $t' = t$, which describes the changes in momentum occupancies due to scattering.

Now let $\hbar \to 0$ with everything else kept the same. This is the WKBJ limit. In that limit, the characteristic decay time of $G_k(t)$ is $\sim \hbar/V_0$, a result which is given correctly by (49). Note that this time depends on the magnitude of the potential rather than the force, in analogy to the dependence of the turbulence Green's function on convecting velocity rather than shear. The decay time of $G_k(t)$ measures a random phase shift in the limit rather than decay of amplitudes in individual realizations, and it is irrelevant to the classical physics. But this decay time also enters the right-hand side of (50) and thereby has a spurious effect on momentum transfer closely analogous to the spurious effect of random convection on turbulent energy transfer. The result is that in the WKBJ limit (50) leads to (47), but with the diffusion coefficient given incorrectly by $\sim \hbar V^{-1} F_0^2$. Thus the scattering goes to zero as $\hbar \to 0$. Again as in the turbulence problem, this phenomenon is repeated in every order of the line- or vertex-renormalized perturbation expansions.

A physically acceptable WKBJ limit can be recovered by reworking the expansions to transform away in proper fashion the unwanted phase shifts, in analogy to the procedure described above for turbulence. The procedure involves the introduction of a generalized Schrödinger field $\psi(\mathbf{x}, t \mid t')$ such that $\psi(\mathbf{x}, t) = \psi(\mathbf{x}, t \mid t)$ obeys the Schrödinger equation as before while for unequal time arguments the field satisfies

$$\hbar \partial \psi(\mathbf{x}, t \mid t')/\partial t = -iv(\mathbf{x}) \psi(\mathbf{x}, t \mid t'). \tag{51}$$

$\psi(\mathbf{x}, 0 \mid t)$ corresponds to an interaction-representation wavefunction where $v(\mathbf{x})$, which is diagonal in the $\mathbf{x}$ representation, is regarded as the free Hamiltonian and $-(\hbar^2/2m)\nabla^2$ as the interaction Hamiltonian. This one-time-argument field is analogous to the ordinary Lagrangian velocity in the turbulence problem.

7. What Can We Calculate?

This question is of traditional importance in assessing the value of theoretical approaches. We shall ask it in two parts. First, what can be calculated in principle from renormalized-perturbation-theory-related approaches, or alternatives? Second, which of the answers accessible in principle are actually in reach? The first part involves convergence properties and related matters. The second part involves the economics of computation.

If the Navier–Stokes system is cut off at a finite $k_{\max}$ (a procedure which seems physically very acceptable if ν is nonzero and $k_{\max}$ is large enough), it becomes a finite set of first-order equations, and this implies that the solution $\mathbf{u}(\mathbf{x}, t)$ in any realization with healthy initial conditions has a nonzero radius of convergence if expanded as a power series either in t or in the ordering parameter λ. Moreover, the conservation of the quadratic energy expression by the nonlinear terms insures that the solution exists for all t.

Analogies with some simpler systems that have the same kind of nonlinearity [21] suggest that the radius of convergence in λ is finite in typical realizations and that it tends to zero as $t \to \infty$ or as the velocity amplitude in the realization tends to infinity. This implies that the radius of convergence in λ of a moment like $U(k, t, t')$ is finite if the initial distribution admits no realizations with unbounded amplitudes, but is zero if the initial distribution is normal or is any distribution that places no bounds on amplitudes.

The divergent power series in t or λ may be used directly to compute approximants to statistical functions by using Padé approximants [22, 23], or other techniques more particularly adopted to the turbulence problem [24]. Good results have been obtained in this way for the turbulent diffusion of marked particles by a random velocity field [24], and it is to be hoped that more problems will be attacked in this direct way.

One might think that the renormalized series, which represent infinite partial summations, would be less divergent than the underlying λ expansion. In fact, this is not so. The reducible diagrams whose explicit appearance is removed by renormalization are only a finite fraction of all diagrams and the eventual rapid growth of terms with increase of order is not reduced by either line- or vertex-renormalization [23]. The situation with the renormalized series is in some respects worse than with the primitive perturbation series. When the radius of convergence is zero, a function is not uniquely determined by its power series. It is ambiguous by any function whose power series vanishes. In the case of the primitive λ expansion, uniqueness can be recovered by using the fact that the functions being calculated are weighted integrals over quantities with nonzero radii of convergence in the individual realizations; this permits the use of suitable limiting procedures. But the renormalized expansions lose this direct contact with the individual realizations. Averaging over the Gaussian distribution of initial values must be carried out in the beginning in order to generate the diagrams. The analyticity properties of the renormalized expansions thus are very much in doubt, and almost nothing is really known about them. It cannot be asserted that the functions $\eta(k, t, t')$ and $S(k, t, t')$ which appear in (34) and (37) are single-valued functionals of G and U [23]. Thus there is the possibility that a sequence of approximants constructed from the renormalized expansions (35) and (38) may converge, if at all, to an unphysical branch. Some interesting results have been obtained by applying Padé-approximant and other techniques to the renormalized expansions [22, 23], but what has been done is wholly heuristic.

The remarks above make it unsurprising that, in general, valid approximations to the turbulence dynamics cannot be obtained by truncating the expansions (35) and (38) and using the truncated series in (34) and (37). The solutions typically blow up spectacularly [16]. This is not the case, however, if the truncation is made at the lowest level, retaining only the terms shown explicitly in (35) and (38). The resulting equations, which have been called the direct-interaction approximation [25], are physically selfconsistent. This approximation succeeds for a reason that is unrelated to perturbation theory. It happens that the direct-interaction approximation for the Navier–Stokes system is an *exact* set of statistical equations for a model dynamical system which has the same energy function and Liouville theorem as the Navier–Stokes system [16].

The direct-interaction equations have been integrated numerically for several problems, including the decay of isotropic turbulence at moderate Reynolds numbers [26], turbulent convection at high Prandtl number [27], and diffusion in a random velocity field [28]. In these cases, there is good quantitative agreement with computer simulations, and the approximation is a definite success. When applied to turbulence at high Reynolds numbers, the approximation exhibits the spurious convection effects discussed in detail in Section 6 and leads to the inertial range (46). The Lagrangian reworking of the approximation, as discussed in Section 6, leads to (16) and gives excellent numerical agreement with spectrum measurements at high Reynolds numbers [20]. A detailed discussion of the direct-interaction approximation and related approximations is given by Leslie [29]. Integrations of the direct-interaction equations for flow in a channel and other shear flows have not yet been carried out, but the results on the problems already treated suggest that good numerical results should be obtained in this case at moderate Reynolds numbers and that the spurious convection effects which lead to (46) should not prevent a good description of large scale excitations and transport properties at high Reynolds numbers.

The inaccessibility of the analyticity properties of the renormalized expansion, of which the direct-interaction approximation represents the first term, suggests that another vehicle be sought for constructing improvements to this approximation. Expansions about the direct-interaction approximation with more transparent properties, and simpler structure, can be constructed [24], and they provide a basis for the possible construction of sequences of convergents. Little has been done so far in this direction.

Now let us turn to questions of computing economy. At the moderate Reynolds numbers where the comparisons have made, integrations of the direct-interaction equations have required roughly two orders of magnitude less computing time than direct integration of realizations of the Navier–Stokes equation, when both are carried out for several characteristic decay times of the total turbulent energy. As Reynolds number rises, this factor of advantage rises also, because the computer simulations of the Navier–Stokes equation must treat all degrees of freedom in detail, while the statistical functions which enter the direct-interaction equations are smooth functions of wavenumber and time and can be represented by relatively few numbers. The factor of advantage goes down with decrease in the symmetries of the flow because such decrease does not much affect the simulation technique but does mean that more numbers are needed to describe the statistical functions. For shear flows at moderate Reynolds numbers, the computing economy of the direct-interaction equations over direct simulation may not be much, and simplified approximations may be more practicable [29].

The question of computing economy becomes a serious one of principle when one considers higher approximations formed as convergents to some expansion about the direct-interaction approximation. The terms in any such expansion will be complicated multiple integrals over wavevectors and times. Except at the lowest orders, calculation of the integrals will be feasible only by Monte Carlo methods. Moreover, the magnitude of the Monte Carlo calculation will grow very rapidly with order and soon dwarf that of direct simulation, even at substantial Reynolds numbers. It seems not unfair to say that any statistical theory whose computation requires more work than direct integration of the primitive equations of motion of the system is not the one we really want. At the present time there exists no practicable way of obtaining accurate numerical values of statistical functions at high Reynolds numbers either by computer experiment or by analytical theory. The direct-interaction approximation and its variants appear to be as complicated an analytical theory as can presently be computed, and higher approximations are not now practicable. It is to be hoped that new approaches, perhaps combining computer experiment and statistical theory in some essential way, will emerge. The statistics of small scales is an outstanding challenge [30].

Acknowledgment

This work was supported by the Fluid Dynamics Branch, Office of Naval Research, under Contract No. N00014-67-C-0284.

References

1. R. H. Kraichnan, J. Fluid Mech. 56 (1972), 287.
2. R. H. Kraichnan, Phys. Rev. 113 (1959), 1181.
3. R. H. Kraichnan, Phys. Fluids 10 (1967), 1417.
4. R. H. Kraichnan, J. Fluid Mech. 67 (1975), 155.
5. R. H. Kraichnan, J. Fluid Mech. 59 (1973), 745.
6. U. Frisch, J. Léorat, A. Mazure, and A. Pouquet, J. Fluid Mech. (1975), to be published.
7. A. N. Kolmogorov, C. R. Acad. Sci. U.S.S.R. 30 (1941), 301; 538.
8. A. N. Kolmogorov, J. Fluid Mech. 13 (1962), 82.
9. A. M. Oboukhov, J. Fluid Mech. 13 (1962), 77.
10. A. Gurvich and A. Yaglom, Phys. Fluids 10 (1967), S59.
11. R. H. Kraichnan, J. Fluid Mech. 62 (1974), 305.
12. S. A. Orszag and G. S. Patterson, Jr., Phys. Rev. Lett. 28 (1972), 76.
13. E. Hopf, J. Rational Mech. Anal. 1 (1952), 87.
14. R. M. Lewis and R. H. Kraichnan, Comm. Pure Appl. Math. 15 (1962), 397.
15. H. W. Wyld, Ann. Phys. 14 (1961), 143.
16. R. H. Kraichnan, J. Math. Phys. 2 (1961), 124.
17. P. C. Martin, E. D. Siggia, and H. A. Rose, Phys. Rev. A 8 (1973), 423.
18. J. R. Dorfman and E. G. D. Cohen, Phys. Rev. Lett. 25 (1970), 1257.
19. R. H. Kraichnan, Phys. Fluids 7 (1964), 1723.
20. R. H. Kraichnan, Phys. Fluids 9 (1966), 1728.
21. R. H. Kraichnan, in "Dynamics of Fluids and Plasmas" (S. I. Pai, Ed.), Academic Press, London, 1966.
22. R. H. Kraichnan, Phys. Rev. 174 (1968), 240.
23. R. H. Kraichnan, in "The Padé Approximant in Theoretical Physics" (G. Baker and J. Gammel, Eds.), Academic Press, London, 1970.
24. R. H. Kraichnan, J. Fluid Mech. 41 (1970), 189.
25. R. H. Kraichnan, J. Fluid Mech. 5 (1959), 497.
26. J. R. Herring and R. H. Kraichnan, in "Statistical Models and Turbulence" (M. Rosenblatt and C. Van Atta, Eds.), Springer-Verlag, Berlin, 1972.

27. J. R. HERRING, *Phys. Fluids* **12** (1969), 39.
28. R. H. KRAICHNAN, *Phys. Fluids* **13** (1970), 22.
29. D. C. LESLIE, "Developments in the Theory of Turbulence," Clarendon Press, Oxford, 1973.
30. R. H. KRAICHNAN, *in* "Statistical Mechanics: New Concepts, New Problems, New Applications" (S. A. Rice, K. F. Freed, and J. C. Light, Eds.), University of Chicago Press, Chicago, 1972.

On an Explicitly Soluble System of Nonlinear Differential Equations Related to Certain Toda Lattices

M. KAC*

The Rockefeller University, New York, New York 10021

AND

PIERRE VAN MOERBEKE

University of Louvain, Louvain, Belgium

DEDICATED TO STAN ULAM

In our paper [1] we introduced a system of nonlinear differential equations which in a certain sense was an analog of the Korteweg-de Vries equation. Our system was discovered by a probabilistic analogy that, in part at least, also explained why exponential unharmonicity is the natural one.

The present note shows how simply the semi-infinite and the finite cases fit into the inverse scattering scheme yielding at the same time alternative (and independently arrived at) derivations of some results recently obtained by J. Moser [2].

At this point we should like to acknowledge our debt to H. Flaschka [3] who first solved the doubly infinite Toda lattice by applying a discrete version of the inverse scattering problem. The strategy we use is essentially that of Flaschka although the details of execution are somewhat different.

If gives us particular pleasure to include this note in a volume dedicated to S. M. Ulam because it is a direct, though by far not the most illustrious, descendant of the classic Fermi, Pasta, Ulam paper.

---

* Supported in part by AFOSR Grant No. 72287.

1

Consider the system of nonlinear differential equations

$$dR_1/dt = -e^{-R_2(t)}, \tag{1.1a}$$

$$dR_n/dt = e^{-R_{n-1}(t)} - e^{-R_{n+1}(t)}, \qquad n \geq 2, \tag{1.1b}$$

and let $Q(t)$ be the (semi-infinite) matrix

$$Q(t) = \begin{pmatrix} 0 & \tfrac{1}{2}e^{-\tfrac{1}{2}R_1(t)} & 0 & 0 & \cdots \\ \tfrac{1}{2}e^{-\tfrac{1}{2}R_1(t)} & 0 & \tfrac{1}{2}e^{-\tfrac{1}{2}R_2(t)} & 0 & \cdots \\ 0 & \tfrac{1}{2}e^{-\tfrac{1}{2}R_2(t)} & 0 & \tfrac{1}{2}e^{-\tfrac{1}{2}R_3(t)} & \cdots \\ \cdots & \cdots & \cdots & \cdots & \end{pmatrix} \tag{1.2}$$

If $B(t)$ is the antisymmetric matrix whose elements are given by the formulas

$$B_{k,k+2} = \tfrac{1}{2}e^{-\tfrac{1}{2}(R_k(t)+R_{k+1}(t))}, \tag{1.3a}$$

$$B_{k,k-2} = -\tfrac{1}{2}e^{-\tfrac{1}{2}(R_{k-2}(t)+R_{k-1}(t))}, \quad \text{and} \tag{1.3b}$$

$$B_{k,l} = 0 \qquad \text{otherwise}, \tag{1.3c}$$

then we can check immediately that

$$dQ/dt = BQ - QB. \tag{1.4}$$

We can thus use the method of Lax [2] and define unitary matrices $V(t)$ by the equation

$$dV(t)/dt = BV, \qquad V(0) = I. \tag{1.5}$$

We then verify directly that

$$V^{-1}(t) Q(t) V(t) = Q(0), \tag{1.6}$$

and thus if $\psi_0(\lambda; n)$ ($n = 1, 2,...$) is such that

$$Q(0)\, \psi_0(\lambda; n) = \lambda \psi_0(\lambda; n), \tag{1.7}$$

then

$$\psi_t(\lambda; n) = \sum_{j=1}^{\infty} V_{nj}(t)\, \psi_0(\lambda; j) \tag{1.8}$$

clearly satisfy
$$Q(t)\,\psi_t(\lambda; n) = \lambda \psi_t(\lambda; n). \tag{1.9}$$

It now follows from (1.8) that

$$\frac{d}{dt}\psi_t(\lambda; n) = \sum_{j=1}^{\infty} B_{nj}\psi_t(\lambda; j) = B_{n,n+2}\psi_t(\lambda; n+2) + B_{n,n-2}\psi_t(\lambda; n-2). \tag{1.10}$$

For large n and $\lambda = \cos\theta$ we have

$$\psi_t(\lambda; n) \sim A(\theta; t)\,e^{in\theta} + B(\theta; t)\,e^{-in\theta}, \tag{1.11}$$

and since we shall be interested only in solutions for which

$$R_t(n) \to 0, \qquad n \to \infty, \tag{1.12}$$

we must have

$$B_{n,n+2} \sim \tfrac{1}{2}, \qquad B_{n-2,n} \sim -\tfrac{1}{2}, \qquad n \to \infty. \tag{1.13}$$

From (1.10), (1.11) and (1.13) it follows that

$$dA/dt = i(\sin 2\theta)A,$$
$$dB/dt = -i(\sin 2\theta)B$$

so that

$$A(\theta; t) = e^{it\sin 2\theta} A(\theta; 0),$$
$$B(\theta; t) = e^{-it\sin 2\theta} B(\theta; 0).$$

It thus follows that the phase shift $\delta_t(\theta)$ of $\psi_t(\cos\theta; n)$ is given by the formula

$$\delta_t(\theta) = \delta_0(\theta) + t \sin 2\theta. \tag{1.14}$$

Assuming (for the sake of simplicity only) that there are no bound states we can use formula (5.17) of [5] to determine the spectral function $\rho_t(\lambda)$ of $Q(t)$ in terms of the spectral function $\rho_0(\lambda)$ of $Q(0)$ and the result is

$$\rho_t(\lambda) = \begin{cases} 0, & \lambda < -1, \\ \dfrac{\int_{-1}^{\lambda} e^{4t\mu^2}\,d\rho_0(\mu)}{\int_{-1}^{1} e^{4t\mu^2}\,d\rho_0(\mu)}, & -1 \leqslant \lambda \leqslant 1, \\ 1, & \lambda > 1. \end{cases} \tag{1.15}$$

We determine orthogonal plynomials $\phi_t(\lambda; n)$ such that

$$\int_{-\infty}^{\infty} \phi_t(\lambda; m)\, \phi_t(\lambda; n)\, d\rho_t(\lambda) = \delta_{mn} \tag{1.16}$$

($\phi_t(\lambda; n)$ is of degree $n - 1$) and obtain the solution to our problem (1.1) in the formulas

$$\tfrac{1}{2} e^{-\tfrac{1}{2} R_k(t)} = \int_{-\infty}^{\infty} \lambda \phi_t(\lambda; k)\, \phi_t(\lambda; k+1)\, d\rho_t(\lambda). \tag{1.17}$$

It is easy to see that, even if there are bound states, formula (1.15) is still valid in the slightly modified form

$$\rho_t(\lambda) = \frac{\int_{-\infty}^{\lambda} e^{4t\mu^2}\, d\rho_0(\mu)}{\int_{-\infty}^{\infty} e^{4t\mu^2}\, d\rho_0(\mu)}. \tag{1.18}$$

2

Having found the solution of the system (1.1) by an application of a reasonably sophisticated method, we may note that a direct verification is extremely simple.

Denoting by $\mu_{2k}(t)$ the even moments of $\rho_t(\lambda)$, i.e.,

$$\mu_{2k}(t) = \int \lambda^{2k}\, d\rho_t(\lambda), \tag{2.1}$$

we find that the first few orthonormal polynomials $\phi_t(\lambda; m)$ are

$$\phi_t(1; \lambda) = 1,$$

$$\phi_t(2; \lambda) = \frac{\lambda}{(\mu^2(t))^{1/2}},$$

$$\phi_t(3; \lambda) = \frac{\lambda^2 - \mu_2}{(\mu_4 - \mu_2^2)^{1/2}},$$

$$\phi_t(4; \lambda) = \frac{(\mu_2)^{1/2}}{(\mu_2 \mu_6 - \mu_4^2)^{1/2}} \left(\lambda^3 - \frac{\mu_4}{\mu_2} \lambda \right)$$

etc.

Thus, e.g.,

$$\tfrac{1}{2}e^{-\tfrac{1}{2}R_1(t)} = \int \lambda \phi_t(1;\lambda)\,\phi_t(2;\lambda)\,d\rho_t(\lambda) = (\mu_2(t))^{1/2}$$

$$\tfrac{1}{2}e^{-\tfrac{1}{2}R_2(t)} = \int \lambda \phi_t(2;\lambda)\,\phi_t(3;\lambda)\,d\rho_t(\lambda) = \frac{(\mu_4(t) - \mu_2^2(t))^{1/2}}{(\mu_2(t))^{1/2}},$$

$$\tfrac{1}{2}e^{-\tfrac{1}{2}R_3(t)} = \int \lambda \phi_t(3;\lambda)\,\phi_t(4;\lambda)\,d\rho_t(\lambda) = \frac{(\mu_2(t)\mu_6(t) - \mu_4^2(t))^{1/2}}{(\mu_2(t))^{1/2}\,(\mu_4(t) - \mu_2^2(t))^{1/2}},$$

and hence

$$e^{-R_1(t)} = 4\mu_2, \qquad e^{-R_3(t)} = 4\,\frac{\mu_2\mu_6 - \mu_4^2}{\mu_2(\mu_4 - \mu_2^2)}$$

$$R_2(t) = -\log 4 - \log(\mu_4 - \mu_2^2) + \log \mu_2.$$

Now,

$$\frac{dR_2(t)}{dt} = -\frac{d\mu_4/dt - 2(d\mu_2/dt)}{\mu_4 - \mu_2^2} + \frac{d\mu_2/dt}{\mu_2}$$

and

$$d\mu_2/dt = 4(\mu_4 - \mu_2^2),$$
$$d\mu_4/dt = 4(\mu_6 - \mu_2\mu_4).$$

It is thus easy to see that

$$dR_2(t)/dt = e^{-R_1(t)} - e^{-R_3(t)}$$

and that to check (1.1) in general one only needs the easily verifiable formula

$$d\mu_{2k}/dt = 4(\mu_{2k+2} - \mu_{2k}\mu_2). \tag{2.2}$$

It now becomes clear that for *every distribution function* $\rho_0(\lambda)$ and a correspondingly defined $\rho_t(\lambda)$ (see formula (1.15)) the formulas (1.17) provide the solution of the system (1.1).

If, in particular, $\rho_0(\lambda)$ is purely discontinuous with jumps $\Delta_1, \Delta_2, ..., \Delta_{N-1}$ at $\lambda_1 < \lambda_2 < \cdots < \lambda_{N-1}$ [1], $(\Delta_1 + \Delta_2 + \cdots + \Delta_{N-1} = 1)$ $\rho_t(\lambda)$ is also purely discontinuous with jumps at the same points (i.e., $\lambda_1, \lambda_2, ..., \lambda_{N-1}$) but with the jump at λ_i given by the expression

$$\Delta_i(t) = \frac{e^{4\lambda_i^2 t}\,\Delta_i}{\sum_{j=1}^{N-1} e^{4\lambda_j^2 t}\,\Delta_j}. \tag{2.3}$$

[1] It should be clear that the λ's come in positive-negative pairs so that $\lambda_1 = -\lambda_N$, $\lambda_2 = -\lambda_{N-1}$, etc.

It is clear that the eigenvalues λ_i are the roots of $\phi_t(\lambda; N)$, i.e.,

$$\phi_t(\lambda_i; N) = 0, \quad i = 1, 2, \ldots, N-1, \tag{2.4}$$

and hence

$$\tfrac{1}{2} e^{-\tfrac{1}{2} R_N(t)} = \int \lambda \phi_t(\lambda; N-1) \phi_t(\lambda; N) \, d\rho_t(\lambda)$$
$$= 0. \tag{2.5}$$

In this way we arrive at the solution of the finite system

$$dR_1/dt = -e^{-R_2(t)}, \tag{2.6a}$$

$$dR_k/dt = e^{-R_{k-1}(t)} - e^{-R_{k+1}(t)} \quad k = 2, \ldots, N-1, \tag{2.6b}$$

$$dR_{N-1}/dt = e^{-R_{N-2}(t)}, \tag{2.6c}$$

which has been suggested by the application of the inverse scattering problem to the solution of the infinite system (1.1).

3

We shall now show how the solution of the system (2.6) yields also the solution to the finite Toda chain with two free ends, a problem that has been recently solved by Moser [2].

Let N be even ($N = 2n$) and set

$$r_k(t) = R_{2k}(t) + R_{2k+1}(t), \quad k = 1, 2, \ldots, n-1, \tag{3.1}$$

$$p_k = -(e^{-R_{2k}(t)} + e^{-R_{2k-1}(t)}) + \alpha, \quad k = 1, 2, \ldots, n-1, \tag{3.2a}$$

$$p_n = -e^{-R_{2n-1}(t)} + \alpha, \tag{3.2b}$$

where α is to be defined later on.

Let us finally set

$$r_k = q_{k+1} - q_k, \quad k = 1, 2, \ldots, n-1, \tag{3.3}$$

and verify at once that

$$dr_k/dt = (dq_{k+1}/dt) - (dq_k/dt) = p_{k+1} - p_k, \quad k = 1, 2, \ldots, n-1. \tag{3.4}$$

We shall also *require* that

$$dq_n/dt = p_n \tag{3.5}$$

which determines $q_n(t)$ once $q_n(0)$ is given and $p_n(t)$ determined.
Equations (3.4) and (3.5) imply at once that

$$dq_k/dt = p_k, \quad k = 1, 2, ..., n. \tag{3.6}$$

Going back to (3.2) and (2.6) we verify that

$$dp_1/dt = -e^{-r_1} = -e^{-(q_2-q_1)},$$

$$dp_k/dt = e^{-r_{k-1}} - e^{-r_k} = e^{-(q_k-q_{k-1})} - e^{-(q_{k+1}-q_k)}, \quad k = 2, ..., n-1, \tag{3.7}$$

$$dp_n/dt = e^{-r_{n-1}} = e^{-(q_n-q_{n-1})},$$

and hence that (3.6) and (3.7) are the Hamilton equations corresponding to the Hamiltonian

$$\mathscr{H} = \frac{1}{2} \sum_{1}^{n} p_k^2 + \sum_{k=1}^{n-1} e^{-(q_{k+1}-q_k)}. \tag{3.8}$$

It remains to show that given $q_k(0)$, $p_k(0)$, $k = 1, 2, ..., n$, one can always determine $R_k(0)$, $k = 1, 2, ..., 2n-1$. In other words, given $r_k(0) = q_{k+1}(0) - q_k(0)$, $k = 1, 2, ..., n-1$ and $p_k(0)$, $k = 1, 2, ..., n$ one can find *real* solutions $R_k(0)$ of the equations

$$r_k(0) = R_{2k}(0) + R_{2k+1}(0), \quad k = 1, 2, ..., n-1, \tag{3.9a}$$

$$p_k(0) = -(e^{-R_{2k}(0)} + e^{-R_{2k-1}(0)}) + \alpha, \quad k = 1, 2, ..., n-1, \tag{3.9b}$$

$$p_n(0) = -e^{-R_{2n-1}(0)} + \alpha. \tag{3.9c}$$

It is immediately clear that since

$$\alpha - p_n(0) = e^{-R_{2n-1}(0)} > 0$$

α must be chosen sufficiently large and what we shall show is that α can be chosen so large as to make the system (3.9) solvable.
Setting

$$\xi_k = e^{-R_{2k}(0)}, \quad \eta_k = e^{-R_{2k+1}(0)}, \quad k = 1, 2, ..., n-1, \tag{3.10}$$

we rewrite the system (3.9) as follows

$$\alpha - p_n(0) = \eta_n,$$

$$e^{-r_k(0)} = \xi_k \eta_{k+1}, \qquad k = 1, 2, ..., n-1,$$

$$\alpha - p_k(0) = \xi_k + \eta_k, \qquad k = 1, 2, ..., n-1,$$

and we note that

$$\xi_{n-1} = \frac{e^{-r_{n-1}(0)}}{\alpha - p_n(0)},$$

$$\eta_{n-1} = \alpha - p_{n-1}(0) - \frac{e^{-r_{n-1}(0)}}{\alpha - p_n(0)},$$

$$\xi_{n-2} = \frac{e^{-r_{n-2}(0)}}{\alpha - p_{n-1}(0) - e^{-r_{n-1}(0)}/(\alpha - p_n(0))},$$

$$\eta_{n-2} = \alpha - p_{n-2}(0) - \frac{e^{-r_{n-2}(0)}}{\alpha - p_{n-1}(0) - e^{-r_{n-1}(0)}/p_n(0)}$$

etc.

It is clear that if α is chosen sufficiently large the ξ's and η's will be positive and hence their logarithms (which are the negatives of $R_k(0)$) real.

It is somewhat curious that while the p's and q's are uniquely determined the R's are not owing to the arbitrariness of α.

4

To appreciate a little better the rather neutral role of α, consider the case $N = 4$ which can be solved in a completely elementary way obtaining

$$R_1(t) = C + \log[D - \sqrt{E} \tanh(t \sqrt{E} + \gamma)], \tag{4.1a}$$

$$R_2(t) = \log[D - \sqrt{E} \tanh(t \sqrt{E} + \gamma)] - \log[E \operatorname{sech}^2(t \sqrt{E} + \gamma)], \tag{4.1b}$$

$$R_3(t) = -\log[D - \sqrt{E} \tanh(t \sqrt{E} + \gamma)], \tag{4.1c}$$

where $D > 0$, $E > 0$ and

$$D^2 = e^{-C} + E. \tag{4.2}$$

The constants E, C, γ are expressible in terms of $R_1(0)$, $R_2(0)$, $R_3(0)$ by the formulas

$$E = \tfrac{1}{4}(e^{-R_1(0)} + e^{-R_2(0)} - e^{-R_3(0)})^2 + e^{-R_2(0)}e^{-R_3(0)}, \tag{4.3a}$$

$$e^{-C} = e^{-R_1}e^{-R_3(0)}, \tag{4.3b}$$

$$\tanh \gamma = \frac{e^{-R_1(0)} + e^{-R_2(0)} - e^{-R_3(0)}}{[(e^{-R_1(0)} + e^{-R_2(0)} - e^{-R_3(0)})^2 + 4e^{-R_2(0)}e^{-R_3(0)}]^{1/2}}, \tag{4.3c}$$

whence by (4.2) it follows that

$$D = \tfrac{1}{2}(e^{-R_1(0)} + e^{-R_2(0)} + e^{-R_3(0)}). \tag{4.3d}$$

Recall now that

$$p_1(t) = \alpha - (e^{-R_1(0)} + e^{-R_2(0)}), \tag{4.4a}$$

$$p_2(t) = \alpha - e^{-R_3(t)}, \tag{4.4b}$$

$$r_1(t) = R_2(t) + R_3(t) = q_2(t) - q_1(t), \tag{4.4c}$$

so that

$$p_1(0) = \alpha - (e^{-R_1(0)} + e^{-R_2(0)}),$$

$$p_2(0) = \alpha - e^{-R_3(0)},$$

$$e^{-r_1(0)} = e^{-R_2(0)}e^{-R_3(0)},$$

and it follows from (4.3) that

$$E = \tfrac{1}{4}(p_1(0) - p_1(0))^2 + e^{-r_1(0)}, \tag{4.5a}$$

$$\tanh \gamma = \frac{p_2(0) - p_1(0)}{[(p_2(0) - p_1(0))^2 + 4e^{-r_1(0)}]^{1/2}}, \tag{4.5b}$$

$$e^{-C} = \alpha - \frac{p_1(0) + p_2(0)^2}{2} - \tfrac{1}{4}(p_1(0) - p_2(0))^2 - e^{-r_1(0)}, \tag{4.5c}$$

$$D = \alpha - \frac{p_1(0) + p_2(0)}{2}. \tag{4.5d}$$

Since D and e^{-C} are to be positive, α has to be chosen sufficiently large.

Once α has been so chosen, we note using (4.4a) and (4.4b) that

$$p_1(t) = \frac{p_1(0) + p_2(0)}{2} - \sqrt{E} \tanh(t \sqrt{E} + \gamma), \qquad (4.6a)$$

$$p_2(t) = \frac{p_1(0) + p_2(0)}{2} + e \tanh(t \sqrt{E} + \gamma), \qquad (4.6b)$$

and by (4.5a), (4.5b) E and γ do not depend on α.

References

1. M. Kac and P. van Moerbeke, Some probabilistic aspects of scattering theory, to appear in the Proceedings of the Conference on Functional Integration and its Applications, held in April 1974 at the Cumberland Lodge near London.
2. J. Moser, Finitely many mass points on the line under the influence of an exponential potential—An integrable system, preprint from the Courant Institute of Mathematical Sciences.
3. H. Flaschka, The Toda lattice II, *Progr. Theoret. Phys.* **51** (1974), 703–716.
4. P. D. Lax, Integrals of nonlinear equations of evolution and solitary waves, *Comm. Pure Appl. Math.* **21** (1968), 467–490.
5. K. M. Case and M. Kac,[1] A discrete version of the inverse scattering problem, *J. Mathematical Phys.* **14** (1973), 594–603.

---

[1] We take this opportunity to correct a number of minor but annoying errors and misprints: 1, In formula (3.13) insert d/dx after c. 2, The right-hand side of (4.22) should be $(2/\pi)(1 - \lambda^2)^{1/2} d\lambda/|A(\cos^{-1} \lambda)|^2$ and correspondingly formula (4.21) and the one which precedes it should be corrected. 3, In the line just above formula (5.11) $z \to 0$ should be replaced by $z \to c$.

Three Integrable Hamiltonian Systems Connected with Isospectral Deformations*

J. Moser

Courant Institute of Mathematical Sciences, New York University, New York, New York 10012

DEDICATED TO STAN ULAM

1. Introduction

(a) *Background.* In the early stages of classical mechanics it was the ultimate goal to integrate the differential equations of motions explicitly or by quadrature. This led to the discovery of various "integrable" systems, such as Euler's two fixed center problems, Jacobi's integration of the geodesics on a three-axial ellipsoid, S. Kovalevski's motion of the top under gravity for special ratios of the principal moments of inertia, to name a few nontrivial examples. These efforts and their climax with the work of Jacobi who applied skillfully the method of separation of variables to partial differential equations, the Hamilton–Jacobi equations associated with the mechanical system, to establish their integrable character.

However, this development took a sharp turn when Poincaré showed that most Hamiltonian systems are not integrable and gave arguments indicating the nonintegrability of the three-body problem. In the same negative direction lies Brun's discovery that the three-body problem has no algebraic integral except for the well-known classical ones and algebraic functions of these. These results express, in other words, that integrability of Hamiltonian systems is not a generic property; it is destroyed under small perturbations of the Hamiltonian.

Therefore it seems an anachronismus to discuss these exceptional

---

* This work was partially supported by the National Science Foundation, Grant No. NSF-GP-42298X.

integrable systems nowadays. However, in recent years various phenomena were discovered which are clearly intimately related to integrable Hamiltonian systems yet they have very different origin. One is related to the discovery by Kruskal and others [6] of so-called solitons for the Korteweg–de Vries equation. These are wave solutions of a nonlinear partial differential equation having a strong stability behavior. Originally these phenomena were brought to light by numerical experiments and later on related to the existence of infinitely many conservation laws that restrict the evolution of the solutions severely. If one interprets the partial differential equation, in this case the Korteweg–de Vries equation, as a Hamiltonian system in an infinite-dimensional function space, with a certain symplectic structure, and the conservation laws as integrals of this system, one can view this as an example of an integrable system of infinitely many degrees of freedom. This was made precise in the work of Zakharov and Faddeev [15].

In an entirely unrelated development Calogero [2, 3] found that the quantum theoretical problem of n mass points on the line interacting under the influence of a potential proportional to the inverse square of the distance can be solved explicitly, and he conjectured that the corresponding classical problem might be integrable. This was established by Marchioro for the "three-body problem" by explicit calculation. Moreover, Calogero used his formula to study the scattering problem associated with the n-particle system in the quantum theoretical framework and found that the scattering is essentially trivial, in the sense that the particles behave asymptotically like elastically reflected mass points.

(b) *Results*. It is our goal to show a close algebraic connection between these so different problems. However, instead of studying the infinite dimensional problems related to the partial differential equation in the one and the quantum theoretical framework in the other case, we will restrict ourselves to finite-dimensional systems. The Korteweg–de Vries equation can be discretized so as to retain the desired integrability, as was shown by Toda [13] and his collaborators. Another discretization leads to the differential equations

$$du_k/dt = \tfrac{1}{2}(e^{u_{k+1}} - e^{u_{k-1}}), \quad (k = 1, 2, ..., n - 1) \qquad (1.1)$$

(where we set formally $e^{u_0} = 0$, $e^{u_n} = 0$) suggested by M. Kac and P. v. Moerbeke [8, 9]. Although this system does not have the appearance of a Hamiltonian system, it can be embedded into one, as was shown in [12]. The remarkable fact is that there are $[n/2] = \nu$ polynomials P_μ of u_k, e^{u_k} which are integrals of the motion, i.e.,

$$dP_\mu/dt = 0 \quad (\mu = 1, 2, ..., \nu)$$

if one inserts a solution of the above differential equations. Moreover, all solutions can be expressed in the form

$$e^{u_k} = R_k(\eta)$$

where R_k are rational functions of

$$\eta = (\eta_1, ..., \eta_\nu) \quad \text{and} \quad \eta_1 = e^{\alpha_1 t}, ..., \eta_\nu = e^{\alpha_\nu t}.$$

These rational functions can, of course, not be explicitly described, but this representation suffices to give a complete description of the scattering problem related to this problem (see Section 7).

Instead of Calogero's quantum theoretical problem we look at the corresponding classical one, described by the equations

$$d^2 x_k/dt^2 = -(\partial U/\partial x_k), \quad (k = 1, 2, ..., n)$$

where

$$U = \sum_{k<l} (x_k - x_l)^{-2}, \quad k, l = 1, 2, ..., n, \tag{1.2}$$

the coordinates x_k of the mass points being distinct real numbers. This system is clearly a Hamiltonian system with

$$\mathscr{H} = \tfrac{1}{2} \sum_{k=1}^{n} y_k^2 + U$$

where y_k are the momenta. We will show that this system is an integrable Hamiltonian system, by which we mean that this system possesses n independent integrals $I_k = I_k(x, y)$, globally defined in the phase space and in involution. In this case these functions are, in fact, polynomials in y_k and $(x_k - x_l)^{-1}$. Using this result it is quite easy to verify Marchioro's conjecture: The particles have an asymptotic velocity $\dot{x}_k(\pm \infty)$ satisfying

$$\dot{x}_k(+\infty) = \dot{x}_{n+1-k}(-\infty).$$

Thus after a fairly complicated interaction the particles emerge as free particles with velocities exchanged, that is, the first particle has for $t \to +\infty$ the velocity of the last for $t \to -\infty$, etc.

As a third example we discuss the equation on the circle

$$d^2 x_k/dt^2 = -(\partial U/\partial x_k)$$

with

$$U = \tfrac{1}{2} \sum_{k \neq l (\mathrm{mod}\, n)} \sin^{-2}(x_k - x_l), \qquad (1.3)$$

Here the x_k are considered mod π as distinct points on a circle. These equations are the classical mechanics analog to those of Sutherland [14]. Also this system will be shown to be an integrable Hamiltonian system with n integrals I_k which are polynomials in y_k and $\cot(x_k - x_l)$.

In contrast to the previous examples the last problem has a compact energy surface. On account of this fact the surfaces $I_k = $ const ($k = 1, 2, ..., n$) are compact and hence, as is well known, tori on which the solutions are quasiperiodic. However, the function theoretical character of these solutions has not yet been satisfactorily described.

(c) *Lax's method.* The common link between these problems is that they can be related to deformations of matrices leaving the eigenvalues fixed, that is, to isospectral deformations. For example, with (1.2) we associate a Hermitean matrix L having y_k as diagonal elements and $i(x_k - x_l)^{-1}$ as elements in the (k, l)-position if $k \neq l$. Then (1.2) gives rise to a differential equation for $L = L(t)$ whose solutions have fixed eigenvalues, i.e., the eigenvalues, and hence their symmetric functions I_k are integrals of the motion. The idea of finding integrals of the motion as eigenvalues of an associated linear operator L was developed by Lax [10] for the Korteweg–de Vries equation, where L is given by the classical Sturm–Liouville operator

$$-(d^2/dx^2) + q$$

and the potential q is to be deformed in such a way that the spectrum is unchanged. This question is intimately related to the inverse problem of determining the spectrum from the potential. Instead of developing these ideas in generality we will illustrate them in the three simple examples mentioned above.

In Section 2 we illustrate this method for Eq. (1.1), although this is in no way new. Indeed Flaschka [4, 5] observed first that this method can be applied to the Toda lattice and this example is only a slight variation on this theme. In Section 3 we put Eq. (1.2) into the same framework and draw the conclusion for the associated scattering problem in

Section 4. The *n*-particle system (1.3) on the circle will be studied in Section 5. Finally in Sections 6 and 7 we discuss the inverse spectrum problem and the scattering problem associated with a special Jacobi matrix. The latter leads to an interesting motion in which particles separate in pairs, each pair having a different asymptotic velocity, while the two particles of one pair have the same asymptotic velocity. The scattering phases can also be determined by relating the differential equations to those for the Toda lattice for finitely many particles.

(d) *General remarks.* These problems have connections with a multitude of topics besides that of dynamical systems. The fact that they are related to isospectral deformation points to the connection with spectral and scattering theory. The function theoretical nature of the solution and the rational character of the integrals relates to functions of complex variables. But also Lie algebras play into the subject; in fact the equations are very similar in nature to those studied by Arnold [1]. Arnold generalized the Euler equation for the rotation of a rigid body to dynamical systems in arbitrary Lie algebra.

Many of these connections are still obscure, and we hope that the study of these simple finite dimensional examples will lead to further investigations clarifying the many questions left open.

I want to express my thanks to H. Flaschka and G. Galavotti for many stimulating discussions in the beginning of this work. I am particularly indebted to Galavotti who pointed out Calogero's work and insisted that the classical analog should be integrable.

2. Isospectral Deformations

We begin with an idea that was introduced by P. D. Lax in a different but closely related connection. Consider a class of matrices, say all Jacobi matrices of the form

$$L = \begin{pmatrix} 0 & a_1 & 0 & & 0 \\ a_1 & 0 & a_2 & & \\ \cdot & \vdots & \vdots & \ddots & \\ & & & \vdots & a_{n-1} \\ 0 & & & a_{n-1} & 0 \end{pmatrix} \quad (2.1)$$

with positive entries $a_1, a_2, ..., a_{n-1}$. Their eigenvalues are real and simple. We ask for all matrices in this class having the same spectrum.

One may expect that there are not enough parameters available, but since

$$K^{-1}LK = -L \quad \text{for} \quad K = \text{diag}(1, -1, +1, \cdots),$$

one has for the characteristic polynomial

$$\Delta_n(\lambda) = \det(\lambda I - L)$$

the relation

$$\Delta_n(\lambda) = (-1)^n \Delta_n(-\lambda). \tag{2.2}$$

Therefore, with λ also $-\lambda$ is an eigenvalue and $\lambda = 0$ is an eigenvalue precisely if n is odd. Thus fixing the eigenvalues amounts to $[n/2]$ conditions and the dimensionality of the isospectral matrices of the form (2.1) is $n - [n/2]$.

To get some isospectral deformations, Lax [10] considered differential equations of the form

$$\frac{d}{dt}L = BL - LB \tag{2.3}$$

where $L = L(t)$, t being the deformation parameter. The matrix B has to be chosen appropriately, so that the commutator $[B, L]$ has zeros except in the two off-diagonals, and those should agree. In this example one finds as one possible choice the skew symmetric matrix

$$B = \begin{bmatrix} 0 & 0 & a_1a_2 & & & & 0 \\ 0 & 0 & 0 & a_2a_3 & & & \\ -a_1a_2 & 0 & \cdot & & \cdot & & \\ & \cdot & & \cdot & & \cdot & \\ & & \cdot & & \cdot & 0 & a_{n-2}a_{n-1} \\ & & & 0 & 0 & 0 & \\ 0 & & & -a_{n-2}a_{n-1} & 0 & 0 \end{bmatrix}$$

for which the differential equation (2.3) takes the form

$$\dot{a}_k = a_k(a_{k+1}^2 - a_{k-1}^2), \quad k = 1, 2, \ldots, n-1 \tag{2.4}$$

where we set formally $a_0 = 0 = a_n$.

It is clear that (2.3) gives rise to isospectral deformations: If we solve the differential equation

$$\frac{d}{dt}U = BU, \quad U(0) = I$$

then (2.3) assures that
$$\frac{d}{dt}(U^{-1}LU) = 0,$$
hence
$$U^{-1}LU = L(0).$$
Thus the eigenvalues of L remain constant under this deformation. Also the coefficients I_k of the characteristic polynomial
$$\Delta_n(\lambda) = \lambda^n + I_1\lambda^{n-1} + \cdots + I_n$$
are integrals of the motion, which are polynomials in $a_1^2, a_2^2,..., a_{n-1}^2$. By (2.2) only $\nu = [n/2]$ of these are not zero, but the remaining I_2, $I_4,..., I_{2\nu}$ are actually independent polynomials.

With
$$a_k = \tfrac{1}{2}e^{\frac{1}{2}u_k}$$
the equations (2.4) take the form
$$\dot{u}_k = \tfrac{1}{2}(e^{u_{k+1}} - e^{u_{k-1}}) \qquad (k = 1, 2,..., n-1) \tag{2.5}$$
where we formally set $u_0 = -\infty$, $u_n = -\infty$. These are the equations which Kac and v. Moerbeke considered in their discretization of the Korteweg–de Vries equation [8].[1] The above derivation is, of course, not new; it is quite analogous to that of Flaschka [4]. But we will use the above representation (2.3) of the differential equation (2.4) to describe its solutions as rational functions of exponentials (Section 6) and to investigate the scattering problem related to (2.5) (Section 7).

Incidentally, the above equations (2.3) do not represent the only deformations of L preserving the spectrum. On the contrary all B giving rise to such deformations form an $(n - [n/2])$-dimensional space [12].

3. The n-Particle System on the Line with the Inverse Square Potential

We consider n particles on the line with coordinates $x_1, x_2,..., x_n$ and define
$$U(x) = \sum_{k<l}(x_k - x_l)^{-2}, \qquad k, l = 1, 2,..., n \tag{3.1}$$

---

[1] As I learned from H. Flaschka, this system (2.5) and its relation to the Toda lattice was already mentioned by M. Hénon in a letter of August 28, 1973.

as their potential so that the equations of motion are given by

$$d^2x_k/dt^2 = -(\partial U/\partial x_k) = 2\sum_{j\neq k}(x_k - x_j)^{-3} \quad (k = 1, 2, ..., n). \quad (3.2)$$

It is remarkable that this system possesses n integrals of the motion which are polynomials in $\dot{x}_k$ and $(x_k - x_l)^{-2}$. This fact can again be derived by considering isospectral deformations of another class of matrices.

The quantum-mechanical analog of (3.2) has been studied by Calogero and Marchioro in a number of papers [2, 3, 11] and Calogero succeeded in determining explicit expressions for the spectrum for this problem. He conjectured from his work that the classical problem, being the limit of the quantum-theoretical one, should be integrable. For $n = 3$ this was already verified by Marchioro [11] but his approach does not lend itself to generalization. In order to introduce the class of matrices adapted to this problem we set

$$z_{kl} = \begin{cases} (x_k - x_l)^{-1} & \text{for } k \neq l \\ 0 & \text{for } k = l \end{cases}$$

and form the matrices

$$Z_\alpha = (z_{kl}^\alpha) \quad \text{for } \alpha = 1, 2,$$
$$Y = \mathrm{diag}\{y_1, ..., y_n\}, \quad (3.3)$$
$$D_\alpha = \mathrm{diag}\left\{\sum_{j=1}^n z_{kj}^\alpha\right\} \quad \text{for } \alpha = 2, 3.$$

Then we define

$$L = Y + iZ_1; \quad B = iD_2 - iZ_2, \quad (3.4)$$

so that $L = L^*$ is Hermitean and B skew Hermitean.

The deformation equations

$$dL/dt = BL - LB \quad (3.5)$$

for this class of matrices can be transformed into the equation of motion (3.2)! This implies by the argument of the previous section that the coefficients I_k of the characteristic polynomial

$$\det(\lambda I - L) = \lambda^n + I_1\lambda^{n-1} + \cdots + I_n$$

are integrals of the differential equations. Moreover, they are rational functions of the coordinates and in involution.

To relate Eqs. (3.2) and (3.5) to each other observe that (3.5) depends only on the $n-1$ differences $x_{k+1} - x_k$ ($k = 1, 2,..., n-1$), while (3.2) involves all n coordinates x_k. Therefore we rewrite (3.2) in terms of the z_{kl} and $y_k = -\dot{x}_k$

$$\dot{y}_k = -\ddot{x}_k = -2 \sum_{j=1}^{n} z_{kj}^3$$
$$\dot{z}_{kl} = z_{kl}^2 (y_k - y_l). \tag{3.6}$$

Of course this system is highly redundant, since only the $n-1$ variables $z_{k,k+1}$ are independent, the other being determined by the relations

$$z_{kl}^{-1} = z_{kr}^{-1} + z_{rl}^{-1} \text{ if } k, l, r \text{ distinct, and } z_{kl} + z_{lk} = 0. \tag{3.7}$$

But one verifies immediately that these relations are consistent with (3.6): If they hold for $t = 0$ then for all t.

Now we identify (3.6) with the deformation equations (3.5). For this purpose we have to compute

$$[B, L] = i[Y, Z_2] - [D_2, Z_1] + [Z_2, Z_1] \tag{3.8}$$

where we used (3.4). The element of $[Z_2, Z_1]$ in the (k, l) position is given by

$$\sum_r (z_{kr}^2 z_{rl} - z_{rl}^2 z_{kr}),$$

hence the corresponding term in $[Z_2, Z_1] - [D_2, Z_1]$ is

$$\sum_r (z_{kr}^2 z_{rl} - z_{rl}^2 z_{kr}) - \sum_r (z_{kr}^2 - z_{rl}^2) z_{kl}.$$

To simplify this expression we use the identities (3.7) as follows: The summands of the sum above can be factored

$$Q_{kl,r} = z_{kr}^2 z_{rl} - z_{rl}^2 z_{kr} - (z_{kr}^2 - z_{rl}^2) z_{kl} = (z_{kr} - z_{rl}) P_{kl,r}$$

with

$$P_{kl,r} = (z_{kr} z_{rl} - (z_{kr} + z_{rl}) z_{kl}).$$

If all k, l, r are distinct, this takes the form

$$P_{kl,r} = z_{kr} z_{rl} z_{kl} \{z_{kl}^{-1} - z_{rl}^{-1} - z_{kr}^{-1}\} = 0$$

on account of (3.7). For $k \neq l$ one gets obviously

$$P_{kl,r} = -z_{kl}^2 \quad \text{if} \quad r = k \quad \text{or} \quad r = l.$$

Thus

$$\sum_{r=1}^{n} Q_{kl,r} = 0 \quad \text{for} \quad k \neq l$$

which shows that $[Z_2, Z_1] - [D_2, Z_1]$ is a diagonal matrix. If one computes the diagonal elements one finds

$$[Z_2, Z_1] - [D_2, Z_1] = -2D_3,$$

with the notation of (3.3). Thus with (3.8) the equations (3.5) take the form

$$dL/dt = i[Y, Z_2] - 2D_3,$$

and, in components,

$$\dot{y}_k = -2 \sum z_{kj}^3,$$

$$\dot{z}_{kl} = (y_k - y_l) z_{kl}^2,$$

in agreement with (3.6).

This establishes the existence of the integrals, as well as their rational character. In Section 4, in which we study the scattering problem for this system, we will find without further calculation that these integrals are in involution.[2]

4. Asymptotic Behavior, Marchioro's Conjecture

The n-particle system of the preceding section has a very simple behavior. Since the particles exert a repelling force on each other they fly apart as $t \to \pm\infty$ and ultimately behave like force particles. From this it is clear that the limits $\lim_{t\to\infty} \dot{x}_k(\pm t) = \dot{x}_k(\pm\infty)$ exist. As a matter of fact, these limit velocities or their symmetric functions can be assigned as integrals to the orbits to which they belong. Thus the *existence* of integrals is no surprise for a system like (3.2). However, the existence of *rational* integrals is remarkable, and it implies that

$$\dot{x}_k(+\infty) = \dot{x}_{n+1-k}(-\infty), \quad k = 1, 2, ..., n, \tag{4.1}$$

[2] Extending this method, M. Adler, a student at New York University, found n rational integrals for $U = \sum_{k<l} \{\alpha(x_k - x_l)^{-2} + \beta(x_k - x_l)^2\}$.

so that the particles simply exchange their velocity. Moreover, the above velocities are distinct and agree with the negative of the eigenvalues of the matrix (3.4) belonging to the orbit considered. This way we will prove the fact that matrices of the form (3.4) always have *simple* eigenvalues. One may ask for the phase shifts δ_k defined by

$$x_k(t) - x_{n-k+1}(-t) - 2\dot{x}_k(\infty)t \to \delta_k$$

for $t \to +\infty$. It is easily verified that $\delta_1 = \delta_2 = 0$ for $n = 2$, and one may conjecture that $\delta_k = 0$ for any $n > 2$, but this we have not been able to establish.[3]

The relations (4.1) have been established by Marchioro [11] for the case $n = 3$ and were conjectured by him for arbitrary n. For the quantum-mechanical problem they were established by Calogero [2].

To prove the above assertion we observe that we may label the particles according to the order

$$x_1 < x_2 < \cdots < x_n.$$

Indeed, since the Hamiltonian of (3.2) is given by

$$\mathcal{H} = \tfrac{1}{2} \sum_{k=1}^{n} y_k^2 + \sum_{k<l} (x_k - x_l)^{-2}, \qquad (4.2)$$

the minimal existence of the particles is bounded away from zero for any solution. Moreover, the velocities $-y_k = \dot{x}_k$ are bounded for all t for every orbit.

Our next goal is to show that

$$\lim_{t \to \infty} y_k(\pm t) = y_k(\pm \infty) \qquad (4.3)$$

exists and that

$$y_1(+\infty) > y_2(+\infty) > \cdots > y_n(+\infty). \qquad (4.4)$$

From

$$\tfrac{1}{2}(\ddot{x}_n - \ddot{x}_1) = \sum_{j<n} (x_n - x_j)^{-3} + \sum_{j>1} (x_j - x_1)^{-3} > 0 \qquad (4.4')$$

and the boundedness of $\dot{x}_k$ we conclude, by integration that

$$\int_{-\infty}^{+\infty} (x_k - x_l)^{-3} dt < \infty \qquad \begin{array}{l} \text{for } k > l = 1 \text{ and} \\ \text{for } l < k = n. \end{array} \qquad (4.5)$$

[3] *Note added in proof.* Meanwhile we have been able to verify that indeed $\delta_k = 0$ for all $n \geqslant q$.

Considering the other differential equations one concludes with a simple induction argument (which we forego) that (4.5) holds for all pairs $k > l$. This, in turn implies from (3.2) that the limits $\lim_{t \to \infty} \dot{x}_k(\pm t)$ exist, proving (4.3). Because of the ordering of the particles we have obviously

$$\dot{x}_1(+\infty) \leqslant \dot{x}_2(+\infty) \leqslant \cdots \leqslant \dot{x}_n(+\infty)$$
$$\dot{x}_1(-\infty) \geqslant \dot{x}_2(-\infty) \geqslant \cdots \geqslant \dot{x}_n(-\infty). \tag{4.6}$$

To prove (4.4) we proceed as follows: Consider first $\phi(t) = x_n - x_1 > 0$ which, by (4.4'), satisfies

$$\tfrac{1}{2}\ddot{\phi} \geqslant 2(x_n - x_1)^{-3} > 0. \tag{4.7}$$

Thus $\dot{\phi}$ is monotone increasing and $\dot{\phi}(+\infty) \geqslant 0$, by (4.6). Were $\dot{\phi}(+\infty) = 0$ then $\dot{\phi}(t) < 0$ and thus ϕ bounded. But then the right-hand side of (4.7) would be bounded away from zero, hence ϕ unbounded. This contradiction shows that

$$\dot{x}_1(+\infty) < \dot{x}_n(+\infty).$$

Thus, in the first row of (4.6) we do not have equality in all places, i.e., there exists an s with

$$\dot{x}_s(+\infty) < \dot{x}_{s+1}(+\infty). \tag{4.8}$$

From this we will show now $\dot{x}_1(+\infty) < \dot{x}_s(+\infty)$ and $\dot{x}_{s+1}(+\infty) < \dot{x}_n(+\infty)$ which implies readily that all velocity are different. It suffices to show $x_1(+\infty) < x_s(+\infty)$, the other case being symmetric to it.

From (4.8) we conclude that $x_j - x_s = 0(t^{-1})$ for $j > s$ and therefore

$$\frac{1}{2} \frac{d^2}{dt^2} (x_s - x_1) = \sum_{j<s} (x_s - x_j)^{-3} - 0(t^{-3}) + \sum_{j>1} (x_j - x_1)^{-3}$$
$$\geqslant 2(x_s - x_1)^{-3} - 0(t^{-3}).$$

Thus $\psi = x_s - x_1 + At^{-1}$ with some positive constant A satisfies

$$\ddot{\psi} \geqslant 4(x_s - x_1)^{-3} \quad \text{for} \quad t > t_0$$

and is bounded from below. Thus $\dot{\psi}$ is increasing and $\dot{\psi}(+\infty) \geqslant 0$. As before we conclude that the assumption $\dot{\psi}(\infty) = 0$ leads to a contradiction. Since $\dot{\psi}(t) < \dot{\psi}(\infty) = 0$ $(t > t_0)$ implies ψ to be bounded for

$t > t_0$ hence $\dot\psi$ would be bounded away from zero, and so ψ unbounded. Thus $\psi(\infty) > 0$ as we wanted to show.

Since $y_k = -\dot x_k$ we have established (4.4). This implies obviously

$$(x_k - x_l)^{-1} = 0(t^{-1}) \quad \text{for } t \to +\infty, \ k \neq l$$

so that we can see that the matrix $L(t)$ has a limit $L(\infty)$ which is a diagonal matrix. Since the eigenvalues λ_k of $L(t)$ are independent of t we have

$$y_k(+\infty) = \lambda_k$$

if we make the convention to order these like

$$\lambda_n < \lambda_{n-1} < \cdots < \lambda_1.$$

For $t \to -\infty$ the matrix L also approaches a diagonal matrix with the same eigenvalues in the diagonal, but, because of (4.6) in reversed order. Thus

$$\dot x_k(+\infty) = -y_k(+\infty) = -\lambda_k; \quad \dot x_{n+1-k}(-\infty) = -y_{n+1-k}(-\infty) = -\lambda_k$$

and (4.1) is proven.

Finally, we observe that the integrals $I_k = I_k(x, y)$ $(k = 1, 2, \ldots, n)$ are in involution. For $x_k - x_{k-1} \to \infty$ these integrals I_k converge with their derivatives to $\sigma_k(y)$, the symmetric functions of y. Thus the Poisson bracket

$$G_{kl} = \sum_{r=1}^{n} \frac{\partial(I_k, I_l)}{\partial(x_r, y_r)} = \{I_k, I_l\}$$

converges to $\{\sigma_k, \sigma_l\} = 0$. Thus, along any solution of our system $G_{kl} \to 0$ as $t \to \infty$. On the other hand, as is well known, G_{kl} are integrals themselves, hence $G_{kl} = 0$ for all x, y.

5. The Periodic Case—Sutherland's Equation

If one wants to study the problems of the previous two sections on the circle it is natural to use the identity

$$\sum_{k=-\infty}^{+\infty} (x - k\pi)^{-2} = \sin^{-2} x$$

as motivation to introduce the potential

$$U(x) = \tfrac{1}{2} \sum_{k \neq l(n)} \alpha^2 \sin^{-2}(\alpha(x_k - x_l)) \qquad (\alpha > 0) \qquad (5.1)$$

where the summation is taken over all distinct pairs $k, l \pmod{n}$. The coordinates x_k of the particles may be defined for all integers k such that

$$x_k = x_l \,(\mathrm{mod}(\pi/\alpha)) \quad \text{if and only if} \quad k = l \,(\mathrm{mod}\, n),$$

so that is suffices to consider x_k for $k = 1, 2, ..., n$. The differential equations take the form

$$d^2 x_k / dt^2 = -(\partial U / \partial x_k) = 2\alpha^3 \sum_{j \neq k(n)} \cot \alpha(x_k - x_j) \sin^{-2}(\alpha(x_k - x_j)) \quad (5.2)$$

which is the classical analog of Sutherland's equation [14]. With $y_k = -\dot{x}_k$ the Hamiltonian is

$$\mathscr{H} = \frac{1}{2} \sum_{k(\mathrm{mod}\, n)} y_k^2 + \frac{\alpha^2}{2} \sum_{k \neq l(n)} \sin^{-2}(\alpha(x_k - x_l))$$

showing that, on an energy surface $\mathscr{H} = \mathrm{const}$, the minimal distance of the particles remains bounded away from zero and the velocities $|y_k|$ bounded away from ∞. Thus the energy surface is compact and most solutions of (5.2) turn out to be quasi-periodic. This will be a consequence of well known facts [1] about integrable Hamiltonian systems if we show that (5.2) has n independent integrals which are in involution.

The construction of these integrals follows the pattern of Section 3. We set

$$z_{kl} = \alpha \cot \alpha(x_k - x_l) \quad \text{if} \quad k \neq l(n),$$
$$z_{kl} = 0 \quad \text{if} \quad k = l(n)$$

and rewrite the system (5.2) in the form

$$\dot{y}_k = U_{x_k} = -2 \sum_{j \neq k(n)} z_{kj}(\alpha^2 + z_{kj}^2)$$
$$\dot{z}_{kl} = (\alpha^2 + z_{kl}^2)(y_k - y_l) \quad \text{for} \quad k \neq l(n). \qquad (5.3)$$

Here the last line follows from the differential equation for $\cot x$.

To put these differential equations in the form (3.5) we introduce the n by n matrices

$$Z_1 = (z_{kl}); \qquad Z_2 = (z_{kl}^2 + \alpha^2)$$

where $k, l = 1, 2, ..., n$. With

$$D_2 = \text{diag}\left\{\sum_{j(\text{mod } n)}(z_{kj}^2 + \alpha^2)\right\}; \quad D_3 = \text{diag}\left\{\sum_{j(\text{mod } n)} z_{kj}(z_{kj}^2 + \alpha^2)\right\}$$

$$Y = \text{diag}\{y_k\}$$

we set

$$L = Y + iZ_1; \quad B = iD_2 - iZ_2. \tag{5.4}$$

Then it is a straightforward, though surprising, calculation that (5.3) can be written in the form

$$dL/dt = BL - LB. \tag{5.5}$$

In fact, for $\alpha \to 0$ the formal identities go over into those of Section 3, except for the boundary conditions.

Thus it follows that the coefficients $I_1, I_2, ..., I_n$

$$\det(\lambda I - L) = \lambda^n + I_1 \lambda^{n-1} + \cdots + I_n$$

are independent integrals of the motion. We will not verify here that they are in involution,[4] but observe that they are rational functions of y_k and $e^{i\alpha(x_k - x_l)}$.

To verify (5.5) one has to use the addition theorem for $\cot x$ which gives, for k, l, r distinct modulo n:

$$z_{kl} = \frac{z_{kr} z_{rl} - \alpha^2}{z_{kr} + z_{rl}}$$

hence, for $k \neq l \pmod{n}$

$$P_{kl,r} = z_{kr} z_{rl} - \alpha^2 - (z_{kr} + z_{rl}) z_{kl} = \begin{cases} 0 & \text{if } r \neq k, l(n) \\ -z_{kl}^2 - \alpha^2 & \text{if } r = k, l(n). \end{cases}$$

This implies for

$$Q_{kl,r} = (z_{kr} - z_{rl}) P_{kl,r} = -z_{kr}(z_{rl}^2 + \alpha^2) + z_{rl}(z_{kr}^2 + \alpha^2) - (z_{kr}^2 - z_{rl}^2) z_{kl}$$

that

$$\sum_r Q_{kl,r} \begin{cases} = 0 & \text{if } k \neq l \pmod{n} \\ = -2 \sum_r (z_{kr}^2 + \alpha^2) z_{kr} & \text{if } k = l \pmod{n} \end{cases}$$

---

[4] This could be done by replacing α by $i\alpha$ and using the same argument as in the previous section.

so that the matrix with the elements $\sum_r Q_{kl,r}$ agrees with the diagonal matrix $-2D_3$.

Now we compute the commutator

$$[Z_2 - D_2, Z_1] = \left(\sum_r Q_{kl,r}\right) = -2D_3$$

and, thus, from (5.4)

$$[B, L] = i[Y, Z_2] - [D_2, Z_1] + [Z_2, Z_1] = i[Y, Z_2] - 2D_3.$$

From this identity one reads off that (5.5) agrees with the equation (5.3). This makes the statement about the I_k being integrals of the motion again obvious.

6. Rational Character of the Solution of (2.4)

We return to the equations (2.4) or (2.5) and investigate their solutions using the fact that these differential equations describe isospectral deformation of Jacobi matrices. We begin with introducing a set of variables r_k on the manifold of Jacobi matrices (2.1) for which the spectrum is fixed. This is the analog of the inverse spectrum problem.

Let

$$R(\lambda) = (\lambda I - L)^{-1}$$

and e_1 be the vector with components $(1, 0, ..., 0)$. We introduce the rational function

$$f(\lambda) = (R(\lambda) e_1, e_1)$$

which has simple poles at $\lambda = \lambda_k$ with a positive residue which we denote by r_k^2, so that

$$f(\lambda) = \sum_{k=1}^n \frac{r_k^2}{\lambda - \lambda_k}.$$

On account of the symmetry property $K^{-1}LK = -L$ derived in Section 2, $f(\lambda)$ is an odd function of λ. Thus, if we order the (always distinct) eigenvalues by

$$\lambda_n > \lambda_{n-1} > \cdots > \lambda_1 \qquad (6.1)$$

we conclude that
$$\lambda_k = -\lambda_{n-k+1}; \quad r_k = r_{n-k+1}$$

and $f(\lambda)$ can be represented by

$$f(\lambda) = \sum_{k=1}^{\nu} \frac{2\lambda r_k^2}{\lambda^2 - \lambda_k^2} + \kappa_n \frac{r_{\nu+1}^2}{\lambda}$$

where $\kappa_n = 1$ for n odd, $\kappa_n = 0$ if n is even and $\nu = [n/2]$. Since $f(\lambda) \sim \lambda^{-1}$ for $|\lambda| \to \infty$ we have

$$\sum_{k=1}^{n} r_k^2 = 1$$

and we prefer to free ourselves from the latter restriction by using the r_k as projective coordinates. Therefore we set

$$f(\lambda) = \frac{\sum_{k=1}^{n} r_k^2/(\lambda - \lambda_k)}{\sum_{k=1}^{n} r_k^2} = \frac{\sum_{k=1}^{\nu} 2\lambda r_k^2/(\lambda^2 - \lambda_k^2) + \kappa_n(r_{\nu+1}^2/\lambda)}{\sum_{k=1}^{\nu} 2r_k^2 + \kappa_n r_{\nu+1}^2}. \quad (6.2)$$

The n variables $r_1, r_2, ..., r_\nu, \kappa_n r_{\nu+1}, \lambda_1, \lambda_2, ..., \lambda_\nu$ can be used to describe the Jacobi matrix (2.1) uniquely up to scaling of the r_k. In fact, the squares a_k^2 ($k = 1, 2, ..., n-1$) of the elements in (2.1) can be expressed rationally in terms of those $r_j, \lambda_j, 1 \leq j \leq \nu$ and $r_{\nu+1}$ if n is odd. The reason for this fact lies in the representation of $f(\lambda)$ as a continued function

$$f(\lambda) = \cfrac{1}{\lambda - \cfrac{a_1^2}{\lambda - \cfrac{a_2^2}{\ddots \cfrac{}{\lambda - \cfrac{a_{n-1}^2}{\lambda}}}}} \quad (6.3)$$

which goes back to Stieltjes (used also in [12]). Since the computation of the continued fraction from the partial fraction expression is a rational process one finds that

$$a_k^2 = R_k(r, \lambda) \quad (6.4)$$

where the R_k are rational functions, homogeneous of degree zero in the r_j and homogeneous of degree two in λ.

Moreover, (6.4) can be viewed as mapping which takes the domain

$$D = \{(\lambda, r), \lambda_1 > \lambda_2 > \cdots > \lambda_\nu \geq 0; r_j > 0 \ (j = 1, 2, \ldots, n - \nu)\}$$

into the domain onto

$$\tilde{D} = \{a_j > 0, j = 1, 2, \ldots, n - 1\}$$

in such a way that the pre-image of each point is precisely one ray $(\rho r, \lambda)$ with a scalar $\rho > 0$.

We will show that in these homogeneous coordinates the differential equations take the simple form

$$\dot{\lambda}_k = 0; \qquad \dot{r}_k = -\lambda_k^2 r_k, \tag{6.5}$$

so that, via (6.4) the a_k^2 appear as rational functions of exponentials $e^{-\lambda_1^2 t}, \ldots, e^{-\lambda_\nu^2 t}$.

To prove this assertion we introduce the eigenvectors $\phi(\lambda_j)$ of L which we normalize by

$$\phi_1(\lambda_j) = (e_1, \phi(\lambda_j)) > 0; \qquad |\phi(\lambda_j)| = 1. \tag{6.6}$$

If L is a solution of (2.3) these eigenvectors become functions of t which evolve according to

$$\phi(\lambda_j, t) = U(t) \phi(\lambda_j, 0)$$

where $U(t)$ is the unitary matrix of Section 2. Thus the eigenvectors satisfy the differential equation

$$\frac{d\phi(\lambda_j, t)}{dt} = -B\phi(\lambda_j, t).$$

We compute the resulting differential equation for the first component $\phi_1 = (e_1, \phi)$

$$\dot{\phi}_1 = -a_1 a_2 \phi_3$$

and use the equations resulting from $(L - \lambda)\phi = 0$

$$-\lambda \phi_1 + a_1 \phi_2 = 0$$
$$a_1 \phi_1 - \lambda \phi_2 + a_2 \phi_3 = 0$$

to express ϕ_3 in terms of ϕ_1. One finds readily

$$a_1 a_2 \phi_3 = (\lambda^2 - a_1^2)\phi_1$$

so that the differential equations for ϕ_1 become

$$\dot\phi_1 = -(\lambda^2 - a_1^2)\phi_1. \tag{6.7}$$

Finally, to show that the $\phi_1(\lambda_k)$ are proportional to the r_k we write the resolvent $R(\lambda)$ in terms of the eigenvectors obtaining

$$f(\lambda) = (R(\lambda) e_1, e_1) = \sum_k \frac{(\phi(\lambda_k), e_1)^2}{\lambda - \lambda_k}$$

so that

$$\phi_1(\lambda_k) = \frac{r_k}{(\sum_{j=1}^n r_j^2)^{1/2}}.$$

Thus the differential equations (6.5) give

$$\dot\phi_1(\lambda_k) = -(\lambda_k^2 - \sum \lambda_j^2 r_j^2) \phi_1(\lambda_k).$$

It is easy to verify that

$$a_1^2 = \sum_j \lambda_j^2 r_j^2 \left(\sum_j r_j^2\right)^{-1}$$

and the second equations of (6.5) have been verified. The first equations of (6.5) are clear from the derivation.

Thus the solutions a_k^2 of (2.4) are rational functions of exponential functions. We describe the solution for $n = 4$. Computing the continued fraction of $f(\lambda)$ explicitly one finds

$$a_1^2 = \frac{\lambda_1^2 r_1^2 + \lambda_2^2 r_2^2}{r_1^2 + r_2^2}, \quad a_2^2 = \frac{(\lambda_2^2 - \lambda_1^2)^2 r_1^2 r_2^2}{(\lambda_1^2 r_1^2 + \lambda_2^2 r_2^2)(r_1^2 + r_2^2)},$$

$$a_3^2 = \frac{\lambda_1^2 \lambda_2^2 (r_1^2 + r_2^2)}{\lambda_1^2 r_1^2 + \lambda_2^2 r_2^2}.$$

Inserting $r_j = r_j(0) e^{-\lambda_j^2 t}$ we obtain the explicit solutions of (2.4).

7. THE SCATTERING PROBLEM ASSOCIATED WITH THE EQUATION OF KAC AND V. MOERBEKE

In order to study the asymptotic behavior of the solution of (2.5) we consider

$$u_k = x_k - x_{k+1}, \qquad k = 1, 2, \ldots, n-1 \tag{7.1}$$

as the difference between the positions x_k of n particles on the line. If the x_k satisfy the differential equations

$$\dot{x}_k = -\tfrac{1}{2}(e^{u_k} + e^{u_{k-1}}), \qquad k = 1, 2, \ldots, n \tag{7.2}$$

where we formally set $e^{u_0} = 0 = e^{u_n}$, or $x_0 = -\infty$, $x_{n+1} = +\infty$ then clearly (2.5) follows. Conversely the x_k are determined only up to translation and for any solution $x_k(t)$ of (7.2) also $x_k(t) + c$ is a solution giving rise to the same solution of (2.5), provided c is a constant. For simplicity we will assume that $n = 2\nu$ is even.

We ask for the asymptotic behavior of the solution of (7.2) for $t \to \pm \infty$ and the relation between the scattering data. We will show that any solution of (7.2) behaves linearly for large t:

$$x_k(\pm t) \sim \pm \alpha_k^{\pm} t + \beta_k^{\pm} \qquad \text{as} \quad t \to +\infty$$

where

$$\alpha_{2j}^+ = \alpha_{2j-1}^+ = \alpha_{n-2j+2}^- = \alpha_{n-2j+1}^-, \qquad j = 1, 2, \ldots, \nu, \tag{7.3}$$

i.e., the particles travel asymptotically in pairs, while the different pairs have negative and different velocities, in fact, it turns out

$$\alpha_{2j}^+ = -2\lambda_j^2, \qquad j = 1, 2, \ldots, \nu \tag{7.3'}$$

where the $\lambda_1 > \lambda_2 > \cdots$ are the eigenvalues of L.

We will also determine the relation between the phases. First of all, for the neighbors we have the asymptotic distances

$$\beta_{2j-1}^+ - \beta_{2j}^+ = \log(-2\alpha_j^+) = \log(4\lambda_j^2)$$

$$\beta_{n-2j+1}^- - \beta_{n-2j+2}^- = \log(-2\alpha_j^+) \tag{7.4}$$

and for the phases of pairs with the same velocities

$$\beta_{2j}^+ - \beta_{n-2j+2}^- = -\sum_{k<j} \log 4(\alpha_{2k}^+ - \alpha_{2j}^+)^2 + \sum_{k>j} \log 4(\alpha_{2k}^+ - \alpha_{2j}^+)^2. \tag{7.5}$$

Thus the particles undergo a scattering in which the pairs behave as if they interacted pairwise at a time.

The results (7.3), (7.3'), (7.4) are easily derived and we begin with their proof. We recall the differential equation (2.4)

$$\dot{a}_k = a_k(a_{k+1}^2 - a_{k-1}^2), \quad k = 1, 2, ..., n-1$$

with $a_0 = 0 = a_n$ from which we see that

$$\sum_{k=1}^{n-1} a_k^2 = \text{const}$$

along solutions. Thus a_k are bounded. Since

$$\frac{d}{dt} \log(a_1 a_3 \cdots a_{2j-1}) = a_{2j}^2$$

we conclude that

$$\int_0^\infty a_{2j}^2 \, dt < \infty.$$

Since $\dot{a}_{2j}$ is bounded this implies that

$$a_{2j}(t) \to 0 \quad \text{as} \quad t \to +\infty. \tag{7.6}$$

Thus, the Jacobi matrix $L(t)$, given by (2.1), is asymptotic to a matrix blocked into two by two matrices with eigenvalues $\pm a_{2j-1}(t), j = 1, 2, ..., \nu$. Since, on the other hand the eigenvalues λ_k are distinct and independent of t it follows that the limits $a_{2j-1}(t) \to a_{2j-1}(\infty)$ exist and agree with these eigenvalues in some order. From the differential equations

$$\frac{\dot{a}_{2j}}{a_{2j}} = a_{2j+1}^2 - a_{2j-1}^2$$

and from (7.6) it follows that

$$a_{2j+1}^2(\infty) < a_{2j-1}^2(\infty)$$

and thus, if we order the eigenvalues λ_k of L according to (6.1) we conclude

$$a_{2j-1}(t) \to \lambda_j, \quad (j = 1, 2, ..., \nu). \tag{7.7}$$

Using the relation

$$4a_k^2 = e^{u_k} = e^{x_k - x_{k+1}} \tag{7.8}$$

we conclude from (7.1), (7.2), (7.7), (7.8) that

$$\dot{x}_{2j}(+\infty) = \dot{x}_{2j-1}(+\infty) = -2\lambda_j^2$$

proving (7.3′) and the first part of (7.3). The other part follows by considering the asymptotic behavior for $t \to -\infty$ analogously.

Moreover, (7.7) and (7.8) implies that

$$x_{2j-1} - x_{2j} \to \log(4\lambda_j^2) \quad \text{for } t \to +\infty$$

proving the first part of (7.4). The second follows similarly.

It remains to prove (7.5). This will be done by relating the first order differential equations (7.2) to a second order system related to the Toda lattice, for which the scattering problem has been solved [12]. We notice that differentiation of (7.2) yields

$$\ddot{x}_k = -\tfrac{1}{2}(e^{u_k}\dot{u}_k + e^{u_{k-1}}\dot{u}_{k-1})$$
$$= -\tfrac{1}{4}\{e^{u_k}(e^{u_{k+1}} - e^{u_{k-1}}) + e^{u_{k-1}}(e^{u_k} - e^{u_{k-2}})\}$$
$$= -\tfrac{1}{4}(e^{x_k - x_{k+2}} - e^{x_{k-2} - x_k})$$

where we set the undefined exponential terms equal to zero. Thus with

$$\xi_j = x_{2j}; \quad \tau = t/2 \tag{7.9}$$

we have

$$\frac{d^2\xi_j}{d\tau^2} = e^{\xi_{j-1} - \xi_j} - e^{\xi_j - \xi_{j+1}} = \frac{\partial U}{\partial \xi_j}, \quad (j = 1, 2, \ldots, \nu) \tag{7.10}$$

where

$$U = \sum_{j=1}^{\nu-1} e^{\xi_j - \xi_{j+1}}.$$

This Hamiltonian system has already been established as an integrable one [13]. For the scattering one has again

$$\xi_j'(+\infty) = \xi_{\nu+1-j}'(-\infty), \quad j = 1, 2, \ldots, \nu$$

which is consistent with (7.3) as $\xi_j'(\pm\infty) = 2\alpha_{2j}^\pm$, and

$$\xi_j(\tau) - \xi_{\nu+1-j}(-\tau) - 2\gamma_j\tau \to \sum_{k \neq j} \delta_{kj} \tag{7.11}$$

where

$$\gamma_j = \xi_j'(+\infty) = 2\alpha_{2j}^+; \qquad \delta_{kj} = \begin{cases} \log(\gamma_k - \gamma_j)^2, & k > j \\ -\log(\gamma_k - \gamma_j)^2, & k < j. \end{cases}$$

With (7.9) the relation (7.11) translates readily into the statement (7.5).

We conclude with a comment on the relation between the differential equation (2.5) by Kac and v. Moerbeke and the equations (7.10) for the the Toda lattice. The first one corresponds to an isospectral deformation of the Jacobi matrix L given by (2.1), with zeros in the diagonal, while the second-order differential equation corresponds to such deformations of such Jacobi matrices with arbitrary diagonal elements (see [4, 12]). To establish the connection between the two we form L^2 which is not any more a tridiagonal matrix, but is similar to one. In fact, with e_α ($\alpha = 1, 2,..., n$) denoting the unit vectors, one finds that L^2 leaves the spaces $E_1 = \text{span}\{e_1, e_3,..., e_{n-1}\}$ and $E_2 = \text{span}\{e_2, e_4,..., e_n\}$ invariant and reduces in each of these spaces to a symmetric Jacobi matrix. This explains why the solutions of (2.4) are rationally expressible in terms of $e^{-\lambda_j^2 t}$ while solutions of the corresponding equations for the Toda lattice are rational in $e^{-\lambda_j t}$. This illustrates in a simple example how the operation $L \to L^2$ and more generally $L \to f(L)$ plays a role in these problems.

References

1. V. I. ARNOLD, Sur la géométrie différentielle des groupes de Lie de dimension infinie et ses applications à l'hydrodynamique, *Ann. Inst. Fourier (Grenoble)* **16** (1966), 319–361.
2. F. CALOGERO, Solution of the one-dimensional n-body problems with quadratic and/or inversely quadratic pair potentials, *J. Mathematical Phys.* **12** (1971), 419–436.
3. F. CALOGERO AND C. MARCHIORO, Exact solution of a one-dimensional three-body scattering problem with two-body and/or three-body inverse square potential, *J. Mathematical Phys.* **15** (1974), 1425–1430.
4. H. FLASCHKA, The Toda lattice, I,[5] *Phys. Rev. B* **9** (1974), 1924–1925.
5. H. FLASCHKA, The Toda lattice, II, *Progr. Theoret. Phys.* **51** (1974), 703–716.
6. C. S. GARDNER, J. M. GREENE, M. D. KRUSKAL, AND R. M. MIURA, Korteweg–de Vries equations and generalizations. VI. Methods for exact solutions, *Comm. Pure Appl. Math.* **27** (1974), 97–133.
7. M. HÉNON, Integrals of the Toda lattice, *Phys. Rev. B* **9** (1974), 1921–1923.
8. M. KAC AND P. VAN MOERBEKE, On an explicitly soluble system of non-linear differential equations related to certain Toda lattices, to appear.

---

[5] The title reads II, apparently a misprint.

9. M. KAC AND P. VAN MOERBEKE, Some probabilistic aspects of scattering theory, to appear.
10. P. D. LAX, Integrals of nonlinear equations of evolution and solitary waves, *Comm. Pure Appl. Math.* **21** (1968), 467–490.
11. C. MARCHIORO, Solution of a three-body scattering problem in one dimension, *J. Mathematical Phys.* **11** (1970), 2193–2196.
12. J. MOSER, Finitely many mass points on the line under the influence of an exponential potential—An integrable system, to appear in Proc. Battelle Rencontres, Lecture Notes in Physics, Springer, 1975.
13. M. TODA, Waves in nonlinear lattice, *Progr. Theoret. Phys. Suppl.* **45** (1970), 174–200.
14. B. SUTHERLAND, Exact results for a quantum many-body problem in one dimension, II, *Phys. Rev. A* **5** (1972), 1372–1376.
15. V. E. ZAKHAROV AND L. D. FADDEEV, Korteweg–de Vries equations: A completely integrable Hamiltonian system, *Funkcional Anal. i Priložen* **5** (1971), 18–27.

Almost Periodic Behavior of Nonlinear Waves*

Peter D. Lax

Courant Institute of Mathematical Sciences, 251 Mercer Street, New York, New York 10012

DEDICATED TO STAN ULAM

1. Introduction

In their very influential paper [3], Fermi, Pasta, and Ulam studied numerically the motion of linearly arranged particles driven by nonlinear forces between nearest neighbors. Contrary to their expectation, the motions were far from being ergodic; on the contrary, each trajectory seemed to occupy only a small portion of phase space; furthermore some of these motions appeared to be almost periodic.

In this talk I shall report briefly on recent theoretical results concerning three nonlinear systems which have a bearing on the questions raised by FPU; here is a brief summary:

Each of the systems discussed has an unusually large number of integrals, i.e., functionals which are conserved during motion; this might explain why some numerically computed trajectories of these systems seem to be confined to such an unexpectedly small portion of phase space. It should be pointed out however, that this cannot be the whole story, since the available part of phase space is still pretty large; in fact other computed trajectories seem to occupy a fairly large portion of phase space. It should be added that it is not known whether the FPU system has any integrals other than total momentum and energy, although the contrary has not been demonstrated either. This shows that there must be an additional mechanism at work; this additional mechanism might very well be the one discovered in low dimensions by Moser, Kolmogoroff, and Arnold.

The first of the systems discussed is Hamiltonian and completely

---

* Results obtained at the Courant Institute of Mathematical Sciences, New York University, under Contract AT(11-1)-3077 with the U.S. Atomic Energy Commission.

integrable; accumulating evidence indicates that so is the second example, with infinitely many degrees of freedom. If so, one might prove the almost periodic behavior of these systems by introducing action and angle variables. Even then it would be desirable to relate the size of almost periods predicted by theory to those observed in calculation.

2. A Method for Constructing Nonlinear Systems with Many Integrals

In [9] a fairly general method was described for constructing nonlinear systems with many integrals. This method has, in the hands of the author and others, led to a number of interesting examples. This section presents very briefly the general method.

Let $L(t)$ be a one-parameter family of operators all of which are similar to each other. That is, we assume that each $L(t)$ can be mapped by a similarity transformation into $L(0)$:

$$U(t)^{-1} L(t)\, U(t) = L(0). \tag{2.1}$$

We assume that both L and U depend differentiably on t, and we introduce the notation

$$U_t U^{-1} = B(t), \tag{2.2}$$

from which we deduce

$$U_t = BU. \tag{2.3}$$

Differentiate (2.1) with respect to t; using

$$\frac{d}{dt} U^{-1} = -U^{-1} U_t U^{-1}$$

and (2.3) we get

$$-U^{-1}BLU + U^{-1}L_t U + U^{-1}LBU = 0$$

which implies

$$L_t = BL - LB. \tag{2.4}$$

Conversely, suppose (2.4) is satisfied and suppose the initial value problem for the differential equation

$$v_t = B(t)v \tag{2.5}$$

can be solved for a sufficiently wide class of initial values $v(0)$. Then the operator

$$U(t): v(0) \to v(t)$$

satisfies (2.1).

Similar operators L have the same spectrum; so it follows from (2.1) that *the eigenvalues $\{\lambda_j\}$ of $L(t)$ are independent of t.*

In any concrete representation the operator L appears as an integral or differential operator, described in terms of coefficients. Relation (2.4) is a nonlinear differential equation for these coefficients. The eigenvalues of L are functionals of the coefficients; being independent of t, they constitute the sought-after integrals.

If the operators L are symmetric or hermitean symmetric then similarity implies unitary equivalence. In fact if the spectrum of L is simple then the operator U appearing in (2.1) must be unitary.

If $U(t)$ is unitary

$$UU^* = I;$$

differentiating with respect to t we get

$$U_t U^* + U U_t^* = 0.$$

The meaning of this equation is that $U_t U^*$ is antisymmetric. Since U is unitary, $U_t U^* = U_t U^{-1}$, the operator denoted in (2.2) as B. So we conclude:

For L hermitean symmetric, B should be chosen antisymmetric:

$$B^* = -B.$$

3. The Toda Lattice

In his recent interesting paper [4], Flaschka has carried out the following construction:

Denote by u a vector:

$$u = (u_1, \ldots, u_N),$$
$$\|u\|^2 = \sum u_j^2. \tag{3.1}$$

Denote by T cyclic translation:

$$(Tu)_j = u_{j-1}, \tag{3.2}$$

where we set

$$u_0 = u_N.$$

Clearly T is a unitary operator:

$$T^* = T^{-1}. \qquad (3.3)$$

Let a be any vector; it is convenient to introduce the abbreviations

$$Ta = a_+, \qquad T^{-1}a = a_-. \qquad (3.4)$$

The following relations are easy to verify:

$$Tau = a_+ Tu, \qquad T^{-1}au = a_- Tu. \qquad (3.5)$$

Define the operator L by

$$L = a_- T^{-1} + c + aT. \qquad (3.6)$$

Using the relation (3.5) we see that L is symmetric:

$$L^* = L.$$

Define B by

$$B = -a_- T^{-1} + aT. \qquad (3.7)$$

Again we see easily that B is antisymmetric:

$$B^* = -B.$$

A simple computation gives

$$BL - LB = a_-(c - c_-) T^{-1} + 2(a^2 - a_-^2) + a(c_+ - c)T. \qquad (3.8)$$

Differentiating (3.6) we get

$$L_t = a_{-t} T^{-1} + c_t + a_t T. \qquad (3.8')$$

Observe that the commutator of B and L belongs to the same class as L_t. Now set, following (2.4)

$$L_t = BL - LB. \qquad (3.9)$$

Equating coefficients we get from (3.8), (3.8')

$$c_t = 2(a^2 - a_-^2), \qquad (3.10c)$$

$$a_t = a(c_+ - c). \qquad (3.10a)$$

As we saw in Section 2, it follows from Eq. (3.9) that the operators L are similar to one another, and therefore their eigenvalues don't change with t.

L is a matrix, and its eigenvalues are rather complicated functions of its entries. The elementary symmetric functions of the eigenvalues however, being the coefficients of the characteristic polynomial of L, are polynomials in the entries of L. Since L is tridiagonal, the first few are easily computed:

$$\det(\lambda I - L) = \lambda^N + I_1 \lambda^{N-1} + \cdots + I_N.$$

Then

$$I_1 = -\sum c_j,$$
$$I_2 = \sum c_j c_k - \sum a_j^2, \qquad (3.11)$$
$$I_3 = \text{cubic, etc.}$$

Using the well known relations

$$\sum \lambda_j = -I_1, \qquad \sum \lambda_j \lambda_k = I_2, \quad \text{etc.}$$

we get from (3.11)

$$\sum \lambda_j = \sum c_j,$$
$$\sum \lambda_j^2 = I_1^2 - 2I_2 = \sum c_j^2 + 2 \sum a_j^2, \qquad (3.12)$$
$$\sum \lambda_j^3 = \text{cubic, etc.}$$

We turn now to lattice vibrations; denote by q_j the lateral displacement of the jth particle from equilibrium; each particle is linked to its two neighbors by identical springs. We denote by $f(s)$ the force exerted by the spring when stretched by the amount s; $f(s)$ is in physically meaningful cases an increasing function of s.

We take the arrangement of the particles to be periodic, i.e.

$$q_{j+N} \equiv q_j.$$

Assume that each particle has unit mass; then the equations of motion are

$$\frac{d^2}{dt^2} q_j = f(q_{j+1} - q_j) - f(q_j - q_{j-1}). \qquad (3.13)$$

This can be written in Hamiltonian form by setting

$$\frac{d}{dt} q_j = p_j, \qquad (3.14)$$

the Hamiltonian being

$$\tfrac{1}{2}\sum p_j^2 + \sum F(q_{j+1} - q_j), \qquad (3.15)$$

where

$$\frac{d}{ds} F(s) = f(s). \qquad (3.16)$$

If $f(s)$ is a linear function of s, the Eqs. (3.13) are linear and analyzable in terms of normal modes. FPU investigated two nonlinear cases, where f was either of the following two forms:

$$f(s) = s + \alpha s^3, \quad \text{or } f(s) \text{ piecewise linear,}$$

Toda, [12], has introduced and studied the lattice where the dependence of f on s is exponential:

$$f(s) = -e^{-s}. \qquad (3.17)$$

The equations of motion are

$$\begin{aligned} dq_j/dt &= p_j, \\ dp_j/dt &= \exp(q_{j-1} - q_j) - \exp(q_j - q_{j+1}). \end{aligned} \qquad (3.18)$$

If one introduces new variables

$$c_j = \tfrac{1}{2} p_j, \qquad a_j = \tfrac{1}{2} \exp(q_{j-1} - q_j)/2 \qquad (3.19)$$

then we can using (3.18) express the derivatives of the new variables as follows:

$$\frac{d}{dt} c_j = \frac{1}{2} \frac{d}{dt} p_j = 2(a_j^2 - a_{j+1}^2), \qquad (3.20c)$$

$$\frac{d}{dt} a_j = \tfrac{1}{2}(p_{j-1} - p_j) = a_j = a_j(c_{j-1} - c_j). \qquad (3.20a)$$

Observe that (3.20c) and (3.20a) are the same as (3.10a) and (3.10c). So one can conclude that the quantities (3.11) are conserved functionals for the Hamiltonian system (3.18). These conserved quantities as well as the transformation (3.19) were originally found by Hénon [7]; he and Flaschka have proved that the quantities I_j are in involution, so that the Hamiltonian system (3.18) is completely integrable.

Note that the first two functionals (3.12) are total momentum and total energy.

4. THE KdV EQUATION

In this application the underlying Hilbert space consists of periodic L_2 functions on the unit interval of the x-axis, and L is the Schroedinger operator

$$L = \partial^2 + u, \quad \partial = d/dx. \tag{4.1}$$

This is the selfadjoint operator, with a discrete spectrum $\{\lambda_j\}$.

For L given by (4.1), $L_t = u_t$ is multiplication by u_t; therefore in order to satisfy Eq. (2.4) we need operators B_j whose commutator with L is multiplication. In [9] the author has shown how to construct a sequence B_j of such operators; these operators have these properties:

(i) B_j is a differential operator of order $2j + 1$.

(ii) B_j is antisymmetric.

(iii) $B_j L - LB_j$ is multiplication by $K_j(u)$; $K_j(u)$ depends in a nonlinear fashion on u and its derivatives up to order $2j + 1$.

Following (2.4) we consider the equations

$$u_t = B_j L - LB_j = K_j(u); \tag{4.2}$$

these equations have the property that for their solutions the spectrum of L defined by (4.1) is independent of t.

The first two of these operators are

$$B_0 = \partial, \quad [B_0, L] = K_0(u) = u_x \tag{4.3_0}$$

and

$$B_1 = \partial^3 + \tfrac{3}{2}u\partial + \tfrac{3}{4}u_x,$$
$$[B_1, L] = K_1(u) = \tfrac{1}{4}u_{xxx} + \tfrac{3}{2}uu_x. \tag{4.3_1}$$

The zeroth equation (4.2_0) is

$$u_t = u_x \qquad (4.2_0)$$

and the first one is

$$u_t = \tfrac{1}{4} u_{xxx} + \tfrac{3}{2} u u_x . \qquad (4.2_1)$$

(4.2_0) describes translation along the x-axis; (4.2_1) is the *KdV* equation, with some inessential rescaling. The general equation (4.2_j) is usually called the *j*th generalized *KdV* equation.

Gardner has shown that the *j*th *KdV* operator K_j has the following structure:

$$K_j(u) = \partial G_j(u),$$

where G_j is the gradient of a functional $F_j(u)$. That is

$$\frac{d}{d\epsilon} F_j(u + \epsilon v) |_{\epsilon=0} = (G_j(u), v).$$

Furthermore Gardner has shown [6] that *each functional $F_m(u)$ is a conserved quantity for each generalized KdV flow* (4.2). From this it is easy to deduce, using the Hamiltonian formalism introduced by Gardner, that *the generalized KdV flows (4.2_j) commute with each other*.

Numerical calculations by Kruskal and Zabusky have indicated an almost periodic behavior of those solutions of *KdV* which are periodic in space. In [10] the author has constructed an abundance of solutions of *KdV* which are periodic in x and almost periodic in t. These solutions can be characterized by a variational problem suggested by Kruskal and Zabusky: minimize $F_N(u)$ subject to the constraints that $F_j(u)$ have prescribed values of $j < N$. The set of solutions of this variational problem consist of smooth N-dimensional tori on which the *KdV* flow—in fact all generalized *KdV* flows—are almost periodic. For details the reader is referred to [10].

I suspect, but cannot prove, that as N tends to ∞ these special solutions become dense among all C^∞ solutions.

5. The Sine-Gordon Equation

In this section we show how to present in the framework of Section 2 a portion of a very interesting theory developed by Ablowitz, Kaup,

Newell, and Segur, [1]. AKNS consider the first order matrix operator

$$L = \begin{pmatrix} -\partial & q \\ r & \partial \end{pmatrix}. \tag{5.1}$$

The analysis of AKNS suggests to seek B of the form

$$B = RL^{-1}, \tag{5.2}$$

where

$$R = \begin{pmatrix} a & b \\ c & d \end{pmatrix}. \tag{5.3}$$

Setting B as given by (5.2) into (2.4) gives

$$L_t = R - LRL^{-1}.$$

Multiplying by L on the right we get

$$L_t L = RL - LR. \tag{5.4}$$

We proceed now to solve this equation for R of form (5.3) when L is of form (5.1).

A straightforward calculation gives

$$RL = \begin{pmatrix} -a\partial + br & aq + b\partial \\ -c\partial + dr & cq + d\partial \end{pmatrix},$$

$$LR = \begin{pmatrix} -\partial a + qc & -\partial b + qd \\ ra + \partial c & rb + \partial d \end{pmatrix}.$$

So

$$RL - LR = \begin{pmatrix} a_x - qc + br & 2b\partial + b_x + aq - qd \\ -2c\partial - c_x + dr - ra & -d_x + cq - rb \end{pmatrix}. \tag{5.5}$$

Differentiating (5.1) we get

$$L_t = \begin{pmatrix} 0 & q_t \\ r_t & 0 \end{pmatrix}.$$

A straightforward calculation gives

$$L_t L = \begin{pmatrix} q_t r & q_t \partial \\ -r_t \partial & r_t q \end{pmatrix}. \tag{5.6}$$

Substituting (5.5) and (5.6) into (5.4) we get 4 sets of relations from the 4 components:

(i) $q_t r = a_x - qc + br$,
(ii) $q_t = 2b$, $b_x + aq - qd = 0$,
(iii) $r_t = 2c$, $-c_x + dr - ra = 0$,
(iv) $r_t q = -d_x + cq - rb$.
$\hfill(5.7)$

Substituting the first relation in (5.7ii) into (5.7i) and the first relation in (5.7iii) into (5.7iv) we get

$$br + qc = a_x \qquad (5.8_1)$$

and

$$qc + rb = -d_x. \qquad (5.8_2)$$

Subtracting these two we get

$$a_x + d_x = 0,$$

which we satisfy by setting $d = -a$. Substituting this into the second relation in (5.7ii) and the second relation in (5.7iii) gives

$$b_x = (d-a)q = -2aq, \qquad (5.9_1)$$
$$c_x = (d-a)r = -2ar. \qquad (5.9_2)$$

Multiply (5.9_1) by c, (5.9_2) by b, and (5.8_1) by $2a$, and add; we get

$$cb_x + bc_x + 2aa_x = 0;$$

from this we conclude that

$$cb + a^2 = \text{const.}$$

We take that constant to be 1; so

$$a = (1 - bc)^{1/2}. \qquad (5.10)$$

Relations (5.9) and (5.10) constitute a system of differential equations for b and c; if initial values are specified, b and c are uniquely determined in terms of q and r. The first relations in (5.7ii) and (5.7iii):

$$q_t = 2b, \qquad r_t = 2c \qquad (5.11)$$

is a system of evolution equations for q and r; the right side is a nonlocal function of q and r.

Equation (5.11) is particularly simple when $q = r$; in this case we choose $b = c$; the resulting system occurs in the theory of self-induced transparency, see [8]. Relation (5.10) suggests the parametrization

$$b = \sin u, \quad a = \cos u. \tag{5.12}$$

Substituting this into (5.9) gives

$$\cos u u_x = -2 \cos u q$$

from which we deduce

$$q = -\tfrac{1}{2} u_x .$$

Substituting this into (5.11) and using (5.12) we get

$$u_{xt} + 4 \sin u = 0, \tag{5.13}$$

the so-called sine-Gordon equation. For application of these ideas to solutions of the sine-Gordon equation we refer the reader to [1].

Acknowledgments

The author is indebted for the shaping of his mathematical taste to Stan Ulam, to whom this article is affectionately dedicated on the occasion of his 65th birthday, and the 20th birthday of his pioneering paper with Enrico Fermi and John Pasta. Many happy recurrences.

References

1. M. J. Ablowitz, D. J. Kaup, A. C. Newell, and H. Segur, Nonlinear evolution equation of physical significance, *Phys. Rev. Letters* **31** (1973), 125.
2. L. Faddeev and V. E. Zakharov, Korteweg-de Vries equation as completely integrable Hamiltonian system, *Funkcional. Anal. i Priložen.* **5** (1971), 18–27.
3. E. Fermi, J. Pasta, and S. Ulam, Studies of nonlinear problems I, Los Alamos Report LA1940 (1955), "Collected papers of Enrico Fermi," Vol. II, p. 978, University of Chicago Press, IL, 1965; "Lectures in Applied Mathematics," Vol. 15, p. 143. *Amer. Math. Soc.* 1974.
4. H. Flaschka, Integrability of the Toda lattice, *Phys. Rev. B* (1974), 703.
5. H. Flaschka, On the Toda lattice, II. Inverse scattering solution, *Phys. Rev. B* **9** (1974), 1924.

6. C. S. GARDNER, Korteweg-de Vries equation and generalizations. IV. The Korteweg-de Vries equation as a Hamiltonian system, *J. Math. Phys.* **12** (1971), 1548–1551.
7. M. HÉNON, Integrals of the Toda lattice, *Phys. Rev. B* **9** (1974), 1921.
8. G. L. LAMB, *Rev. Mod. Phys.* **43** (1971), 99.
9. P. D. LAX, Integrals of nonlinear equations of evolution and solitary waves, *Comm. Pure Appl. Math.* **21** (1968), 467–490.
10. P. D. LAX, Periodic solution of the *KdV* equation, *Comm. Pure Appl. Math.* **28** (1975).
11. J. MOSER, On invariant curves of area-preserving mappings of an annulus, *Nachr. Akad. Wiss. Göttingen Math.-Phys. Kl. II* No. 1 (1962).
12. M. TODA, Vibrations of a nonlinear chain, *J. Phys. Soc. Japan* **22** (1967), 431.
13. M. TODA, Studies on a nonlinear lattice, *Ark. Fys. Sem. i. Trondheim.* No. 2 (1974).
14. V. E. ZAKHAROV AND A. B. SHABAT, *Soviet Phys. JETP* **34** (1972), 62.

The Real Numbers as a Wreath Product

F. Faltin

Cornell University, Ithaca, New York 14850

N. Metropolis

Los Alamos Scientific Laboratory, Los Alamos, New Mexico 87544

B. Ross and G.-C. Rota

Massachusetts Institute of Technology, Cambridge, Massachusetts 02139

DEDICATED TO S. M. ULAM

1. Introduction

Few mathematical structures have undergone as many revisions or have been presented in as many guises as the real numbers. Every generation re-examines the reals in the light of its values and mathematical objectives.

Our present purpose is one more such re-examination. We are motivated largely by the constructive views of mathematics which are making headway under the aegis of the computer.

It is often deplored that the field of real numbers is not constructive in any of the currently accepted meanings of the word. How then do we propose to adhere to the seemingly impossible objective of making the real numbers conform to the credo of constructivity?

A way out can be gleaned from a neglected corner of algebra, the theory of local fields. Few people would admit that much of the analytic simplicity and the explicit computability of the p-adic fields could be carried over to the real numbers by a simple algorithmic device. Yet, this is precisely what we attempt to do in this paper.

Our idea consists in obtaining the real numbers as the quotient, by a maximal ideal, of a specified subset of the ring of formal Laurent series with integer coefficients. In other words, a real number turns out to be an equivalence class of such formal Laurent series, or *strings* as we call them, and arithmetic operations are performed on these equivalence classes. In this way, addition and multiplication become carryless.

The carry operation of ordinary arithmetic is incorporated in the description of the equivalence classes, that is, in the ideal. Thus, in contrast with classical constructions, such as by Dedekind cuts or other topological devices, the field operations are explicitly performable and defined at the outset. On the other hand, a real number turns out to be represented by infinitely many strings of digits, any two equivalent strings being transformable into each other by a sequence of carries. In particular, the ordinary binary representation of a real number is obtained by performing a sequence of carries that we call clearing.

This procedure can be used to construct several other fields, for example the p-adics, and leads to the general notion of digital representation of fields, briefly developed at the end. The difference between p-adics and real numbers appears, as it must, in a boundedness restriction on the strings used in the construction of the real number field. Aside from this restriction, our construction is completely algorithmic, and leads, we would like to believe, to a revised concept of the real number system.

2. Synopsis

In this paper we shall give a construction of the real numbers that differs from those in current usage. We could construct the real numbers by algorithmically describing the operations of binary addition, multiplication, and division on infinite strings of zeroes and ones. It turns out, however, that an explicit description of the algorithms for the elementary operations is very cumbersome, owing to the presence of carries. A Boolean description of the carry operations involved in addition and multiplication was given in a previous paper in this series (Metropolis and Rota [4]).

In this paper we bypass the difficulty of an explicit description of carries by an extremely simple idea. We devise a notation that allows carries to remain unperformed. The digits of a string which is to represent a real number are allowed to be arbitrary integers. Two strings that differ only in that a carry has been performed on one of them are decreed equivalent, and a real number is an equivalence class under iterated performances of carry operations.

More precisely, the binary carry operation consists in removing two units from the nth digit and adding one unit to the $(n-1)$th.

The enormous advantage of the present notation is that addition, multiplication, and division become carryless operations, as they are simply operations on formal Laurent series. The price we pay is that a real number can be represented by infinitely many strings. But the nonuniqueness of representation of real numbers is not a novel phenomenon, even in ordinary arithmetic.

When arbitrary integer digits are allowed in the representation of real numbers, the definition of strings as well as the characterization of the equivalence relation defined by the carry operation become more delicate. From a pedestrian point of view, what we are doing is simply representing a real number in the form

$$\sum_n a_n/2^n,$$

where the a_n's are integers and where only a finite number of negative values of n appear. Two such representations are equivalent when the sums are the same. Abstracting from this well-known representation we are led to define a bounded string **A** as a sequence of integers a_n for which

$$\sum_n |a_n|/2^n$$

is finite. A carry string **C** is defined as a string for which

$$\sum_n |c_n - 2c_{n-1}|/2^n$$

is finite, and moreover $c_n/2^n$ tends to zero. Two bounded strings **A** and **B** are said to be equivalent when $\mathbf{A} = \mathbf{B} + \mathbf{KC}$, where **K** is the string 1. $-2000\cdots$, or digitwise when $a_n = b_n + c_n - 2c_{n-1}$; here **C** is a carry string, and multiplication is defined as in ordinary Laurent series. We then prove that in each equivalence class there is a unique clear string, that is, essentially, a unique binary real number. Thus, to recover ordinary addition and multiplication in the present context, one adds or multiplies clear strings and then clears the result.

The p-adic numbers can be treated along the same lines, in fact, even more simply, and so can several other fields of common occurrence in algebra.

3. Strings

Let R be a commutative ring with identity. A *string* with values in R is a function from the integers to the ring R, which may be written $\mathsf{A} = \{a_i, i \in Z\}$, with the property that $a_i = 0$ for all sufficiently small i. The value a_i is called the *i-th digit*.

Two notations will be used for strings. The first is intended to suggest the operations of ordinary arithmetic:

$$\mathsf{A} = \cdots a_{-n} a_{-n+1} \cdots a_0 \cdot a_1 a_2 a_3 \cdots;$$

the least integer i for which $a_i \neq 0$ is called the *leading digit*. The radix point is placed between the zeroth and first digits. A string whose leading digit is the zeroth digit is denoted by $\mathsf{A} = a_0 \cdot a_1 a_2 \cdots$. A string with finitely many nonzero digits is called a *finite string*. We sometimes use the notation $(\mathsf{A})_i$ to indicate the ith digit of the string A.

The second suggests the operations on strings:

$$\mathsf{A} = \sum_{n \in Z} a_n x^n,$$

where there are only a finite number of negative powers of x. Addition and multiplication of strings A and $\mathsf{B} = \{b_i, i \in Z\}$ are defined by the corresponding operations on formal Laurent series. We summarize the well-known results:

Addition $\mathsf{A} + \mathsf{B} = \mathsf{C}$ is defined elementarily by $a_i + b_i = c_i$.

Multiplication $\mathsf{AB} = \mathsf{C}$ is defined by convolution:

$$c_i = \sum_{n \in Z} a_n b_{i-n}.$$

Under these definitions the set of strings with digits in the ring R becomes a commutative ring with identity, written $\sum (R)$.

If R is a field, then the ring $\sum (R)$ is a field, and division of strings is defined by ordinary division of formal Laurent series.

The identity I of the ring $\sum (R)$ is the string with the zeroth digit equal to the multiplicative identity of R and all other digits equal to the zero element. The zero of $\Sigma(R)$ is the string O all of whose digits are the zero element of R.

A topology can be given to $\Sigma(R)$; define a neighborhood $N(D, A)$ of a string A, determined by a finite set D of digits, to be the set of all strings B such that $b_i = a_i$ for all i in D. The collection of all such $N(D, A)$ for all strings A in $\Sigma(R)$ is the basis for a topology under which the ring of strings $\Sigma(R)$ becomes a topological ring. Although topological considerations motivate some of the results derived in the sequel, they are not required for the reading of this work.

4. Equivalence of Strings

From now on, the ring R will be the ordinary integers, and the ring of strings with integer digits will be written $\Sigma(Z)$.

We next introduce an equivalence relation on the ring $\Sigma(Z)$ which corresponds to the intuitive notion of carrying in ordinary arithmetic.

For two strings A and B in $\Sigma(Z)$, we say B *can be carried* into A whenever $A = B + KC$, for some finite string C in $\Sigma(Z)$, where K is the constant string with $k_0 = 1$, $k_1 = -2$, and $k_i = 0$, for $i \neq 0$ or 1. We call C the carry of B into A. The string K is fixed throughout; it is called the *carry constant*.

Equivalently, for A, B in $\Sigma(Z)$, B can be carried into A whenever for some finite C in $\Sigma(Z)$,

$$a_i = b_i + c_i - 2c_{i-1}, \tag{*}$$

for all i in Z.

The correspondence to the notion of carrying is evident when one considers the case $c_i = 1$ and $c_j = 0$, for $j \neq i$. This carries any B into A where

$$a_i = b_i + 1,$$
$$a_{i+1} = b_{i+1} - 2, \quad \text{and}$$
$$a_j = b_j, \quad \text{for} \quad j \neq i, i+1.$$

We have removed two units from the $(i+1)$th digit and added one unit to the ith digit. Every carry (*) may be obtained by iteration of carries of this simple form.

One may easily show that this notion of carrying defines an equivalence relation on $\sum(Z)$.

We wish to extend the preceding definition to allow infinite carries **C**. However, any two strings turn out to be equivalent when arbitrary infinite carries are allowed. To obviate this difficulty we are led to the following definition:

A string **A** with integer digits is *bounded* when a positive integer z exists with the property that for all nonnegative integers n

$$\sum_{i \leqslant n} |a_i| \, 2^{n-i} \leqslant z 2^n,$$

where $|a_i|$ is the absolute value of the digit a_i.

The set of bounded strings will be denoted by $\sum_2(Z)$.

PROPOSITION 1. *The set $\sum_2(Z)$ of all bounded strings is a subring with identity of the ring $\sum(Z)$ of all strings with integer digits.*

The easy proof is omitted.

We can now extend the notion of carry so as to include certain infinite strings. The motivation for the following definition is deferred to the following section. A string **C** is a *carry string* if:

(1) **KC** is bounded, that is, there exists a positive integer z such that

$$\sum_{i \leqslant n} 2^{n-i} | c_i - 2c_{i-1} | \leqslant z 2^n,$$

for all nonnegative integers n, and

(2) For every positive integer z there is an integer $k \geqslant 0$ such that

$$z \, | c_j | \leqslant 2^j \quad \text{for all} \quad j > k.$$

The set of all carry strings is denoted by Φ.

We next verify some elementary properties of carry strings.

PROPOSITION 2.

(a) *The sum of carry strings is a carry string.*

(b) *The product of a bounded string and a carry string is a carry string.*

Proof.

(a) Consider the carry strings **A** and **B** in Φ. Since **K(A + B)** =

KA + KB, property (1) above is immediate. Note that $(A + B)_j = a_j + b_j$. Given a positive integer z, choose $k \geqslant 1$ so large that $(2z)|a_j| \leqslant 2^j$ for all $j > k$, and similarly for b_j. Then for $j > k$,

$$z|(A+B)_j| \leqslant z|a_j| + z|b_j| \leqslant 2^{j-1} + 2^{j-1} = 2^j,$$

as desired.

(b) Now let $A = BC$, where B is a bounded string and C is a carry string. Since $KA = (KC)B$ it follows that the string KA is bounded.

We must show that for each positive integer z, we can find an integer $j \geqslant 0$ such that $z|a_n| \leqslant 2^n$ for $n > j$. Taking b_0 and c_0 to be the leading digits of B and C we have

$$a_n = \sum_{i=0}^{n} b_i c_{n-i},$$

hence

$$z|a_n| \leqslant z \sum_{i=0}^{n} |b_i||c_{n-i}|.$$

Let m be the greatest integer for which $2m \leqslant n$; then for $n \geqslant 0$, omitting the absolute value signs for simplicity,

$$za_n \leqslant z \sum_{i=0}^{m} b_i c_{n-i} + z \sum_{i=m+1}^{n} b_i c_{n-i},$$

$$za_n 2^n \leqslant z \sum_{i=0}^{m} (b_i 2^{n-i})(c_{n-i} 2^i) + z \sum_{i=m+1}^{n} (b_i 2^{n-i})(c_{n-i} 2^i),$$

$$za_n 2^n \leqslant z \max_{i=0}^{m} [c_{n-i} 2^i] \sum_{i=0}^{m} b_i 2^{n-i} + z \max_{i=m+1}^{n} [c_{n-i} 2^i] \sum_{i=m+1}^{n} b_i 2^{n-1}.$$

A change of indices yields

$$za_n 2^n \leqslant z \max_{i=n-m}^{n} [c_i 2^{n-i}] \sum_{i=0}^{m} b_i 2^{n-i} + z \max_{i=0}^{n-m-1} [c_i 2^{n-i}] \sum_{i=m+1}^{n} b_i 2^{n-i}.$$

By the definition of a bounded string, there is some integer $k > 0$ such that

$$\sum_{i \leqslant n} b_i 2^{n-i} \leqslant k 2^n \quad \text{for all} \quad n \geqslant 0.$$

By the definition of a carry string, we can choose a $w \geqslant 0$ such that $|c_i| \leqslant 2^i$, for all $i > w$. Let v be the least positive integer such that $|c_i| \leqslant v 2^i$ for $0 \leqslant i \leqslant w$. Then for all $i \geqslant 0$, we have $|c_i| \leqslant v 2^i$. Thus the sequence c_i grows at most exponentially. Therefore

$$z a_n 2^n \leqslant z k 2^n \max_{i=n-m}^{n} [c_i 2^{n-i}] + v z 2^n \sum_{i=m+1}^{n} b_i 2^{n-i}.$$

Now choose $j_1 \geqslant 0$ so large that $2zk |c_i| \leqslant 2^i$, for $i > j_1$. Also choose $j_2 \geqslant 0$ such that for integers $x \geqslant y > j_2$,

$$2vz \sum_{i=y}^{x} b_i 2^{x-i} \leqslant 2^x \quad \text{(cf. Appendix).}$$

Let $j = 2\max(j_1, j_2)$. Then for $n > j$,

$$z a_n 2^n \leqslant 2^n 2^{n-1} + 2^n 2^{n-1} = 2^{2n}.$$

Thus $z a_n \leqslant 2^n$ for $n > j$. Q.E.D.

We can now extend the definition of equivalence of (bounded) strings to include (infinite) carry strings.

For bounded strings A *and* B, *we say that* A *is equivalent to* B, *in symbols* A $\sim$ B, *whenever there exists a carry string* C *such that* A $=$ B $+$ KC.

An important fact about equivalence is:

THEOREM 1. *Equivalence* ($\sim$) *is an equivalence relation on the ring of bounded strings.*

Proof. Reflexivity is immediate since O is a carry string.

Symmetry is also clear, since whenever C is a carry string, $-$C is also a carry string.

To establish transitivity, suppose A $=$ B $+$ KC and B $=$ F $+$ KG, where A, B, F are bounded strings and C, G are carry strings. Then A $=$ F $+$ KG $+$ KC $=$ F $+$ K(C $+$ G). It follows from Proposition 2(a) that A is equivalent to F.

This equivalence relation on bounded strings is nontrivial; we verify that O is not equivalent to I. Indeed, if O = I + KC, then C must be the string C = $-1 \cdot -2 -4 \cdots$; but the string C is not a carry string.

An equivalence class of bounded strings shall be called a *Real Number*.

5. Carries[1]

In the previous section we introduced two notions of carrying. The first defines an equivalence relation on the ring $\sum(Z)$ of all strings, but at the cost of restricting the carries to be finite strings. The second is obtained by choosing the subring of bounded strings and prescribing a significantly larger set of carries, which nonetheless defines a nontrivial equivalence relation on this ring. We shall now justify our definition of carry strings.

Let Λ be a subring of $\sum(Z)$ with a topology T making it into a topological ring. The *carry set* of Λ is the set of all strings C in $\sum(Z)$ such that:

(1) (closure) KC belongs to Λ,

(2) (truncation) let C_n be the string with digits c_j, for $j \leqslant n$, and 0 for $j > n$. If KC belongs to Λ, then KC_n is in Λ for all n in Z, and

(3) (convergence) the sequence KC_n converges to KC in Λ.

For example, the carry set for the entirety of $\sum(Z)$ under the topology of Section 3 is all of $\sum(Z)$ itself.

As a second example, the definition of bounded strings suggests a topology T_2 for the ring $\sum_2(Z)$ of bounded strings. A basis for T_2 is the set of all neighborhoods $N(A, z)$ of bounded strings A, where z is any positive integer, consisting of all strings B such that

$$z \sum_{i \leqslant p} 2^{p-i} \mid a_i - b_i \mid \leqslant 2^p,$$

for all nonnegative integers p.

The following fact is of limited importance, so we state it without proof.

[1] This section may be omitted at first reading without loss of continuity.

PROPOSITION 1. *Under the topology* T_2, *the ring* $\Sigma_2(Z)$ *becomes a complete topological ring.*

With a topology thus defined on $\Sigma_2(Z)$, we now inquire what the carry set of $\Sigma_2(Z)$ is. The answer is provided by:

PROPOSITION 2. *A string belongs to the carry set of* $\Sigma_2(Z)$ *if and only if it is a carry string in the sense of Section* 4.

Proof. Condition 1 of Section 4 for carry strings (KC is bounded) is clearly necessary. We next verify that Condition 2 of Section 4 is also necessary. We must show that $z \mid c_n \mid \leqslant 2^n$ for n sufficiently large and for any positive integer z. Let C_n be as in the definition of truncation above. By Condition (3) above, the sequence KC_n converges to KC. Therefore $K(C_n - C_{n-1})$ converges to the zero string.

Thus for every positive integer z, there must exist a positive integer k, such that for $n > k$,

$$z \sum_{i \leqslant p} 2^{p-i} \mid (K(C_n - C_{n-1}))_i \mid \leqslant 2^p$$

for all $p \geqslant 0$. In particular, for $p = n + 1$ the nonvanishing terms are

$$z[2^{n+1-(n+1)} \mid -2c_n \mid + 2^{n+1-n} \mid c_n \mid] \leqslant 2^{n+1}$$

whence

$$z[2 \mid c_n \mid + 2 \mid c_n \mid] \leqslant 2^{n+1},$$

and, lastly,

$$z \mid c_n \mid \leqslant 2^{n-1} < 2^n,$$

the desired conclusion.

We now establish the sufficiency of the conditions in Section 4. The boundedness of KC yields closure at once. Further, since for any carry string C, KC_n is a bounded string for all integers n, truncation is also immediate. Thus, we need only show that the sequence KC_n converges to KC for any carry string C. Thus, we must show that for any positive integer z there exists a k such that for $n > k$,

$$z \sum_{i \leqslant p} 2^{p-i} \mid (K(C - C_n))_i \mid \leqslant 2^p$$

for all $p \geqslant 0$. For $p \geqslant n + 1 \geqslant 0$, the left side equals

$$z \left[2^{p-n-1} \mid c_{n+1} \mid + \sum_{i=n+2}^{p} 2^{p-i} \mid c_i - 2c_{i-1} \mid \right].$$

By Property 2 of carry strings, we may choose $k_1 \geqslant 0$ so large that $(2z) \mid c_{n+1} \mid \leqslant 2^{n+1}$ for $n > k_1$. Also, using the boundedness of $K(C - C_n)$, choose (cf. Appendix) $k_2 \geqslant 0$ sufficiently large that

$$2z \sum_{i=n+2}^{p} 2^{p-i} \mid c_i - 2c_{i-1} \mid \leqslant 2^p$$

for all $p - 2 \geqslant n > k_2$. Then for all $p - 2 \geqslant n > k = \max(k_1, k_2)$ the above sum equals

$$z 2^{p-n-1} \mid c_{n+1} \mid + z \sum_{i=n+2}^{p} 2^{p-i} \mid c_i - 2c_{i-1} \mid \leqslant 2^{p-n-1} 2^n + 2^{p-1} = 2^p. \quad \text{Q.E.D.}$$

Thus we see that the definition of carry strings in Section 4 conforms to minimal conditions that assure reasonable behavior of the carries on the topological ring $\Sigma_2(Z)$. In the ensuing sections these facts will be used to establish the central result, that the equivalence classes of bounded strings form a field isomorphic to the reals.

6. Clearing

A *clear string* is a bounded string with the following properties:

(a) The leading digit equals 1 or -1.

(b) If the leading digit is -1, then the following digit is 0.

(c) All digits to the right of the leading digit are 0 or 1.

(d) There is no integer j such that all the digits beyond the jth equal 1.

The reason for condition (b) on the leading digit is that

$$-1 \cdot 1 a_2 a_3 a_4 \cdots \sim 0 \cdot -1 a_2 a_3 a_4 \cdots.$$

Our present purpose is to show that every bounded string is equivalent to a clear string. To this end, some preliminary results will be derived.

PROPOSITION 1. *Let* A *be a finite string with* n *the greatest integer for which* $a_n \neq 0$. *Then for any integer* j, *there exists a finite string* C *such that* $(A + KC)_i = 0$ *or* 1 *for all* $i > j$ *and*

(1) $\sum_{i=j}^{n-1} |2c_i - c_{i+1}| 2^{n-1-i} \leqslant \sum_{i=j+1}^{n} |a_i| 2^{n-i} + 2^n$,

(2) $|c_i| 2^{n-i} \leqslant |a_{i+1}| 2^{n-i-1} + |a_{i+2}| 2^{n-i-2} + \cdots + |a_n| + 2^{n-i}$

for $i \geqslant j$.

Proof. The conclusion is trivial if $j \geqslant n$. Otherwise we may assume $j = 0$. There exists a unique sequence ϵ_i, defined for $1 \leqslant i \leqslant n$, where $\epsilon_i = 0$ or 1 such that the equations

$$2c_{n-1} = a_n - \epsilon_n,$$
$$2c_{n-2} - c_{n-1} = a_{n-1} - \epsilon_{n-1}, \qquad (*)$$
$$\cdots$$
$$2c_0 - c_1 = a_1 - \epsilon_1$$

have a solution in integers c_{i-1}. All other digits of C are defined to be zero.

Taking absolute values, multiplying by suitable powers of two, and summing, we obtain

$$\sum_{i=0}^{n-1} |2c_i - c_{i+1}| 2^{n-1-i} \leqslant \sum_{i=1}^{n} |a_i| 2^{n-i} + \sum_{i=1}^{n} \epsilon_i 2^{n-i}.$$

This completes the proof of (1).

To prove (2), note that for $0 \leqslant i < n$ the explicit solution of (*) satisfies

$$2^{n-i} c_i = a_{i+1} 2^{n-i-1} + a_{i+2} 2^{n-i-2} + \cdots + a_{n-1} 2^1 + a_n 2^0$$
$$- (\epsilon_{i+1} 2^{n-i-1} + \epsilon_{i+2} 2^{n-i-2} + \cdots + \epsilon_{n-1} 2^1 + \epsilon_n 2^0). \qquad (**)$$

Take absolute values and remark that

$$\epsilon_{i+1} 2^{n-i-1} + \epsilon_{i+2} 2^{n-i-2} + \cdots + \epsilon_{n-1} 2^1 + \epsilon_n 2^0$$
$$\leqslant 2^{n-i-1} + 2^{n-i-2} + \cdots + 2^1 + 2^0 < 2^{n-i}.$$

PROPOSITION 2. *Let* A *be a bounded string with nonnegative digits. Then there is a carry string* B *such that* $(A + KB)_i = 0$ *or* 1 *for all* i.

Proof. Assume that the leading digit of A is a_0. If A is not finite, define $\mathbf{A}_n$ by $(\mathbf{A}_n)_i = a_i$ for $i \leqslant n$ and $(\mathbf{A}_n)_i = 0$ otherwise. Let $\mathbf{C}_n$ be related to $\mathbf{A}_n$ as in Proposition 1 with $j = 0$.

$\mathbf{A} - \mathbf{A}_n$ is a bounded string. Therefore there exist positive integers h and k such that

$$\sum_{i=n+1}^{m} a_i 2^{m-i} \leqslant h2^m = h2^n 2^{m-n} = k 2^{m-n}$$

for all $m \geqslant n \geqslant 0$. Thus for each such n there exists a least integer k_n such that $a_{n+1} 2^{m-n-1} + a_{n+2} 2^{m-n-2} + \cdots + a_m 2^0 \leqslant k_n 2^{m-n}$. Then there is some integer r, such that for all $s \geqslant r$,

$$a_{n+1} 2^{s-n-1} + a_{n+2} 2^{s-n-2} + \cdots + a_{s-1} 2^1 + a_s 2^0 > (k_n - 1) 2^{s-n},$$

since k_n is minimal.

From (**) we see that $(\mathbf{C}_m)_n$ is the greatest integer such that $2^{m-n}(\mathbf{C}_m)_n \leqslant a_{n+1} 2^{m-n-1} + a_{n+2} 2^{m-n-2} + \cdots + a_m 2^0$. Since all digits of $\mathbf{A}$ are nonnegative, the sequence $(\mathbf{C}_m)_n$, $(\mathbf{C}_{m+1})_n$,... is nondecreasing. But then $k_n - 1 = (\mathbf{C}_r)_n = (\mathbf{C}_{r+1})_n = \cdots$. Let $\mathbf{C}$ be the string for which $c_n = 0$ for $n < 0$ and $c_n = k_n - 1$ for $n \geqslant 0$. Since the $\mathbf{C}_m$'s obey the inequalities of Proposition 1, $\mathbf{C}$ is a carry string by the Tail Sum Lemma (cf. Appendix).

If $\mathbf{A}$ is finite, let the carry string $\mathbf{C}$ be as in Proposition 1 with $j = 0$.

In either case $(\mathbf{A} + K\mathbf{C})_i = 0$ or 1 for $i > 0$. Denote $(\mathbf{A} + K\mathbf{C})_0$ by p. Let q be the least integer such that $2^q \geqslant p$. For $j = -q$ let the carry string $\mathbf{D}$ be related to the bounded string $\cdots 0p \cdot 0 \cdots$ as in Proposition 1. Then $\mathbf{B} = \mathbf{C} + \mathbf{D}$ is the desired carry string.

PROPOSITION 3. *Let $\mathbf{A}$ be any bounded string. Then there is a bounded string $\mathbf{D}$ equivalent to $\mathbf{A}$ which has digits 0 or 1, except, perhaps, for the leading digit, which may be $1, -1,$ or -2.*

Proof. Let $\mathbf{A}'$ be the string for which $a_i' = -a_i$ if $a_i < 0$, and $a_i' = 0$ otherwise. Also define $\mathbf{A}'' = \mathbf{A} + \mathbf{A}'$. Then $\mathbf{A}'$ and $\mathbf{A}''$ have nonnegative digits, and thus there exist carry strings $\mathbf{C}'$ and $\mathbf{C}''$ as in Proposition 2. Let $\mathbf{B} = (\mathbf{A}'' + K\mathbf{C}'') - (\mathbf{A}' + K\mathbf{C}')$; then $\mathbf{B} \sim \mathbf{A}$ and $|b_i| \leqslant 1$ for all i.

Assume the leading digit of $\mathbf{B}$ is b_0. Let $0 < k_1 < k_2 < \cdots$ be all positive integers such that $b_{k_i} = -1$. For each such k_i, let $n_i \geqslant 0$ be the greatest integer such that $n_i < k_i$ and $b_{n_i} \neq 0$.

Then define a carry string C by

$$C = \begin{cases} c_j = -1 & \text{if } n_i \leqslant j < k_i \text{ for some } i, \\ c_j = 0 & \text{otherwise.} \end{cases}$$

We claim $D = B + KC$ is the desired string.

For $j > 0$, suppose $b_p = 0$ for all $p > j$. Then $d_j = b_j$ for $b_j = 0$ or 1; if $b_j = -1$, $d_j = b_j + c_j - 2c_{j-1} = -1 + 0 - 2(-1) = 1$.

Otherwise let m be the least integer greater than j such that $b_m \neq 0$. Suppose $b_j = 1$. Then if $b_m = -1$, we have $d_j = 1 + (-1) - 2(0) = 0$; if $b_m = 1$, $c_j = c_{j-1} = 0$, so $d_j = b_j = 1$.

Suppose $b_j = 0$. For $b_m = -1$, $d_j = 0 + (-1) - 2(-1) = 1$; for $b_m = 1$, $d_j = b_j$ as before.

Finally, suppose $b_j = -1$. If $b_m = -1$, $d_j = -1 + (-1) - 2(-1) = 0$; if $b_m = 1$, $d_j = -1 + 0 - 2(-1) = 1$.

One sees easily that the leading digit of D is d_0 or d_1. The simple verification that $-2 \leqslant d_0 \leqslant 1$ then completes the proof.

PROPOSITION 4. *Let A be a bounded string. Suppose there exists an integer j, such that $a_i = 1$ for all $i > j$. Then A is equivalent to a finite string.*

Proof. We may assume $j = 0$. Set

$$c_i = \begin{cases} 0 & \text{for } i < 0, \\ 1 & \text{for } i \geqslant 0. \end{cases}$$

This defines a carry string C. Let $B = A + KC$. Then B is a finite string. Q.E.D.

This result shows that the string constructed in Proposition 3 may easily be cleared, that is, it is equivalent to a clear string. We have thus proved the result announced at the beginning of the section:

THEOREM 2. *Let A be a bounded string. Then there exists a carry string C which clears A.*

THEOREM 3. *Two clear strings are equivalent only if they are equal.*

Proof. Let A and B be distinct clear strings such that $D = A - B = KC$, for some carry string C. We shall establish a contradiction.

Assume that the leading digit of D is $d_0 > 0$. Since A and B are clear, we then have $d_0 = 1$ or 2 and $d_n = -1, 0,$ or 1 for $n \geqslant 1$.

Solving for C using $c_0 = d_0$ and $c_n - 2c_{n-1} = d_n$ for $n \geq 1$, we find inductively that

$$c_n = \sum_{i=0}^{n} 2^{n-i} d_i = 2^n d_0 + \sum_{i=1}^{n} 2^{n-i} d_i.$$

Suppose not all d_i are negative for $i \geq 1$. Then for all large n we have the inequality

$$c_n > 2^{n-i}$$

for some i, and C cannot be a carry string, as Condition (2) in the definition is violated.

We infer that $d_i = -1$ for all $i \geq 1$. But then $c_n = 2^n(d_0 - 1) + 1$ for $n \geq 0$. If $d_0 = 2$, Condition (2) is again violated; otherwise, $c_n = 1$ for all $n \geq 0$. Then $b_i = 1$ for all $i > 1$, and B is not clear.   Q.E.D.

These two theorems establish that there is a one-to-one correspondence between clear strings and Real Numbers.

7. Addition and Multiplication

An arbitrary equivalence relation $*$ is *compatible* with addition and multiplication if whenever $A * B$ and $D * E$, then $A + D * B + E$ and $AD * BE$.

PROPOSITION 1. *Equivalence of bounded strings* $(\sim)$ *is compatible with string addition and string multiplication.*

Proof. Let $A = B + KC$ and $D = E + KF$. Then $A + D = B + KC + E + KF = B + E + K(C + F)$. Since the sum of two carry strings is a carry string, $A + D \sim B + E$.

Similarly, $AD = (B + KC)(E + KF) = BE + K(CE + BF + KCF)$. We have shown in Proposition 2(b) of Section 4 that the product of a bounded string, and a carry string is a carry string; hence $AD \sim BE$.
     Q.E.D.

Recall that $O = 0.000...$, $I = 1.000...$, and let

$$-A = \cdots - a_{-1} - a_0 \cdot -a_1 a_2 \cdots.$$

Clearly $A + O = A$, $AO = O$, $AI = A$, and $A + (-A) = O$. Thus, using the results of Section 3 and the fact that string addition and multiplication are compatible with equivalence, we have

THEOREM 4. *The set of equivalence classes of bounded strings under string addition and multiplication form a commutative ring with identity.*

8. DIVISION

As is shown by the example $\mathsf{A} = 1 \cdot 1 -1\, 1 -1 \cdots$, which has inverse $\mathsf{B} = 1 \cdot -1\, 2 -4\, 8 \cdots$, the inverse of a bounded string is not always bounded. It turns out, nevertheless, that the inverse of a *clear* string is always bounded. This will lead to a proof of the fact that the Real Numbers form a field, and moreover provide an algorithm for division that appears to be new and that is in some ways preferable to conventional long division. The algorithm can be summarized as follows: to find the quotient of two real numbers (in binary form), first find the quotient of the numbers considered as strings, and then clear the resulting string.

Let $\mathsf{A} = a_0 \cdot a_1 a_2 a_3 \cdots$ be a clear string whose leading digit is $a_0 = \pm 1$. The string B inverse to A, that is, the string for which $\mathsf{AB} = \mathsf{I}$, where, as usual, $\mathsf{I} = 1.000 \cdots$, is determined recursively by the equations

$$b_0 = a_0,$$
$$b_n = -a_0(b_{n-1}a_1 + b_{n-2}a_2 + \cdots + b_0 a_n), \quad \text{for} \quad n \geq 1. \qquad (*)$$

LEMMA 1. *If b_n is determined by* (*), *then for all n,*

$$|b_n| \leq c_n,$$

where c_n is the Fibonacci sequence of integers determined recursively by the equations

$$c_0 = c_1 = c_2 = 1,$$
$$c_n = c_{n-1} + c_{n-2}, \quad \text{for} \quad n \geq 3. \qquad (**)$$

Proof.

Case 1. $a_1 = 0$. Then for $n > 1$ we have, recalling that $a_i = 0$ or 1 for $i \geq 1$,

$$|b_n| \leq |b_{n-2}| + |b_{n-3}| + \cdots + |b_0|$$

since $b_0 = a_0$ and $b_1 = 0$.

Case 2. $a_1 = 1$. Since **A** is clear, we must have $a_0 = 1$. Substituting on the right side of (*) the recursive equation for b_{n-1}, we find again that

$$|b_n| \leqslant |b_{n-2}| + |b_{n-3}| + \cdots + |b_0|,$$

since

$$b_n = b_{n-2}(a_1 - a_2) + b_{n-3}(a_2 - a_3) + \cdots + b_0(a_{n-1} - a_n)$$

and $|a_i - a_{i-1}| \leqslant 1$ for all i. Thus, $|b_2| \leqslant |b_0| = 1$.

In either case, $|b_n| \leqslant \sum_{i=2}^{n} |b_{n-i}|$. Now let $c_0 = c_1 = c_2 = 1$ and

$$c_n = \sum_{i=2}^{n} c_{n-i} \quad \text{for} \quad n \geqslant 3.$$

Then it is obvious that $|b_n| \leqslant c_n$ for all $n \geqslant 0$. Moreover, from the above equality we infer that for $n \geqslant 3$

$$c_n = c_{n-2} + \sum_{i=2}^{n-1} c_{n-1-i} = c_{n-2} + c_{n-1},$$

as desired.

LEMMA 2. *For the sequence c_n defined in the preceding lemma the following inequalities hold*:

$$4c_n \leqslant 7c_{n-1}, \quad \text{for} \quad n \geqslant 4.$$

Proof. We shall proceed by induction, proving the above inequality simultaneously with the inequality

$$3c_n \geqslant 4c_{n-1}, \quad \text{for} \quad n \geqslant 4.$$

For $n = 4$ we have $c_4 = 3$ and $c_3 = 2$ and the verification is immediate. Assuming now $4c_{n-1} \leqslant 7c_{n-2}$ and $4c_{n-2} \leqslant 3c_{n-1}$, we find, from (**), the inequality $7c_n \geqslant 7c_{n-1} + 4c_{n-1} = 11c_{n-1}$, which trivially implies $3c_n \geqslant 4c_{n-1}$.

Similarly, the identity $4c_n = 4c_{n-1} + 4c_{n-2}$ together with the second induction hypothesis gives $4c_n \leqslant 4c_{n-1} + 3c_{n-1} = 7c_{n-1}$, as desired. Moreover, from $c_0 = c_1 = c_2 = 1$ and $c_3 = 2$, we have $4^i c_i \leqslant 7^i$ for all $i \geqslant 0$.

Using the preceding lemmas, we can now establish our main result.

THEOREM 5. *If A is a clear string, then the string B for which* $AB = I$ *is a bounded string.*

Proof. Assume as in the lemmas that a_0 is the leading digit of A. Since $|b_n| \leqslant c_n$, we need only show that $\sum_{i=0}^{n} 2^{n-i} c_i \leqslant k2^n$, for all $n \geqslant 0$ and some integer k. Indeed, it suffices to show that $\sum_{i=0}^{n} 7^i 8^{n-i} \leqslant 8 \cdot 8^n$ for all n. For, from the preceding lemma, we know that $4^i c_i \leqslant 7^i$, so that $\sum_{i=0}^{n} 4^i c_i 8^{n-i} \leqslant 8 \cdot 8^n$; consequently, $\sum_{i=0}^{n} c_i 2^{n-i} \leqslant 8 \cdot 2^n$. The proof is completed by remarking that

$$\sum_{i=0}^{n} 7^i 8^{n-i} = (8-7) \sum_{i=0}^{n} 7^i 8^{n-i} = 8^{n+1} - 7^{n+1} = 8 \cdot 8^n - 7^{n+1} \leqslant 8 \cdot 8^n,$$

as desired.

We can now establish the validity of the algorithm proposed at the beginning of the section.

COROLLARY. *If A is a clear string and F is a bounded string, there exists a bounded string G such that* $G = F/A$, *that is,* $AG = F$.

Proof. Let $G = BF$, where B is the inverse of A as above.

We thus have a new algorithm for long division of real numbers. In particular, let A, B be ordinary positive real numbers in binary notation. Considering A and B as bounded strings, we may take the quotient A/B in the ring of bounded strings. The resulting string can be cleared to give the ordinary quotient of A and B.

9. THE REAL NUMBERS

We have shown that equivalence classes of strings can be added, subtracted, multiplied, and divided. It is an easy matter to show that such equivalence classes, or Real Numbers, as we have called them, form a field. To complete such a verification, we need only show that Real Numbers form an integral domain under addition and multiplication. But this is immediate from Theorem 5. Indeed, suppose A and B are strings and $AB \sim O$. We may assume A and B are clear. If A is not equivalent to O, then multiply both sides by the inverse of A, A^{-1}; we then see that $A^{-1}AB \sim A^{-1}O$, or $B \sim O$, as desired. In conclusion we have

THEOREM 6. *Equivalence classes of strings (Real Numbers) form a field.*

We prove next that this field is the field of ordinary real numbers. To this end, it suffices to verify that it is a complete ordered field.

Say that a Real Number is positive, if, when cleared, its leading digit is 1, and is negative if, when cleared, its leading digit is -1. The sum and product of two positive real numbers is positive. It follows that the set of positive strings defines a linear order. It is easy to reconstruct the explicit description of this order relation: it is the usual "lexicographic" ordering of clear strings.

Finally, the verification that every bounded set of positive real numbers has a least upper bound is carried out much like in ordinary arithmetic, digit by digit. Thus we have

THEOREM 7. *The Real Numbers are a complete ordered field.*

10. DIGITAL REPRESENTATION OF FIELDS

We briefly summarize the construction of the real number field given in the preceding sections, with a view towards extending the construction to other fields.

Starting with a naive notion of carry, we were led to introduce the carry constant $K = 1 \cdot -2000 \cdots$ and thereby to formalize the iteration of carries in the language of strings. This suggests an investigation of the fields that can be obtained by varying the carry constant. In general, the digits of the string need not even be integers, but can belong to an arbitrary commutative ring, say R, with identity. This leads to the following definition:

Two strings A and B are said to be *equivalent relative to a string* K whenever

$$A = B + KC$$

for some string C.

It is immediate that this is an equivalence relation.

In the language of ring theory, this equivalence relation is simply the equivalence relation defined by the principal ideal I(K) generated by K. Thus, the set of equivalence classes is isomorphic to the quotient ring $\sum (R)/\text{I}(K)$.

A field F is said to be *digitally represented by the ideal* I *of the ring*

$\sum(R)$ whenever an isomorphism is given between F and the quotient ring $\sum(R)/I$. Intuitively, the ideal I describes the carry rules.

In this section, we briefly consider the case where the digits are integers. As we shall see, it then suffices to take principal ideals $I = I(K)$, whose generators (arbitrarily chosen) will again be called carry constants.

Recall that a Euclidean domain is a ring in which the division algorithm holds.

PROPOSITION 1. *The ring of strings $\sum(Z)$ is a Euclidean domain.*

Proof. If S is a nonzero string, define $d(S)$ to be the absolute value of the leading digit. Then for any nonzero string T, $d(S) \leqslant d(ST)$ and it is easily verified that the division algorithm works. That is, given S and T, there is an R such that $S = QT + R$, with either $R = O$ or $d(R) < d(T)$. We omit the routine details.

PROPOSITION 2. *A string K is a unit (that is, an invertible element) of the ring $\sum(Z)$ if and only if its leading digit is ± 1.*

The easy proof is omitted.

Recall that a string K is said to be irreducible when K is not a unit and $K = AB$ only when either A or B is a unit.

PROPOSITION 3.

(a) *A string K in $\sum(Z)$ is irreducible when the leading digit is a prime.*

(b) *If a string K is irreducible then the leading digit of K is a power of a prime.*

Proof.

(a) We may assume that the leading digit of each of the following strings is the zeroth digit. Thus, if k_0 is a prime and $AB = K$, then $a_0 b_0 = k_0$ and either a_0 or b_0 is 1. The conclusion thus follows from the preceding proposition.

(b) If k_0 is not a power of a prime, then we can find relatively prime integers $a_0, b_0 > 1$ such that $a_0 b_0 = k_0$. The equation

$$a_0 b_n + b_0 a_n = k_n - \sum_{j=1}^{n-1} a_j b_{n-j}$$

can be recursively solved for a_n and b_n in view of the fact that a_0 and b_0 are relatively prime. In this way, two strings A and B may be defined, for which AB = K. Neither A nor B is a unit. The proof is thus concluded.

Note that the converse of part (a) fails, as is shown by the string $p^2 \cdot -1000 \cdots$ for prime p.

Suppose now that the field F has a digital representation with integer digits and with carry rules I. Then the ideal I must be maximal. In virtue of Proposition 1, every ideal in $\sum(Z)$ is principal; thus, I is generated by a string K, that is I = I(K). Consequently the carry rules are described by a single carry constant K, much as in ordinary and p-adic arithmetic.

The problem of determining all fields having integer digital representations is thus reduced by Proposition 1 to the problem of determining all irreducible strings in $\sum(Z)$. These problems will be dealt with in a succeeding paper of this series; we only sketch an elementary result.

PROPOSITION 4. *The only field of characteristic $p \neq 0$ which is digitally representable with integer digits is the field $\sum(Z_p)$ of strings with digits in the prime field Z_p with p elements. In other words, a carry constant K generates a field of characteristic p if and only if K is an associate of $\cdots 0p \cdot 000 \cdots$.*

Proof. Let P be the string $p \cdot 000 \cdots$. The field $\sum(Z)/I(K)$ has characteristic p if and only if the sum of p terms each equal to the identity string, that is, the string P, satisfies

$$P = O + KC$$

for some string C. This implies $k_0 c_0 = p$. Since K is not a unit, we must have $k_0 = p$ and $c_0 = 1$. Thus, the string P is also a generator of the ideal I(K), and it is easily verified that the quotient $\sum(Z)/I(K)$ is the desired field.

In closing, we recall that our construction of the reals called for a restriction of a topological nature in the definitions of string and of carry. We leave to another occasion the development of topological digital representations, indispensable for a concept of digital representation of global fields.

In the next section we show that the p-adic fields allow an explicit digital representation, constructed as a quotient field of $\sum(Z)$.

11. THE p-ADIC FIELDS

We shall outline a construction of the p-adic fields in the spirit of the preceding section.

Let P be the string $\cdots 00p \cdot -1000 \cdots$, where p is a prime. By Proposition 3, P is an irreducible string, and therefore the quotient of the ring of strings $\Sigma(Z)$ by the principal ideal I(P) generated by the string P is a field by Proposition 1 of the preceding section. We shall call this field the field of p-adic numbers, denoted by Q_p.

The structure of the field Q_p is easily determined. In analogy with our construction of the real number field, we shall say a string S is p-clear if all its digits satisfy $0 \leqslant s_i < p$. Recall from the preceding section that two strings S and T are p-equivalent, that is, their difference belongs to the ideal I(P), whenever $S = T + PC$ for some string C.

The p-adic analogue of clearing is

PROPOSITION 1. *For each string S there exists a p-clear string T that is p-equivalent to S.*

Proof. Say s_0 is the first digit of S.

For any integer j, define $G_p(j)$ to be the greatest integer k such that $kp \leqslant j$.

Now let C be the string with digits

$$c_n = \begin{cases} 0, & \text{if } n < 0 \\ -G_p(s_n - c_{n-1}), & \text{if } n \geqslant 0. \end{cases}$$

Let $T = S + PC$, so that

$$t_n = \begin{cases} 0, & \text{if } n < 0 \\ s_n + pc_n - c_{n-1}, & \text{if } n \geqslant 0. \end{cases}$$

Now $s_n + pc_n - c_{n-1} = (s_n - c_{n-1}) - pG_p(s_n - c_{n-1})$. Thus $t_n \geqslant 0$ for all integers n; moreover, since $G_p(j)$ is the greatest integer such that $pG_p(j) \leqslant j$, it is also clear that $t_n < p$ for all n.

Therefore, T is a p-clear string p-equivalent to S, as desired.

The analog of uniqueness for the reals (Theorem 2) is:

PROPOSITION 2. *No two distinct p-clear strings are p-equivalent.*

Proof. Let S, T be distinct p-clear strings, and suppose $S - T = PC$ for some string C. The leading digit of $S - T$, say $s_k - t_k$, must be divisible by p. But $0 \leqslant s_k, t_k \leqslant p - 1$, hence $-(p - 1) \leqslant s_k - t_k \leqslant p - 1$. If p is to divide $s_k - t_k$, we must then have $s_k - t_k = 0$, but then $s_k - t_k$ is not the leading digit of $\mathsf{S} - \mathsf{T}$.

Thus, every string is p-equivalent to exactly one p-clear string.

We conclude by indicating how the well-known valuation on p-adic fields is constructed in the present context. If S is a p-clear string whose leading digit is the nth, set $|\,\mathsf{S}\,|_p = 1/p^n$. It is easily verified (Bachman [3, pp. 2–4]) that this is indeed the p-adic valuation. Thus Q_p is indeed a complete valued field of characteristic 0.

We thus have obtained the p-adic fields Q_p by means of our concept of a digital representation. To be sure, the present representation does not greatly differ from the classical construction of Hensel [8]. The novelty consists in identifying an element of the field, not with a single string, as is done classically, but with an equivalence class of strings containing infinitely many elements. That this approach permits a simpler and more lucid expression of the carries is the major idea of the present paper.

It is to be noted that several other representations of the p-adic fields Q_p are available. While we leave a complete investigation of such representations to another occasion (see however Tanny's thesis [10]), it may be worth noting that the present idea leads to a pleasing simplification in the construction of the classical Witt fields.

12. Further Work

We sketch a few of the lines of further development and applications of the idea of digital representation. Some of these are proposed as open problems, and some are given by way of motivation.

1. *Algorithms.* The device we used for division can be considerably generalized to obtain algorithms for various algebraic operations. The idea is to extend the recursive procedure used in Hensel's Lemma for p-adic fields. The advantage of the present development of the reals is that an algorithm reminiscent of Hensel's Lemma can often be made to work. Suppose, for example, that we wish to find a root of the equation $f(x) = 0$ in either real or p-adic fields. We may proceed in two steps. First, we try to find a string which satisfies the equation in the ring of strings; in the case of the reals, we then verify that such a string is bounded. To this end we establish a procedure whereby the digits

of the string are recursively determined. With the computation of a new digit, one carries in such a way that the string is still bounded. Carries may have to be performed at each stage in order to ensure the solvability of the equation (a common occurrence in p-adic fields). It would be interesting, for example, to give a proof of the fundamental theorem of algebra along these lines. Or, more modestly, one might seek to develop efficient algorithms for ordinary algebraic operations such as roots, logarithms, and exponentials.

2. *Complexity of algebraic operations.* The *dependence function* $d(A) = S$ of a bounded string A is a function whose value is a string S defined as follows:

Let x_n be the greatest integer such that $x_n < \sum_{j=n+1}^{\infty} |a_j|/2^{j-n}$. Then s_n is the least nonnegative integer for which $\sum_{j=n+1}^{n+s_n} |a_j|/2^{j-n} \geqslant x_n$.

In other words, the nth digit of A becomes in some sense *stable* after clearing the succeeding s_n digits. The string cannot be said to be constructively given unless its dependence function is specified.

The dependence function can be used to classify the complexity of algebraic operations. For example, the sum of 2^k clear strings is a string whose dependence function has, for large n, the constant value k. Similarly, one expects that the dependence function of the product of N clear strings is of the order of n^N, and that the dependence function of the inverse of a string is of the order of $(1 \pm \sqrt{5})/2)^n$. We surmise that these facts can be used to derive Tarski's decision procedure for real fields, and that similar considerations for p-adic fields lead to Cohen's decision criteria. Knowledge of the dependence function allows us to determine the value of the string to within the desired accuracy, and can be used to investigate the propagation of errors.

3. *Asymptotic strings.* An unbounded string may be obtained by formally solving an algebraic or transcendental equation in the ring of strings. For example, unbounded strings may sometimes be obtained by inverting bounded strings. Strictly speaking, such strings do not represent real numbers. Nevertheless, in several cases the string obtained by disregarding all but a finite number of digits gives a good approximation to the real solution. The situation is similar to that which obtains in ordinary differential equations, where a divergent series which formally satisfies the equation turns out to be asymptotic to a real solution. We are led to conjecture that a notion of asymptotic strings can be developed along these lines.

4. Digital representation.

Among the applications of the concept of digital representation used in this paper, one of the most fruitful is to Witt vectors. It turns out that Witt's scheme [7] for the construction of local fields with given residue class field can be greatly simplified in the language of carrying. This program was begun in Tanny's thesis [10] and will be expanded elsewhere.

Appendix

We shall prove below a useful property of bounded strings which is needed in establishing the results of Sections 4, 5, and 6. This lemma is the analog in integer terminology of the familiar fact that

$$\lim_{j \to \infty} \sum_{i=j}^{\infty} |a_i|/2^i = 0$$

for a sequence of reals $\{a_i\}$ for which

$$\sum_{i=0}^{\infty} |a_i|/2^i < \infty.$$

THE TAIL SUM LEMMA. *Let* $A = \cdots a_{-1}a_0 \cdot a_1a_2 \cdots$ *be a bounded string. Then for any positive integer z there exists a nonnegative integer j such that*

$$z \sum_{i=n}^{m} 2^{m-i} |a_i| \leqslant 2^m$$

for all integers $m \geqslant n > j$.

Proof. Suppose the contrary. Then for some positive integer p there exists for every nonnegative integer j, integers $m \geqslant n > j$ for which

$$p \sum_{i=n}^{m} 2^{m-i} |a_i| > 2^m.$$

Since A is bounded, there exists some nonnegative integer k such that

$$\sum_{i \leqslant s} 2^{s-i} |a_i| \leqslant k2^s.$$

for all nonnegative integers s. Fix an integer $q \geqslant 0$. Then let v be the least integer such that

$$v > k2^q - \sum_{i \leqslant q} 2^{q-i} |a_i| \geqslant 0.$$

Let $w = pv$. Pick $n_w > n_{w-1} > \cdots > n_1 > q$ and $m_w > m_{w-1} > \cdots > m_1 \geqslant n_1$ such that

$$m_h < n_{h+1} \leqslant m_{h+1} \quad \text{for} \quad 1 \leqslant h \leqslant w-1,$$

and

$$p \sum_{i=n_h}^{m_h} 2^{m_h-i} |a_i| > 2^{m_h}$$

for each h, $1 \leqslant h \leqslant w$.

Then

$$p \sum_{i=n_h}^{m_h} 2^{m_w-i} |a_i| > 2^{m_w}$$

for each such h; hence

$$p \sum_{i \leqslant m_w} 2^{m_w-i} |a_i| = p \sum_{i \leqslant q} 2^{m_w-i} |a_i| + p \sum_{i=q+1}^{m_w} 2^{m_w-i} |a_i|$$

$$\geqslant p 2^{m_w-q} \sum_{i \leqslant q} 2^{q-i} |a_i| + \sum_{h=1}^{w} p \sum_{i=n_h}^{m_h} 2^{m_w-i} |a_i|$$

$$\geqslant p 2^{m_w-q} \sum_{i \leqslant q} 2^{q-i} |a_i| + vp 2^{m_w}.$$

Hence

$$\sum_{i \leqslant m_w} 2^{m_w-i} |a_i| \geqslant 2^{m_w-q} \sum_{i \leqslant q} 2^{q-i} |a_i| + v 2^{m_w-q}$$

$$= 2^{m_w-q} \left(\sum_{i \leqslant q} 2^{q-i} |a_i| + v \right)$$

$$> k 2^{m_w}.$$

But this contradicts our definition of k. Q.E.D.

References

1. N. Metropolis, G.-C. Rota, and S. Tanny, Significance arithmetic: The carrying algorithm, *J. Combinatorial Theory* **14** (1973), 386–421.
2. O. Zariski and P. Samuel, "Commutative Algebra," Vol. 1, Van Nostrand, Princeton, NJ, 1958.
3. G. Bachman, "Introduction to p-adic Numbers and Valuation Theory," Academic Press, New York, 1964.
4. N. Metropolis and G.-C. Rota, Significance arithmetic: On the algebra of binary strings, published in the volume dedicated to Cornelius Lanczos, Academic Press, London, 1973.
5. S. Lang, "Algebra," Addison-Wesley, Reading, MA, 1965.
6. N. Metropolis, Algorithms in unnormalized arithmetic. I. Recurrence relations, *Numer. Math.* **7** (1965), 104–112.
7. E. Witt, Zyklische Körper und Algebren der Charakteristik p vom Grad p^n, *Journal Reine Angew. Math.* **176** (1936–37), 126–140.
8. K. Hensel, "Zahlentheorie," Göschen, Berlin and Leipzig, 1913.
9. K. Mahler, "Introduction to p-adic Numbers and Their Functions," Cambridge University Press, Cambridge, 1973.
10. S. Tanny, Studies in significance arithmetic, Doctoral dissertation, Massachusetts Institute of Technology, 1969.
11. S. Eilenberg, "Automata, Languages, and Machines," Vol. A, Academic Press, New York, 1974.

QA
1
L588
1974

AUG 4 1977